Centrifugal Materials
Processing

Centrifugal Materials Processing

Edited by

Liya L. Regel and
William R. Wilcox

Clarkson University
Potsdam, New York

Springer Science+Business Media, LLC

Library of Congress Cataloging-in-Publication Data

Centrifugal materials processing / edited by Liya L. Regel and William
R. Wilcox.
    p.   cm.
   "Proceedings of the Third International Workshop on Materials
Processing at High Gravity, held June 2-7, 1996, at Clarkson
University, Potsdam, New York"
   Includes bibliographical references and index.
   ISBN 978-1-4613-7722-1   ISBN 978-1-4615-5941-2 (eBook)
   DOI 10.1007/978-1-4615-5941-2
   1. Materials--Effect of high gravity on--Congresses. 2. Crystal
growth--Congresses. 3. Thin films--Congresses. 4. Flow
visualization--Congresses. 5. Solidification--Congresses.
I. Regel', L. L. II. Wilcox, William R. III. International
Workshop on Materials Processing at High Gravity (3rd : 1996 :
Clarkson University)
TA417.7.H53C46 1997
660'.2842--dc21   97-13653
                                                 CIP

Proceedings of the Third International Workshop on Materials Processing at High Gravity, held June 2–7, 1996, at Clarkson University, Potsdam, New York

ISBN 978-1-4613-7722-1

© 1997 Springer Science+Business Media New York
Originally published by Plenum Press, New York in 1997
Softcover reprint of the hardcover 1st edition 1997

http://www.plenum.com

PREFACE

This volume constitutes the proceedings of the Third International Workshop on Materials Processing at High Gravity. It offers the latest results in a new field with immense potential for commercialization, making this book a vital resource for research and development professionals in industry, academia and government. We have titled the proceedings *Centrifugal Materials Processing* to emphasize that centrifugation causes more than an increase in acceleration. It also introduces the Coriolis force and a gradient of acceleration, both of which have been discovered to play important roles in materials processing.

The workshop was held June 2-8, 1996 on the campus of Clarkson University in Potsdam, New York, under the sponsorship of Corning Corporation and the International Center for Gravity Materials Science and Applications. The meeting was very productive and exciting, with energetic discussions of the latest discoveries in centrifugal materials processing, continuing the atmosphere of the first workshop held in 1991 at Dubna (Russia) and the second workshop held in 1993 in Potsdam, New York. Results and research plans were presented for a wide variety of centrifugal materials processing, including directional solidification of semiconductors, crystallization of high Tc superconductors, growth of diamond thin films, welding, alloy casting, solution behavior and growth, protein crystal growth, polymerization, and flow behavior. Also described were several centrifuge facilities that have been constructed for research, with costs beginning at below $1000. Unlike the small bench-top, high-acceleration machines of the past, these centrifuges can accommodate usual laboratory equipment and provide an acceleration of up to 20 times earth's gravity.

During the 15 years that have passed since the first crystal growth experiment on a large centrifuge, the science and practice of centrifugal materials processing have become better understood and more widely applied to a variety of techniques. Thanks to all of the creative and inquiring researchers in this field around the world.

To my consternation, I discovered that it is impossible to recall all the occasions when graduate students and postdoctoral fellows, colleagues and visitors gave me advice and criticism of inestimable value. My heartfelt thanks to all of those who so contributed, not only during the last five years at Clarkson, but also during my life at the Space Research Institute (IKI) in Moscow and during extended stays at CNRS in Meudon, France. I'm particularly indebted to my dearest friends and colleagues, Huguette and Michel Rodot. I wish also to acknowledge the help and professionalism of co-workers in my former IKI laboratory. It is a pleasure to thank all of the contributors to the three international meetings and the resulting proceedings devoted to centrifugal materials processing.

I believe only one person has influenced almost every session of the three workshops and every paper of the last two proceedings --- Bill Wilcox, inspired scientist and wit. He read each abstract and followed through on each presentation and paper. He immeasurably improved these proceedings with his passion for exactitude and with an astounding ability to sum up his own experience in seemingly spontaneous perfect sentences.

My heartfelt appreciation to Bill Wilcox, Bill Arnold, and Ramnath Derebail, who transformed my dream into reality and built the 6-ton HIRB centrifuge facility -- the first in the world dedicated to materials processing research and related flow visualization.

I'm most grateful to the National Science Foundation, the Russian Academy of Sciences, and the French Ministry of Research for their valuable support at critical stages in the continued development of this field.

Finally, I wish to express my deep thanks to my mother for her support and understanding throughout my life, and for introducing me to the fascinating subject of the science of materials. The value of her help and contributions far exceed what these words of appreciation can express.

Liya L. Regel, Chair of Workshops and Co-Editor

November 1996
International Center for Gravity Materials Science and Applications
Clarkson University
Potsdam, New York

CONTENTS

CENTRIFUGAL MATERIALS PROCESSING

Liya L. Regel and William R. Wilcox

International Center for Gravity Materials Science and Applications
Clarkson University, Potsdam, NY 13699-5814

SUMMARY

This is a brief review of the influence of centrifugation on materials processing. The emphasis in this summary is on papers presented at the Third International Workshop on Materials Processing at High Gravity. This highly successful meeting was held June 2-8, 1996 on the campus of Clarkson University in Potsdam, New York, under the sponsorship of Corning Corporation and the International Center for Gravity Materials Science and Applications. The present volume constitutes the proceedings of this workshop.

The workshop began with our discussion of the influence of centrifugation on transport phenomena. We pointed out that centrifugation not only increases the acceleration, but also introduces the Coriolis force and a spatial dependence for the acceleration vector. Increasing the acceleration to a few g_e greatly increases sedimentation of second-phase particles, but should have little effect on sedimentation of the components in ordinary solutions or on the pressure in a liquid. Much of this material is in the present paper.

The primary stimulus for increased activity in centrifugal materials processing was the 1988 report by Liya Regel and Huguette Rodot of uniform Ag doping in PbTe grown by a gradient freeze technique. This observation indicated diffusion-controlled segregation at a particular acceleration -- an unexpected and surprising result. At the Second International Workshop in 1993, numerical modeling indicated that a sharp minimum in convection should occur at the rotation rate where the acceleration vector is nearly perpendicular to the solid-liquid interface. At the present Workshop, Friedrich and Müller confirmed the Regel-Rodot results, both experimentally and theoretically. They obtained a maximum in the effective distribution coefficient for Ga in Ge at a particular rotation rate. They used a scaling analysis, in which they equated the axial and radial buoyancy terms to their corresponding Coriolis terms, in order to predict a minimum in convection versus rotation rate for a concave interface shape. Their three-dimensional numerical model predicted that this convection should be steady, with circumferential rotational features. These features were, in turn, confirmed by the flow visualization experiments of Skudarnov, Regel and Wilcox.

At the 1996 Workshop, Arnold and Regel showed numerical results for rotation of a cylinder about its own axis. They predicted a sharp minimum in buoyancy-driven convection for the gradient freeze configuration using this simple technique. They confirmed this prediction by flow visualization experiments.

Everyone agreed that large cross-sectional variations in doping may occur even when the axial composition is constant because the convection is very weak.

Parfeniev, Shamshur and Regel found that the effective distribution coefficient for Tl in PbTe became unity at 10 g_e, although there was much scatter in the composition data. Likewise, the axial variation in superconducting transition temperature decreased as g was increased. Ladeira, Shen, Wilcox and Regel showed that centrifugation had no effect on the dislocation density of Zn-doped CdTe grown by the gradient freeze technique, but greatly diminished the number of precipitates. Grain size was a maximum at a particular rotation rate. It was suggested that centrifugation floats away particles that cause nucleation at the ampoule walls and at the freezing interface.

Bandeira, An, Franzan and Salgado determined the compositional homogeneity of PbSnTe and HgCdTe solidified at 1g_e and at 7g_e in normal and inverted geometries. Glazov, Pavlova and Stanuks reported on vertical stratification in molten HgCdTe, and attributed it to sedimentation of clusters.

Inatomi, Kuribayashi, Kitajima and Huang mounted a microscopic interferometer on a rotating table to measure temperature fields during solidification of organic compounds. With salol, no change with acceleration was observed. For mixtures, the interface position and shape depended on g. Temperature fluctuations of about 0.1°C were observed at 1 g_e, and 1°C at 10 g_e. Bergelin and Chevy measured temperatures in molten tin held in a "destabilizing" temperature gradient. The axial and radial temperature differences were measured versus Rayleigh number. The transition to fluctuating temperatures was measured versus acceleration, temperature gradient and mass of the melt. The temperature fluctuations at different locations were correlated in order to obtain the flow pattern. A scaling analysis was used in which the buoyancy term was equated to the Coriolis term.

When Volkov, Melekh, Kartenko, Zamorjanskaja and Regel crystallized a Bi-based, high-Tc superconductor mixture in a centrifuge, it stratified to form different compositions and crystal structures versus height.

Gurin grew small crystals of KCl and KBr from aqueous solutions by cooling in a bench-top centrifuge. Remarkably, the Knoop microhardness of KBr grown at 12 g_e was 140% higher than that grown at 1g_e, while KCl was 27% harder. The size of the crystals was diminished and the habit changed by centrifugation. Morgunova, Blagova, Smirnova, Mikhailov, Armstrong, Mao and Ealick crystallized proteins in space, on earth, and in a centrifuge at 70,000 g_e. Supersaturated solutions were used on earth and in microgravity, while undersaturated solutions were used in the ultracentrifuge experiments. The resolution of the crystal structure was best in the space-grown crystals, and poorest in those grown at earth's gravity. However, reproducibility was greater and the ease of growth was much easier at high g than at low g. Lanzerotti, Autera, Pinto and Sharma reported that ultracentrifugation allowed single crystals of TNT, RDX and TNAZ to be grown without inclusions of particles, bubbles, or solution. The stability of these energetic materials is thought to be degraded by the presence of such inclusions.

Izmailov and Myerson described the surprising sedimentation behavior of supersaturated aqueous solutions at 1g_e. The traditional theory for settling of nano-clusters predicts a sedimentation time of hundreds of years, while the experiments yielded a stable stratification in days. A new theory was developed that predicts unusual behavior at high accelerations, such as spatially periodic nucleation and a critical radius that is a function of g.

Riahi and Baker demonstrated theoretically that rotation can be used to reduce or eliminate freckles in castings. Aidun and Domey reported on their welding experiments. The depth to width ratio of the weld fusion zone decreased with increasing g for 304 and 316 stainless steel, while the reverse behavior was observed for commercially pure nickel.

Briskman, Kostarev and Moshev produced a spatial variation of properties in plastics by centrifugation during thermal- and photo-polymerization. Karimov, Akhmedov and Achourov greatly increased the electrical conductivity of polymer films for solar cells by using an impeller rotating at 1400 rpm during precipitation.

Folkersma, Van Diemen and Stein studied the agglomeration of dispersions of 2 micron polystyrene particles in space and at 1, 2, 4 and 7 g_e. The minimum coagulation rate was at 2 g_e and the maximum was in microgravity experiments. Simmons *et al.* produced a sticky, porous film of polytetrafluoroethylene from a colloidal dispersion by use of a table-top centrifuge.

CENTRIFUGES

In some sense, centrifugal materials processing is an old field. Bench top centrifuges have been used for decades to separate colloidal particles from liquids. Ultracentrifuges have been used to separate heavy molecules from solutions, and occasionally for crystal growth.[1-4] Large industrial centrifuges are used to separate crystals from their mother liquor following crystallization operations. Centrifugal casting is used commercially for metal alloys.

In the last 20 years, centrifugal materials processing research has been done on a variety of techniques and types of materials. This increasing level of activity has resulted in international meetings being held in 1991,[5] 1993,[6] and 1996 (present volume). What has differed from older work is that large centrifuges are being used to carry laboratory equipment to moderately high accelerations. Some of these centrifuges had been built commercially for use in civil engineering modeling or for testing of human beings. These machines were neither under the control of materials scientists nor designed for materials research. Furthermore, they were expensive, ranging from over $100,000 to approximately $100 million. As a consequence, materials scientists have been constructing their own centrifuges, with very modest budgets. Table 1 summarizes the large or specialized centrifuges that have been used for materials processing research. Several of these are described later in the present volume.

FLUID PHYSICS OF CENTRIFUGATION

When a fluid is placed in a centrifuge rotating at angular velocity ω, the acceleration is increased by the centrifugal acceleration $g_a = \omega^2 r$, where r is the distance from the axis of rotation. When centrifugation is carried out on earth, the total acceleration g is the vector sum of the acceleration due to earth's gravity, g_e (9.8m/s^2), and g_a. Because of the r-dependence during centrifugation, g varies throughout the fluid, both in magnitude and in direction. We call this r-variation the "acceleration gradient." This variation is normally small compared to the total acceleration because the dimensions of the container of the fluid are small compared to the radius of the centrifuge arm. Nevertheless, there are materials processing situations in which the acceleration gradient plays an important role in buoyancy-driven convection, as we discuss later.

Centrifugation also introduces the Coriolis acceleration, $2\omega \times V$, into the equations of motion. Here, $\times$ represents the vector cross product and V is the local fluid velocity in the rotating frame. The effect is to deflect the flow and create circulation.

Now we examine the consequences of centrifugation on various phenomena. To illustrate these, we occasionally refer to a sealed ampoule containing a liquid and its vapor, such as might be used in directional solidification. We find that some phenomena are so strongly influenced by centrifugation that interesting changes occur at a few g_e. Other phenomena are observable only at very high accelerations. To put this into perspective, large centrifuges for testing humans operate up to about 20g_e. Some of the centrifuges used by civil engineers for testing model structures can operate to about 100g_e with moderately large loads. For higher accelerations, one must use small laboratory machines. These ultracentrifuges are capable of several $100,000g_e$, but only with small samples. Thus, most materials processing experiments, and the potential for centrifugal manufacturing, are limited to moderate g.

Table 1. Large or specialized centrifuges utilized for materials processing research. **Bold type indicates centrifuges constructed specifically for materials processing research.** *Italics indicates those believed to be no longer used for such research.*

Authors	Location	Applications	Refs.
Regel et al.	*Star City, Russia*	*Gradient freeze crystallization, vapor transport*	*7-26*
Müller et al.	**Erlangen, Germany**	**Gradient freeze solidification, zone melting, temperature measurements**	**27-35**
Barta, Triska	*Prague, Czech Rep.*	*Gradient freeze solidification*	*36,37*
Sokolowski, Glicksman	*Troy, NY, USA*	*Eutectic alloy solidification*	*38-40*
Grugel	*Nashville, TN, USA*	*Dendrite solidification and flow visualization*	*41*
Rodot, Chevy et al.	Nantes, France	Gradient freeze, temperature measurements, chemical vapor deposition, combustion expts.	8,18,20, 42-49
Bandeira et al.	**Saõ José dos Campos, Brazil**	**Gradient freeze semiconductor alloys**	**50-52**
Lin, Ma	**Beijing, China**	**Horizontal Bridgman, temperature measure.**	**53-55**
Hibiya	*Tsukuba, Japan*	*Heat transfer studies*	*56*
Regel, Wilcox et al.	**Potsdam, NY, USA**	**Gradient freeze solidification, flow visualization, diamond films, coagulation of dispersions**	**20,57-62**
Aidun et al.	**Potsdam, NY, USA**	**Welding**	**63,64**
Stein, Folkersma	**Amsterdam, Neth.**	**Coagulation of colloidal dispersions**	**65**
Inotami	**Tokyo, Japan**	**Observation of solidification**	**66**
Abe	**Tsukuba, Japan**	**Diamond film deposition**	**67**

Hydrostatic Pressure

The rate of increase of hydrostatic pressure p with distance h below the surface of a stationary fluid of density ρ is given by:

$$\frac{\partial p}{\partial h} = \rho g = \rho(g_e + \omega^2 r) \tag{1}$$

Again, r is the distance from the center of rotation of the centrifuge. If $\omega^2 r \gg g_e$ (large acceleration), then h is practically in the same direction as r, and we can write $(\partial p/\partial h) \approx (dp/dr)$ and integrate accordingly to find p versus r and h. On the other hand, if $h \ll r$, then the net g varies little throughout the material and we can write $p = p_0 + \rho gh$, where p_0 is the pressure at the top of the fluid. The relative pressure change, $(p/p_0 - 1) = \rho gh/p_0$, is small for gases because ρ is very small. Thus if there is a gas above a liquid, the pressure in this gas is essentially uniform and the pressure at the surface of the liquid is the same as in the gas. The relative pressure change is also small for liquids, unless p_0 is very small, h is large, or g is large. For example, at $1g_e$ the pressure only increases by 0.097 atm for each meter of water. If, for example, one wanted to increase the pressure at the bottom of a 10 cm column of water to 100 atm, the acceleration would have to be over 10,000 g_e.

The thermodynamic properties of condensed phases are not strongly influenced by pressure, and, therefore, by gravity. We conclude from the foregoing that appreciable changes in thermodynamic properties, such as solubility, would be observed only at high g, i.e. in ultracentrifuges.

Sedimentation in Solutions

The components in a mixture tend to separate in an accelerational field because of differences in their molecular weights. For a binary mixture, the sedimentation flux J_{sed} (mol/m^2.s) can be given by:

$$J_{sed} = \frac{DC(M - \rho\overline{V})g}{\nu RT\left(1 + \frac{\partial \ln \gamma}{\partial \ln C}\right)} \qquad (2)$$

where D is the diffusion coefficient (m^2/s) in the fluid, C is the concentration (mol/m^3) of the sedimenting component, M is its molecular weight, $\overline{V}$ is its partial molar volume (m^3/mol), ν is the number of ions into which its molecule dissociates (e.g., 2 for NaCl), and γ is its activity coefficient. The average velocity of sedimentation of these molecules is J_{sed}/C. As noted later, in an ultracentrifuge the sedimentation of dissolved components can be sufficient to cause crystallization, even from a solution that is initially undersaturated. In the absence of such crystallization, a concentration gradient develops that causes molecular diffusion in the opposite direction. Eventually an equilibrium is reached, for which the vertical concentration gradient is given by:

$$\frac{\partial \ln C}{\partial h} = \frac{(M - \rho\overline{V})g}{\nu RT\left(1 + \frac{\partial \ln \gamma}{\partial \ln C}\right)} \qquad (3)$$

Substitution of numbers into these equations reveals that sedimentation is unobservable unless the system is very tall (as in the earth's atmosphere), or the acceleration is very large. We would not expect to observe sedimentation of dissolved constituents at the accelerations used in materials processing experiments, unless an ultracentrifuge is used. Settling of second-phase particles or bubbles is another matter (see below).

On the other hand, supersaturated aqueous solutions have been observed to sediment at $1g_e$.[68-78] After several hours, the solute concentration at the bottom of a column of solution appears to become slightly larger than the concentration at the top. Although it is known that clusters form in supersaturated solutions, these clusters are far from being large enough to settle as particles at 1 g_e ; Brownian motion would be sufficient to keep them suspended. Using the observed steady state concentration difference together with equation 3, the cluster size has been estimated and found to be reasonable. But with this cluster size, the time required to reach steady state would be on the order of centuries rather than a few days as observed experimentally. Thus, the true nature of this phenomenon remains a mystery. Centrifugal experiments on supersaturated solutions would be very helpful in determining the mechanism behind this phenomenon. Since nucleation and crystal growth take place in supersaturated solutions, such experiments would also improve our understanding of these processes.

Sedimentation has also been observed at 1 g_c in metal alloys,[e.g. 79-81] and has been called "liquation." Proposed explanations include sedimentation of clusters, preferential melting of one phase combined with settling of the other phase, and segregation during solidification combined with buoyancy-driven convection. Again, centrifugal experiments would be helpful in understanding and controlling such behavior.

Sedimentation of Second-Phase Particles and Gas Bubbles

At earth's gravity, we observe settling of heavier particles and the rise of gas bubbles in a liquid. The terminal settling velocity V_t of a small, isolated spherical particle of density ρ_p and diameter d_p in a stagnant liquid of viscosity μ is given by Stokes law:

$$V_t = \frac{g(\rho_p - \rho)d_p^2}{18\mu} \tag{4}$$

Individual colloidal particles substantially below 1 μm in size do not settle at 1 g_c, because of Brownian motion. This is particular true of nano-sized particles, which are of great interest today. Centrifugation, sometimes at very high g, can be useful for collecting colloidal particles.

The settling of slurries, flocculation of suspensions, and aggregation of particles depends not only on the size distribution of the particles but also on their electric charge. Under some conditions, gravity would influence the kinetics of these processes, even though current theory rarely includes a gravity dependence. Centrifugal experiments would be useful in improving our understanding of colloidal behavior. Centrifugation could also influence the structure resulting from settling, flocculation and aggregation, in turn altering the properties of the consolidated material or even making it possible to produce new types of materials. Two papers in this volume deal with the influence of centrifugation on coagulation of colloidal dispersions.[62,82]

Gas bubbles detach more readily from surfaces in the centrifuge, and should result in fewer voids being incorporated in solids formed during centrifugation.[e.g.4]

Buoyancy-Driven Natural Convection

Even small rotation rates can influence buoyancy-driven convection. If g were simply increased, the convective velocity would be proportional to $g^{1/2}$, and the heat and mass transfer rates would increase with $g^{1/4}$. However, not only is g increased by centrifugation, but it also introduces the Coriolis acceleration and the acceleration gradient (see above).

The Coriolis effect deflects the flow, causing a secondary circulation or vortices. The present volume reports on the first truly successful flow visualization experiments in a simulated gradient freeze geometry,[58] confirming the theoretically predicted circulation patterns.[83-86] It proved necessary to view the ampoule from the bottom in order to see the circumferential circulation, and not to rely on the usual side view with light sheet illumination.

The Coriolis effect also influences the stability of buoyancy-driven convection and can introduce new flow modes.[e.g.32,33] It can influence the tendency of the flow to become oscillatory or turbulent, either by causing a transition to a more stable flow mode or by stabilizing the current mode. Thus, for example, an increase in rotation rate led to a cessation of fluctuations in convection, even though the strength of the convection was increased.[27-35,42]

Under some conditions, centrifugation can even reduce convection velocities, due to the spatial gradient of acceleration. This can occur when the density decreases with height in the fluid, with weak buoyancy-driven convection due to a horizontal variation in density. Under some conditions, centrifugation causes the acceleration vector to become more aligned with the density gradient, thereby reducing the driving force for convection.[83-85,87-90] A sharp minimum in convection is produced at a particular rotation rate, with a reversal in flow direction as the rotation rate exceeds this value. Such behavior offers the potential for fine-tuning both the convection velocity and direction.

Forced Convection

Fluid also may be circulated by mechanical means. This is called "forced convection." The Coriolis effect causes secondary circulation and thereby influences heat and mass transfer, changes the pressure drop, and alters the transition to turbulent flow. The present volume contains two papers dealing with this subject.[91,92]

MATERIALS PROCESSING

Directional Solidification

Much of the recent research on centrifugation has involved the gradient freeze technique at moderate accelerations. In this technique, an ampoule containing a melt is placed in a furnace with a temperature gradient along its length. The temperature is slowly reduced, causing directional solidification of the melt.

The increasing interest in centrifugal materials processing stems largely from the surprising results of Rodot and Regel.[7,8] In a series of experiments spanning approximately ten years, they discovered that the axial distribution of Ag in PbTe became uniform when the material was solidified in a centrifuge at a particular rotation rate. Such results should occur only when the buoyancy-driven convection is negligible compared to the freezing velocity. On the other hand, since centrifugation increases the acceleration g in all of the equations of hydrodynamics, one would have expected the convection to increase monotonically with increasing acceleration. Consequently, the results of Rodot and Regel were greeted with considerable skepticism by both the crystal growth and the hydrodynamic communities. By now, however, their results have been explained theoretically [83-85,87-90] and confirmed experimentally in the present volume.[35] We return to this subject later. For now, we discuss more generally the influence of centrifugation on directional solidification.

Increased heat transfer in the melt during directional solidification by the gradient freeze technique can lower the temperature gradient in the melt, increase the freezing rate, and lead to morphological breakdown of the solid-liquid interface. The position and shape of the solid-liquid interface can be changed, thereby influencing impurity incorporation, thermal stress in the solid, and propagation of dislocations and grain boundaries.

Increasing the net acceleration also increases the settling of second phase material, such as particles and bubbles. To the extent that particles cause nucleation of new grains, enhanced settling of foreign particles can alter the grain size of solidified materials[e.g.,60] and decrease the spurious nucleation of new crystals during vapor and solution growth. Rise of bubbles during solidification would reduce the incorporation of gas bubbles throughout the solid and at the ampoule walls.

The "weight" of a material is increased by centrifugation. This can cause more intimate contact of a melt and its resulting solid with the surface of its container, especially if the surface of the ampoule is rough. A solid may be plastically deformed by the force arising from its own weight. However, as noted earlier, the pressure exerted is small unless the acceleration is large. Thus, theoretical calculations have shown that the stress due to differential thermal expansion between the material and the ampoule is much larger than that due to hydrostatic pressure.[93,94]

Experiments on Zn-doped CdTe reported in this volume showed no influence of centrifugation on dislocation etch pit density.[60]

Inverted Gradient Freeze Technique. Müller and coworkers performed an extensive set of solidification experiments on InSb, temperature measurements in liquid metals, and theoretical modeling for the inverted gradient freeze technique.[27-34] The ampoule containing the material was placed in a furnace with the temperature decreasing with height. Solidification was made to proceed downward by slowly decreasing the furnace temperature. The furnace was attached to the centrifuge with a fixed angle, so that the resultant acceleration vector was aligned with the ampoule axis at only one rotation rate.

As the rotation rate was increased, the axial temperature gradient in the melt declined, indicating increased buoyancy-driven convection. Temperature fluctuations began to occur in the melt. Etching revealed impurity striations in the resulting crystals, indicating a fluctuating freezing rate. As the rotation rate was increased farther, the temperature fluctuations increased in magnitude until a critical rate was reached at which the temperature again became steady and the crystals no longer had striations. Müller and coworkers showed that this transition corresponds to a change in flow direction due to the Coriolis acceleration. Below the critical rotation rate, a single flow cell circulates in one direction and is irregular. Above the critical rotation rate, a single flow cell circulates in the opposite direction. Temperature fluctuations did not recur when the rotation rate was decreased, i.e. there was hysteresis in the flow behavior.

Chevy et al.[12] performed similar experiments, except that the furnace was attached to the centrifuge by a hinge, so that the net acceleration was always aligned with the ampoule axis. In this case, there was no hysteresis in flow behavior. As the rotation rate was increased, temperature fluctuations stopped at a critical rate, and resumed below this value when rotation was decreased. The acceleration required to stop the temperature fluctuations increased rapidly as the ampoule diameter was increased. The axial temperature gradient in the melt decreased dramatically as acceleration was increased, and was much less than the gradient in the furnace. This decrease in axial temperature gradient indicates vigorous convection. A continuation of this work is reported in the present volume.[47]

Horizontal Gradient Freeze Technique. Gallium arsenide was solidified by the gradient freeze technique in a horizontal boat, attached to a centrifuge by a hinge so that the resultant acceleration was always normal to the ampoule axis and the surface of the melt.[53] Impurity striations were found in the resulting crystals, indicating fluctuations in freezing rate due to fluctuations in the convection. These striations diminished as the acceleration was increased.

Temperature measurements were made in liquid tin in the horizontal boat configuration with a temperature gradient down the furnace.[53,54] As the acceleration was increased, the axial temperature gradient in the molten tin decreased, indicating increased convection. Temperature fluctuations were observed in the melt. These temperature fluctuations decreased considerably with increasing acceleration if the centrifuge rotation was in the same sense as the convection roll in the melt. If the centrifuge rotation was in the opposite direction, then the temperature fluctuations increased with increasing rotation rate. A numerical model[55] of buoyancy-driven convection in the melt agreed with the experimental results. It appears that when the Coriolis force pushes the top and bottom streamlines closer together, then the flow is destabilized. When the Coriolis force pushes the streamlines apart, then the flow is stabilized and the temperature fluctuations should decrease with increasing rotation rate.

Vertical Gradient Freeze Technique. Regel and Rodot pioneered centrifugal directional solidification in the normal gradient freeze orientation.[7,8] In the horizontal boat method and the inverted gradient freeze technique, significant buoyancy-driven convection is expected. In the normal gradient freeze orientation at 1 g_e there would be no convection at all if there were no horizontal temperature gradients. However, some radial temperature gradients are inevitable, and so gentle convection does occur.[95-100]

As mentioned earlier, Regel and Rodot directionally solidified Ag-doped PbTe in a gradient freeze furnace.[7,8] The ampoule axis was aligned with the net acceleration vector. The Ag concentration was measured along the centerline of each ingot, but not near the ends. Ingots solidified in the 18 m arm centrifuge at the Gagarin Cosmonaut Training Center outside Moscow had an uniform Ag concentration when the net acceleration was 5.2 g_e. As the acceleration deviated more and more from 5.2 g_e, the Ag concentration became less and less constant. Similar results were obtained for PbTe in the 5.5 m arm centrifuge at Nantes in France, except that an uniform Ag concentration was obtained at about 2 g_e. The rotation rate for an uniform concentration was approximately the same in the Russian and French centrifuges.

In the present volume, Friedrich and Müller also report an apparent minimum in convection for gradient freeze solidification of germanium during centrifugation.[35]

Other materials solidified in the Soviet centrifuge exhibited behavior expected for reduced convection as the rotation rate was increased.[9-16] Recent theoretical work has provided considerable insight into convection in the vertical gradient freeze technique and an explanation for these experimental results.[83-85,87-89] The following is based on the papers by Arnold[84] and by Urpin.[89]

If density ρ decreases with height ($g \cdot \nabla\rho < 0$), buoyancy-driven convection occurs only when the acceleration vector is not perfectly aligned with the density gradient ($g \times \nabla\rho \neq 0$). When the acceleration vector is parallel to the ampoule axis, as in a vertical ampoule on earth, then convection occurs whenever a horizontal density gradient is present. In the absence of concentration gradients, the freezing interface is an isotherm, and the density gradient is perpendicular to the interface. Thus the driving force for convection near the freezing interface is directly related to the curvature of this interface. Indeed, from his numerical simulation of gradient freeze growth of germanium, Motakef found that the maximum velocity in the melt is proportional to the interface deflection and to the axial temperature gradient along the wall.[100] In the gradient freeze technique, the interface is predicted to be concave,[e.g.,95] and so convection is always expected on earth.

In a centrifuge, the acceleration vector is no longer parallel everywhere to the ampoule axis. Under some conditions, the acceleration can become normal to the concave interface in a gradient freeze experiment. When this happens, $g \times \nabla\rho = 0$, and there is no driving force for convection in the neighborhood of the interface.[83-85,89] We believe this is what happened in the gradient freeze centrifuge experiments described above. As the rotation rate was increased, the acceleration vector became more and more perpendicular to the concave freezing interface. For Ag-doped PbTe, we believe it became very nearly perpendicular at one particular rotation rate, and then deviated from this condition as the rotation was increased farther. Similar results were obtained by Friedrich and Müller using a scaling analysis, in which they equated the axial and radial buoyancy terms to the corresponding Coriolis terms.[85]

Convection could not have been completely absent when an uniform Ag concentration was obtained in PbTe, because the acceleration could not have been precisely aligned with the density gradient everywhere in the melt. Nevertheless, at the moderately high freezing rate used for the PbTe experiments, an uniform axial doping profile can be obtained. When the ratio of freezing rate V to diffusion coefficient D is sufficiently high, we can obtain a concentration profile corresponding to the absence of convection, even though some gentle convection is present.[e.g.,100] Such conditions can yield an uniform axial concentration profile, while at the same time producing substantial radial variations in doping. In the PbTe experiments, the Ag concentration was measured only along the centerline, and so cross sectional variations could not have been observed. Similarly, the interface shape and the freezing rate were not measured, making comparison with theory impossible. Until now, no one has measured experimentally the interface shape, freezing rate, and cross-sectional variations in doping during solidification in a centrifuge.

No impurity striations could be found in InSb[18-20], GaSb[14] or Te[22] ingots prepared by the gradient freeze technique using either the 18 m Star City centrifuge or the 5.5 m Nantes centrifuge. These negative results indicate that the freezing rate did not fluctuate and that the convection was steady. Temperature measurements in molten Sn, 70%Sb-30%Bi, Ge and Al revealed no fluctuations and no change in axial temperature gradient as the acceleration was increased.[42]

The primary dendrite arm spacing in Pb-Sn alloys decreased significantly as acceleration was increased.[41] Such behavior corresponds to reduced convection. The axial variation in charge carrier concentration in tellurium crystals corresponded to the presence of some convection.[22] This variation reached a maximum as the acceleration was increased and then decreased at higher g. This behavior was attributed to the peculiar properties of molten Te, which shows a minimum in density as temperature is increased beyond the melting point.

$Pb_{0.8}Sn_{0.2}Te$ was directionally solidified in several configurations.[23,50,52] Solidification upward at 1 g produced compositional profiles corresponding to those expected in the presence of moderately strong convection. Concentration profiles in crystals solidification downward by the inverted Bridgman method indicated that convection was reduced. This was explained by the dominance of solutal effects, i.e. the density of the melt was influenced more by composition gradients than by temperature gradients. Gradient freeze growth in centrifuges yielded complex composition profiles, in both the inverted and normal orientations. This behavior may indicate a strong variation in freezing rate and/or convection during solidification in the centrifuge.

Influence of Centrifugation on Microstructure and Perfection. We have concentrated thus far on the influence of centrifugation on compositional variations in the resulting materials due to changes in convection in the melt during solidification. We have gained a fairly good understanding of the influence of centrifugation on buoyancy-driven convection and on compositional homogeneity. Centrifugation has also been found to influence the microstructure and perfection of directionally solidified materials. In most cases, we do not know why.

There are at least three possible explanations for an improvement of microstructure. First, foreign particles may sediment out during centrifugation prior to solidification, so that they cannot nucleate new grains and twins during solidification. Second, by altering the heat transfer in the system, centrifugation may cause the interface to become more favorable for grain selection, i.e. less concave. Third, centrifugation causes gas bubbles in the melt to float away from the freezing interface. (Many fewer gas bubbles are present on the surface and in the interior of materials solidified in the centrifuge.) It is possible that gas bubbles at the freezing interface cause nucleation of new grains or twins. For example, a bubble may suddenly move, altering the heat transfer in the neighborhood of the interface and causing very rapid solidification at that location.

Following are some examples of the influence of centrifugation at moderate accelerations on microstructure and perfection. In their gradient freeze experiments on Ag-doped PbTe, Rodot and Regel found that only the ingot solidified at 5.2 g was a high quality single crystal.[7,8] Experiments on gradient freeze solidification of germanium in the centrifuge also showed improved grain size as the rotation rate was increased.[43,44] The solid-liquid interface of GaSb solidified in the Star City centrifuge was flatter than when GaSb was solidified in space or at 1 g_e.[14] The number of gas bubbles in InSb decreased as the acceleration increased.[18-21,26] Although there were differences in the mobility and numbers of grain and twin boundaries between the InSb ingots, no trend with acceleration could be discerned. The mobility of holes in Te was relatively constant when it was prepared by the gradient freeze technique at 1 g_e.[22] Te solidified at 5 and 10 g_e had a significantly lower mobility that increased as one moved down the ingots. This behavior was attributed to increased convection in the melt during solidification that increased the disorder in the resulting crystals.

As noted earlier, in the solidification of PbTe-SnTe alloys, convection is greater in the normal orientation than in the inverted orientation.[50] Increasing acceleration in the normal gradient freeze orientation caused formation of a cellular structure, metallic inclusions, and nucleation of new grains.[23] This was attributed to increased convection, leading to a reduced temperature gradient and an increased freezing rate, both of which increase the likelihood of morphological instability.[23]

It has been speculated that increased acceleration might increase the sticking of crystals to the ampoule wall, thereby increasing plastic deformation due to the difference in thermal expansion coefficient between the crystal and the ampoule.[94] Acceleration was observed to increase the dislocation density of GaAs grown by the horizontal gradient freeze technique in sand-blasted silica boats.[53] It was suggested that the increased weight forced the melt into the pits and valleys of the boat wall, thereby increasing the adhesion of the resulting solid. One might also expect increased weight of the melt and the solid to force the solid to deform plastically in order to remain in contact as it cools from the melting point. (Normally the ampoule is coated with a non-stick coating to assist the solid in breaking away from the wall during cooling.)

Directional Crystallization of High TC Superconductors

A high TC superconductor mixture of Bi-Sr-Ca-Cu-O was crystallized in the vertical gradient freeze orientation.[24] The superconducting transition temperature increased as one moved down the crystallized mixture, reaching as high as 130 K. Apparently crystals settled as crystallization proceeded.

Welding and Casting

In a numerical study of welding, increased acceleration was predicted to enhance the convection in the weld pool, thereby influencing the heat transfer, the depth and width of the two phase region, and the pool depth-to-width ratio.[101] The depth to width ratio of the weld fusion zone decreased with increasing g for 304 and 316 stainless steel, while the reverse behavior was observed for commercially pure nickel.[64] In this volume, theoretical modeling is presented that predicts decreased freckling of metal alloy castings by use rotation during solidification.[102,103]

Solution Crystal Growth

By increasing buoyancy-driven convection, centrifugation should increase the kinetics of solution crystal growth when stirring is not used. Centrifugation should also influence the formation of solution inclusions, which are the most troublesome defect resulting from solution crystal growth. Because inclusion formation is a poorly understood process, the exact influence of centrifugation needs to be determined experimentally. For example, the direction of the convection over the growing surface influences the stability of a train of growth steps. Gas bubbles are implicated in formation of some inclusions, and centrifugation could cause them to float away before inclusions form. We noted earlier the strange properties of supersaturated solutions, from which crystals grow. Sedimentation was observed even at 1 g_e. Centrifugation could cause surprising results because of this.

Interesting results have been achieved using ultracentrifuges, in which the acceleration can reach several hundred thousand times g_e. As pointed out earlier, sedimentation of dissolved constituents can become appreciable.[e.g.104,105] This phenomenon has been used to move solvent inclusions through crystals[106-108] and to cause crystals to grow from solution.[1-4,109] Centrifugation caused crystals of the explosive materials TNT, RDX and TNAZ to grow free of inclusions, and hopefully be more stable.[3,4] Resolution of the crystal structure was best in protein crystals grown in space, and poorest in those grown at earth's gravity, while reproducibility was increased by use of ultracentrifugation.[109]

Centrifugation at modest g modified the morphology of small KCl and KBr crystals and increased their hardness.[110]

Sedimentation of the gel phase has proved helpful in the precipitation of zeolites.[111-113]

Polymerization

Experiments and theory were described for gel polymerization by light and heat in a centrifuge.[114,115] As g was increased, the radial variation in properties of the polymer increased, including Young's modulus and pore size. The effect depended on when centrifugation was begun during the polymerization. Very high acceleration caused mechanical destruction of the polymer.

A sticky, porous form of polytetrafluoroethylene was produced from a colloidal dispersion by use of a bench-top centrifuge.[62] Rapid stirring yielded improved electrical conductivity in polymer films for solar cells.[116]

Vapor Transport and Deposition

GeSe crystals were grown by the vapor transport technique in the normal and inverted gradient freeze orientations using the Star City centrifuge.[25] At $1 g_e$, the net transport rate was only slightly higher for the inverted orientation, which would be expected to produce more vigorous convection in the gas. There were fewer but larger crystals from the normal orientation. Centrifugation at $10 g_e$ in the normal orientation increased the transport rate by about 35%. In the inverted orientation, the transport rate was roughly proportional to g. The deposition pattern in both orientations was nearly axisymmetric at $1 g_e$, and become very asymmetric at $10 g_e$. The largest crystals, with well defined Laue patterns, were obtained at $10 g_e$ in the inverted orientation.

GaAs was chemically vapor deposited onto a GaAs hemispherical substrate.[45] The growth rate was proportional to $g^{1/4}$ from 1 to $10 g_e$, with more spurious nucleation on the ampoule wall as g was increased. A numerical model was developed to explain the results.

The flame height of a fire[46] varied as $g^{-1/3}$, the fluctuation frequency as $g^{1/2}$, and the total radiant power as $g^{-0.3}$. The experimental results were correlated using the predictions of numerical models and scaling analysis.

Two groups are now studying the influence of centrifugation on diamond film deposition.[61,67] Preliminary results show an increase in growth rate and more uniform deposition over a larger area.[61]

FUTURE RESEARCH

There are two primary motives for studying materials processing in centrifuges. First, such research improves our understanding of the influence of acceleration and convection on materials processing. Second, there are commercial opportunities for production of unique and improved materials that cannot be prepared under normal earth conditions or in space.

Additional careful experiments are needed in gradient freeze solidification, with materials of various types, including semiconductors, metal alloys, organic compounds and oxides. A wide range of freezing rates, temperature gradients, and compositions should be investigated. When possible, seeding should be used to provide single crystals of desired orientations. Interface demarcation should be used so that the interface shape and freezing rate versus position are known. The resulting ingots should be characterized more thoroughly than in the past, including, for example, analysis of impurity concentration to the ends of the crystals and over entire cross sections, microstructure, dislocation content, inclusions, and electrical properties.

Similarly, additional research should be performed on solution crystal growth, vapor transport, chemical vapor deposition, polymerization, and welding. Other opportunities exist in Bridgman-Stockbarger solidification, electrodeposition, fabrication and joining of composite materials, and fine particle processing.

Flow visualization and temperature measurements should be performed and compared with theoretical predictions.

Acknowledgment

We are grateful for the continuing support of the National Science Foundation under Grant DMR-9414304.

REFERENCES

1. P.J. Shlichta, *J. Crystal Growth* 119:1 (1992).
2. P.J. Shlichta and R.E. Knox, *J. Crystal Growth* 3/4:808 (1968).
3. M.Y.D. Lanzerotti, J. Autera, J. Pinto, and J. Sharma, *in ref.* 6.
4. M.Y.D. Lanzerotti, J. Autera, L. Borne, and J. Sharma, *in present volume.*
5. L.L. Regel, M. Rodot, and W.R. Wilcox, editors, "Material Processing in High Gravity, Proceedings of the First International Workshop on Material Processing in High Gravity," North-Holland, Amsterdam (1992). Also, volume 119 of the *Journal of Crystal Growth.*
6. L.L. Regel and W.R. Wilcox, editors, "Materials Processing in High Gravity," Plenum Press (1994).
7. H. Rodot, L.L. Regel, G.V. Sarafanov, H. Hamidi, I.V. Videskii, and A.M. Turchaninov, *J. Crystal Growth* 79:77 (1986).
8. H. Rodot, L.L. Regel, and A.M. Turtchaninov, *J. Crystal Growth* 104:280 (1990).
9. L.L. Regel *et al.*, *Fiz. Khim. Obrab. Mater.* 45 (1989).
10. L.L. Regel, A.M. Turchaninov, R.V. Parfeniev, I. Farbshtein, N.K. Shulga, S.V. Nikitin, and S.V. Yakimov, "Electrofizicheskie Svoistva Monokristallov Tellura i Splava $Te_{1-x}Se_x$, Poluchennikh v Usloviyakh pri Vishennoi Gravitatsii (5 g_o i 10 g_o)," USSR Space Research Institute, Moscow (July 1989).
11. L.L. Regel, I.V. Videnskii, V.V. Zubenko, I.M. Cafonova, and I.V. Telegina, *Fizika i Chimiya Obrabotki Materialov* 23:45 (1989).
12. P. Bartsi, L.L. Regel and I. Solyom, *in:* "Proceedings of the 4th Intercosmos Seminar on Cosmic Materials and Technologies," Bucharest (1989) pp 117-137.
13. B.V. Burdin, L.L. Regel, A.M. Turchaninov, and O.V. Shumaev, *J. Crystal Growth* 119:61 (1992).
14. L.L. Regel and O.V. Shumaev, *J. Crystal Growth* 119:70 (1992).
15. P. Barczy, J. Solyom, and L.L. Regel, *J. Crystal Growth* 119:160 (1992).
16. L.L. Regel et al., *J. Phys. France* 2:373 (1992).
17. M.A. Fikri, G. Labrosse, and M. Betrouni, *J. Crystal Growth* 119:41-60 (1992).
18. R. Derebail, W.R. Wilcox and L.L. Regel, *J. Spacecraft & Rockets* 30:202 (1993).
19. R. Derebail, "Study of Directional Solidification of InSb under Low, Normal, and High Gravity," M.S. Thesis, Clarkson University (1990).
20. R. Derebail, "Directional Solidification of Indium Antimonide under High Gravity in Large Centrifuges," PhD Thesis, Clarkson University (1994).
21. L.I. Farbshtein, R.V. Parfeniev, S.V. Yakimov, L.L. Regel, R. Derebail, and W.R. Wilcox, *in ref.* 6.
22. L.I. Farbshtein, R.V. Parfeniev, N.K. Shulga and L.L. Regel, *in ref.* 6.
23. L.L. Regel, A.M. Turchaninov, O.V. Shumaev, I.N. Bandeira, C.Y. An, and P.H.O. Rappl, *J. Crystal Growth* 119:94 (1992).
24. M.P. Volkov, B.T. Melekh, R.V. Parfeniev, N.F. Kartenko, and L.L. Regel, *J. Crystal Growth* 119:122 (1992).
25. H. Wiedemeier, L.L. Regel, and W. Palosz, *J. Crystal Growth* 119:79 (1992).
26. R. Derebail, W.R. Wilcox, and L.L. Regel, *J. Crystal Growth* 119:98 (1992).
27. G. Müller, *in:* ESA Special Publication No. 114, European Space Agency, Paris (1980) pp 213-216.

28. G. Müller and G. Neumann, *J. Crystal Growth* 59:548 (1982).
29. G. Müller, *in:* "Convective Transport and Instability Phenomena," J. Zierep and H. Oertel, Jr., eds., Braun Verlag, Karlsruhe (1982).
30. G. Müller, *in:* Selisch Fachbuch-Verlag, Langensendelbach (1986) pp 151-165.
31. G. Müller, *J. Crystal Growth* 99:1242 (1990).
32. W. Weber, G. Neumann, and G. Müller, *J. Crystal Growth* 100:100 (1990).
33. G. Müller, G. Neumann, and W. Weber, *J. Crystal Growth* 119:8 (1992).
34. G. Müller, E. Schmidt, and P. Kyr, *J. Crystal Growth* 49:387 (1980).
35. J. Friedrich and G. Müller, *in present volume.*
36. Z. Chvoj and C. Barta, *Czech. J. Phys. B* 36:868 (1986).
37. C. Barta, F. Fendrych, E. Krcova, and A. Triska, *Adv. Space Res.* 8:167 (1988). *Also in:* "Proceedings of the 4th Intercosmos Seminar on Cosmic Materials and Technologies," V. Lupei and D. Toma, eds., Rumanian Academy of Science, Bucharest (1989).
38. R.S. Sokolowski, Ph.D. Thesis, Rensselaer Polytechnic Institute, Troy, NY (1981).
39. M.E. Glicksman and R.S. Sokolowski, *Adv. Space Res.* 3:129 (1983).
40. R.S. Sokolowski and M.E. Glicksman, *J. Crystal Growth* 119:126 (1992).
41. R.N. Grugel, A.B. Hmelo, C.C. Battaile, and T.G. Wang, *in ref. 6.*
42. A. Chevy, P. Williams, M. Rodot, and G. Labrosse, *in ref. 6.*
43. A. Chevy, *Compte Rendue Acad. Sci. Paris* 307:1147 (1988).
44. A. Chevy, Private Communication, Universite Pierre et Marie Curie, Paris, France (1990).
45. J.C. Launay, S. Bouchet, A. Randriamampianina, P. Bontoux, and P. Gibart, *in ref. 6.*
46. J. Chen, J.M. Most, P. Joulain, and D. Durox, *in ref. 6.*
47. L. Bergelin and A. Chevy, *in present volume.*
48. R. Parfeniev, D. Shamshur, L.L. Regel, and S. Nemov, *in present volume.*
49. J. Garnier and L.M. Cottineau, *J. Crystal Growth* 119:66 (1992).
50. Y.A. Chen, I.N. Bandeira, A.H. Franzan, S. Eleutério Filho and M.R. Slomka, *in ref. 6.*
51. Y.A. Chen, L.C. Russo, M.F. Ribeiro, and I.N. Bandeira, *in present volume.*
52. Y.A. Chen, E.G. Salgado, C.R.M. Silva, and I.N. Bandeira, *in present volume.*
53. B. Zhou, F. Cao, L. Lin, W. Ma, Y. Zheng, F. Tao, and M. Xue, *in ref. 6.*
54. W.J. Ma, F. Tao, Y. Zheng, M.L. Xue, B.J. Zhou, and L.Y. Lin, *in ref. 6.*
55. F. Tao, Y. Zheng, W.J. Ma, and M.L. Xue, *in ref. 6.*
56. T. Hibiya, S. Nakamura, K.W. Yi, and K. Kakimoto, *in ref. 6.*
57. R. Derebail, W.A. Arnold, G.J. Rosen, W.R. Wilcox, and L.L. Regel, *in ref. 6.*
58. P.V. Skudarnov, L.L. Regel, and W.R. Wilcox, *in present volume.*
59. I. Moskowitz, L.L. Regel, and W.R. Wilcox, *in present volume.*
60. L.O. Ladeira, J. Shen, L.L. Regel, and W.R. Wilcox, *in present volume.*
61. Y. Takagi, L.L. Regel, and W.R. Wilcox, *in present volume.*
62. J. Simmons, L.L. Regel, W.R. Wilcox, and R. Partch, *in present volume.*
63. D. Aidun, *in present volume.*
64. D. Aidun, *private communication,* Clarkson University, Potsdam, NY (1995).
65. R. Folkersma, A.J.G. van Diemen, J. Laven, and H.N. Stein, *in present volume.*
66. Y. Inatomi, O. Kitajima, W. Huang, and K. Kuribayashi, *in present volume.*
67. Y. Abe, G. Maizza, H. Rouch, N. Sone, and Y. Nagasaka, *in present volume.*
68. J.W. Mullin and C.L. Leci, *Phil. Mag.* 19:1075-1077 (1969).
69. A.T. Allen, M.P. McDonald, W.M. Nicol, and R.M. Wood, *in:* "Particle Growth in Suspensions," A.L. Smith, ed., Academic Press, London (1973) pp 239-246.
70. A.T. Allen, M.P. McDonald, W.M. Nicol, and R.M. Wood, *Nature Phys. Sci.* 235:36-37 (1972).
71. M.A. Larson and J. Garside, *Chem. Eng. Sci.* 41:1285-1289 (1986).
72. I.T. Rusli and M.A. Larson, *in:* "Crystallization and Precipitation," G.L. Strathdee, M.O. Klein, and L.A. Melis, eds., Pergamon, Oxford (1987) pp 71-77.
73. A.S. Myerson and P.Y. Lo, *J. Crystal Growth* 99:1048-1052 (1990).
74. V. Veverka, O. Söhnel, P. Bennema, and J. Garside, *AIChE J.* 37:490-498 (1991).
75. A.S. Myerson and P.Y. Lo, *J. Crystal Growth* 110:26-33 (1991).
76. K. Ohgaki, Y. Makihara, M. Morishita, M. Ueda, and N. Hirokawa, *Chem. Eng. Sci.* 46:3283-3287 (1991).
77. R.M. Ginde and A.S. Myerson, *J. Crystal Growth* 116:41-47 (1992).
78. A.F. Izmailov and A.S. Myerson, *Physica* A224:503-532 (1996).
79. R. Derebail and J.N. Koster, *Metallurgical & Mat. Trans.* 27B:1-4 (1996).
80. A.P. Mohanty, "Determination of the Soret Coefficient of Mn-Bi Melts," M.S. Thesis, Clarkson University, Potsdam (1990).
81. V.M. Glazov, L.M. Pavlova, and S.V. Stankus, *in present volume.*

82. R. Folkersma, A.J.G. van Diemen, J. Laven, and H.N. Stein, *in present volume.*
83. W. Arnold, "Numerical Modeling of Directional Solidification in a Centrifuge," PhD Thesis, Clarkson University (1993).
84. W.A. Arnold and L.L. Regel, *in ref. 6.*
85. J. Friedrich and G. Müller, *in present volume*
86. Sh. Mavlonov, *J. Crystal Growth* 119:167-175 (1992).
87. W. Arnold, W.R. Wilcox, F. Carlson, A. Chait, and L.L. Regel, *J. Crystal Growth* 119:24 (1992).
88. W. Arnold, W. Wilcox, F. Carlson, A. Chait, and L. Regel, *in:* "Proceedings of the Society of Engineering Science," Gainesville (November 1991).
89. V.A. Urpin, *in ref. 6.*
90. W.R. Arnold and L.L. Regel, *in present volume.*
91. S.Ya. Gertsenshtein, N.V. Nikitin, and A.N. Sukhorukov, *in present volume.*
92. D.T. Valentine and C.C. Jahnke, *in present volume.*
93. T.P. Lee, "Finite Element Analysis of the Thermal and Stress Fields during Directional Solidification of Cadmium Telluride," Ph.D. Thesis, Clarkson University (1996).
94. T. Lee, J.C. Moosbrugger, F.M. Carlson, and D.J. Larson, Jr., *in ref. 6.*
95. C.E. Chang, V.F.S. Yip, and W.R. Wilcox, *J. Crystal Growth* 22:247 (1974).
96. C.E. Chang and W.R. Wilcox, *J. Crystal Growth* 21:135 (1974).
97. S. Sen and W.R. Wilcox, *J. Crystal Growth* 28:36 (1975).
98. T.W. Fu and W.R. Wilcox, *J. Crystal Growth* 48:416 (1980).
99. G.T. Neugebauer and W.R. Wilcox, *J. Crystal Growth* 89:143 (1988).
100. S. Motakef, *J. Crystal Growth* 102:197 (1990).
101. J. Domey, D.K. Aidun, G. Ahmadi, L.L. Regel, and W.R. Wilcox, *in ref. 6.*
102. C.F. Baker and D.N. Riahi, *in present volume.*
103. D.N. Riahi, *in present volume.*
104. K.O. Pedersen, *Z. Phys. Chem.* A170:41 (1934).
105. D.J. Cox, *Arch. Biochem. Biophys.* 119:230 (1967).
106. W.R. Wilcox and P. Shlichta, *J. Appl. Phys.* 42:1823 (1971).
107. W.R. Wilcox, *J. Crystal Growth* 13/14:787 (1972).
108. T.R. Anthony and H.E. Cline, *Phil. Mag.* 22:893 (1970).
109. E. Blagova, E. Morgunova, E. Smirnova, A. Mikhailov, S. Armstrong, C. Mao, and S. Ealick, *in present volume.*
110. V.N. Gurin, S.P. Nikanorov, L.L. Regel and L.I. Derkachenko, *in present volume.*
111. D.T. Hayhurst, P.J. Melling, W.J. Kim, and W. Bibbey, *in:* "Zeolite Synthesis," M.L. Occelli and H.E. Robson, eds., American Chemical Society (1989) ch 17.
112. W.J. Kim, Ph.D. Thesis, Cleveland State University, Cleveland, Ohio (1989); through *Chem. Abstr.* 112:219459 (1990).
113. D.T. Hayhurst, W.J. Kim, and P.J. Melling, US Patent Application 233,287 (1988); PCT Int. Appl. WO 90 02,221 (1990); through *Chem. Abstr.* 113:32438 (1990).
114. V.A. Briskman, K.G. Kostarev and T.P. Lyubimova, *in ref. 6.*
115. V. Briskman, K. Kostarev, and T. Yudina, *in present volume.*
116. Kh.S. Karimov, Kh.M. Akhmedov, and A.M. Achourov, *in present volume.*

CONVECTION IN CRYSTAL GROWTH
UNDER HIGH GRAVITY ON A CENTRIFUGE

J. Friedrich and G. Müller

Institut für Werkstoffwissenschaften
Universität Erlangen - Nürnberg
91058 Erlangen, Germany

ABSTRACT

We have theoretically analyzed buoyancy-driven convection in directional solidification configurations under the influence of additional forces acting on flow on a centrifuge, the centrifugal and the Coriolis forces. The influence of centrifugation on buoyancy-driven convection in the melt depends on the following conditions: the geometrical orientation of the melt cylinder on the centrifuge, the presence of radial temperature gradients in the melt (curved crystal-melt interface), the centrifuge radius, and the rotation rate of the centrifuge. The behaviour of convection depends on the complex interaction of buoyancy and Coriolis forces. It is demonstrated by theoretical considerations and experiments that a suppression of the vigor of convection up to one order of magnitude is possible under certain conditions on the centrifuge.

INTRODUCTION

Materials processing under high gravity using a centrifuge has been performed only in a few places. Some important results concerning this topic were obtained by the groups of Müller[1] and Rodot and Regel.[2] In their experiments, a furnace with a growth ampoule was pivoted on a centrifuge so that the symmetry axis of the furnace was parallel or antiparallel to the resultant acceleration **b** (vectorial sum of the centrifugal acceleration and the buoyancy acceleration) when the centrifuge rotated. Müller et al.[1] showed theoretically and experimentally that in 'top-seeded' vertical Bridgman growth, the Coriolis force acts on the fluid flow in a centrifuge and is responsible for a transition from an oscillatory convection regime to a stationary one. Rodot and Regel[2] performed solidification experiments with PbTe:Ag on two different centrifuges using a "stabilizing temperature gradient." They observed a uniform dopant distribution in crystals grown at a certain rotation rate, while experiments performed without rotation or at other rotation rates resulted in non-uniform concentration profiles. Such homogeneous profiles are expected only when the intensity of convection is weak compared to the freezing rate. Such growth conditions were achieved, for example, during growth under microgravity conditions, where the driving force for convection, i.e. the buoyancy force, is very small. In contrast to growth under microgravity conditions, one would actually expect that on a centrifuge the vigor of convection should be increased because the buoyancy force is increased by centrifugation. Up to now, only a few attempts have been made to explain the surprising results of Rodot and Regel.[3,4] In this paper, we systematically analyze the influence of centrifugation on convection in cylindrical metal and semiconductor melts.

THEORETICAL MODEL

Description of the Idealized Growth Configuration

The idealized growth configuration used in our theoretical analysis is shown in Figure 1. The analysis is restricted to a cylindrical melt region with radius r and height h.[a] For simplification of the heat transfer in the furnace, we assume a constant temperature gradient $\Delta T/h$ along the inner wall of the melt's container. In general, a radial temperature gradient also exists near the solidification front in real gradient-freeze configurations, resulting in a concave liquid-solid interface. This effect is considered in our model by using an isothermal paraboloidal as a curved bottom for the melt volume. It is characterized by the maximum interface deflection Δx_i. The top is assumed to be flat and isothermal.

The growth furnace is pivoted on the centrifuge arm, so that the symmetry axis of the furnace is parallel to the resulting gravity at its centre of mass while the centrifuge rotates around its vertical axis with angular velocity ω. The distance between the rotation axis and the bearing point of the furnace is the centrifuge radius R_a. The deviation of the furnace from its vertical position without rotation ($\omega = 0$) is described by the swinging angle θ, which depends on the rotation rate ω, on the centrifuge radius R_a, and on the distance between the bearing point of the furnace and its centre of mass x_s. From geometrical considerations (see Figure 1), the swinging angle θ is given by:

$$R \cdot \omega^2 = g \cdot \tan\theta \tag{1}$$

where

$$R = R_a + x_S \cdot \sin \theta$$

In order to calculate the resulting gravity field **b** in the melt volume, we used the Cartesian coordinate system x, y z shown in Figure 1. The origin of this coordinate system is the centre of mass of the furnace. Thus the acceleration vector is:

$$\mathbf{b} = \mathbf{g} - \left(\omega \times (\omega \times (\mathbf{r} + \mathbf{R}))\right) \tag{2}$$

where $\mathbf{g} = g \cdot [-\cos(\theta), -\sin(\theta), 0]$ is the vector of the normal earth acceleration, the vector $\mathbf{R} = R \cdot [-\sin(\theta), \cos(\theta), 0]$ is oriented from the rotation axis to the origin of the coordinate system, $\omega = \omega \cdot [\cos(\theta), \sin(\theta), 0]$ is the vector of the angular velocity, and $\mathbf{r} = [x, y, z]$ is the vector from the origin to any position in the melt.

---

[a] All symbols and characteristic numbers are defined in Reference 14, in the present volume.

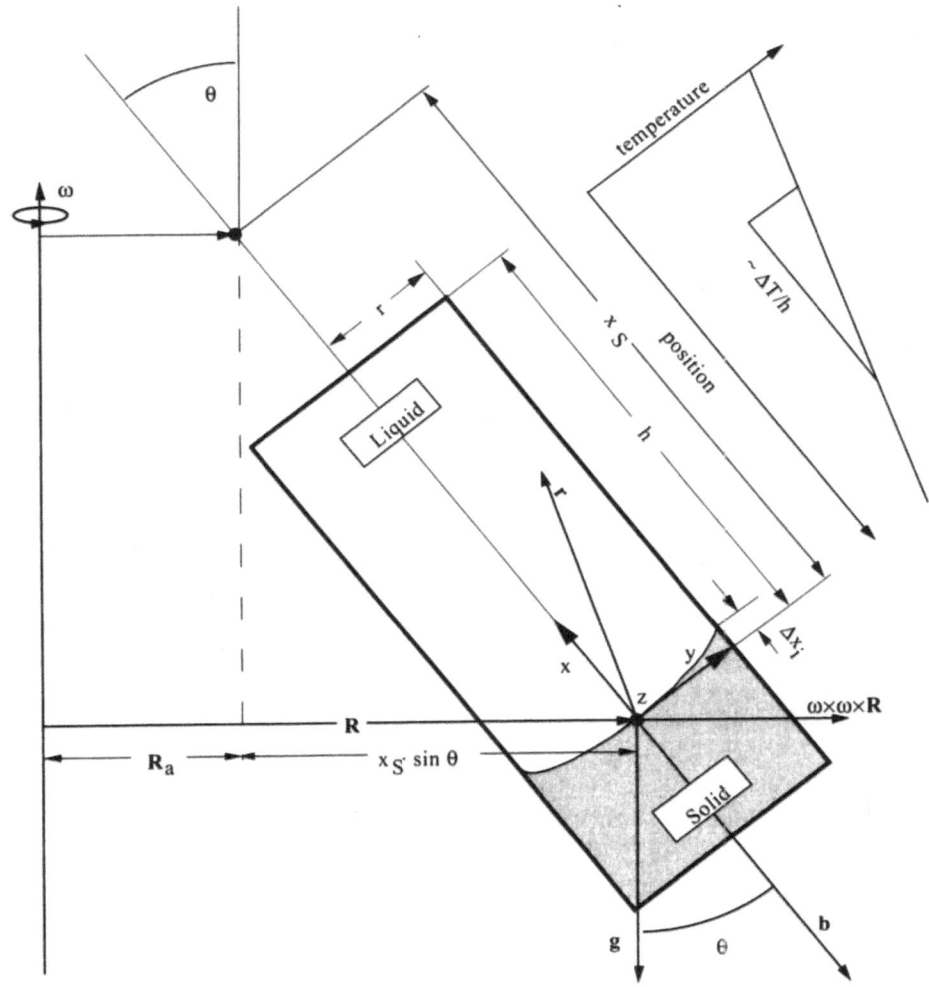

Figure 1. Idealized growth configuration and coordinate system.

Taking into account the geometrical relations between **g**, **R**, and ω with respect to the coordinate system, the components of the resulting acceleration at position **r** can be written:

$$b_x = \omega^2 \cdot x \cdot (\sin\theta)^2 - \omega^2 \cdot y \cdot \sin\theta \cdot \cos\theta - \omega^2 \cdot R \cdot \sin\theta - g \cdot \cos\theta \qquad (3a)$$

$$b_y = \omega^2 \cdot y \cdot (\cos\theta)^2 - \omega^2 \cdot x \cdot \sin\theta \cdot \cos\theta + \omega^2 \cdot R \cdot \cos\theta - g \cdot \sin\theta \qquad (3b)$$

$$b_z = \omega^2 \cdot z \qquad (3c)$$

The absolute value of the resultant acceleration N·g at the centre of mass ($x = y = z = 0$) is then simply given by:

$$N \cdot g = R\,\omega^2 \sin(\theta) + g\cos(\theta) \qquad (4)$$

19

Considering the acceleration field (Equation 3) and the Coriolis acceleration ($2\omega \times \mathbf{v}$), the equations governing mass, momentum and heat transport in a fluid on a centrifuge can be written, while applying the usual classical assumptions (Newtonian fluid, Boussinesq approximation, etc.):[1,3,5]

$$\nabla \cdot \mathbf{v} = 0 \tag{5}$$

$$-(1/\rho_0)\nabla P - (\mathbf{v} \cdot \nabla)\mathbf{v} + \nu \nabla^2 \mathbf{v} - \beta \cdot (T-T_0) \cdot [\mathbf{g} - (\omega \times (\omega \times (\mathbf{r} + \mathbf{R})))] - 2 \cdot (\omega \times \mathbf{v}) = \partial \mathbf{v}/\partial t \tag{6}$$

$$-(\mathbf{v} \cdot \nabla)T + \kappa \nabla^2 T = \partial T/\partial t \tag{7}$$

where $\mathbf{v}$ is the fluid velocity vector, P the pressure, T the temperature, t the time, ρ_0 the fluid density at reference temperature T_0, β the thermal expansion coefficient, ν the kinematic viscosity, κ the thermal diffusivity and ∇ the differential operator.

We used the following two methods to calculate the influence of centrifugation on buoyancy-driven convection.

Numerical Method

First we solved numerically the transport equations with the code FASTEST developed at the Institute of Fluid Mechanics of our university. FASTEST is a finite volume code using a non-orthogonal grid and a pressure correction with the SIMPLE algorithm. Numerical details of FASTEST are explained elsewhere.[6,7]

For our three-dimensional, time-dependent calculations, in which the no-slip condition for the flow velocity was applied at all boundaries, we used a grid size of 52 x 18 x 36 control volumes in the axial, radial and azimuthal directions of the melt cylinder. Simulations calculated with a finer grid size (78 x 28 x 58 control volumes) for some specific cases showed no significant differences in comparison to results obtained with the coarser grid. These numerical solutions were performed on a CRAY YMP and always resulted in a steady flow.

Order of Magnitude Analysis

In the literature, especially in the publications of Camel and Favier,[8] the influence of the buoyancy force on convection has been studied by order of magnitude analyses. In such analyses it is usually assumed that in the case of metal and semiconductor fluids the temperature field and the velocity field can be treated separately because the heat transport is mainly conductive. To estimate the influence of centrifugation on convection with an order of magnitude analysis, our approach is slightly modified in comparison to Camel and Favier, where the centrifugal force and the Coriolis force were not taken into account.

First we apply the curl operator to the stationary momentum equation (Equation 6). Noting that $\nabla \times (\nabla P) = 0$ and $\nabla \times (\nabla^2 \mathbf{v}) = -\nabla \times (\nabla \times (\nabla \times \mathbf{v}))$:[9]

$$-\nabla \times ((\mathbf{v}\nabla)\mathbf{v}) - \nu \nabla \times (\nabla \times (\nabla \times \mathbf{v})) + \beta [\mathbf{b} \times \nabla T] - 2\nabla \times (\omega \times \mathbf{v}) = 0 \tag{7}$$

Next we normalize Equation 7 with a proper set of reference quantities. We introduce the radius r of the ampoule and the *a priori* unknown velocity v_{ref} as the characteristic length and velocity in Equation 7. The vector angular velocity ω in the Coriolis term and the buoyancy term are normalized by the absolute values, $|\omega| = \omega$ and $|b \times \nabla T|$. The relations between the dimensional and non-dimensional variables (marked with a superscript °) are:

$$\mathbf{r} = r \cdot \mathbf{r}^\circ \ , \ \mathbf{v} = v_{ref} \cdot \mathbf{v}^\circ \ , \ \mathbf{b} \times \nabla T = |\mathbf{b} \times \nabla T| \cdot (\mathbf{b} \times \nabla T)^\circ \ , \ \boldsymbol{\omega} = \omega \cdot \boldsymbol{\omega}^\circ \tag{8}$$

With this normalization, Equation 7 can be transformed to:

$$\frac{v_{ref}^2}{r^2} \nabla^\circ \times \left((\mathbf{v}^\circ \nabla^\circ) \mathbf{v}^\circ \right) + \frac{v v_{ref}}{r^3} \nabla^\circ \times \left(\nabla^\circ \times (\nabla^\circ \times \mathbf{v}^\circ) \right) + \frac{2\omega v_{ref}}{r} \nabla \times (\boldsymbol{\omega}^\circ \times \mathbf{v}^\circ) = \beta \left| \mathbf{b} \times \nabla T \right| \left(\mathbf{b} \times \nabla T \right)^\circ \tag{9}$$

The relative importance of the forces on the characteristic flow velocity v_{ref} is given by the dimensional coefficients in front of each non-dimensional term, which should be on the order of one.[8,10] For this purpose, it is assumed that the only driving action is the buoyancy force.[3,8,10]

$$\left(\frac{v_{ref} \ r}{v} \right)^2 + \left(\frac{v_{ref} \ r}{v} \right) + \left(\frac{v_{ref} \ r}{v} \right) \cdot \left(\frac{2 \omega r^2}{v} \right) = \frac{\beta \left| \mathbf{b} \times \nabla T \right| r^4}{v^2} \tag{10}$$

Next, we introduce the non-dimensional Reynolds, Grashof and Taylor numbers:

$$Re = \frac{v_{ref} \cdot r}{v} \qquad Gr = \frac{\beta \ (\Delta T/h) (N \ g) \ r^4}{v^2} \qquad Ta = \frac{4 \cdot r^4 \cdot \omega^2}{v^2} \tag{11}$$

where $\Delta T/h$ is the axial temperature gradient. It follows with these relations, from Eq. 10:

$$Re^2 + Re(1 + Ta^{0.5}) = Gr \frac{|\mathbf{b} \times \nabla T|}{(Ng)(\Delta T / h)} \tag{12}$$

With the requirement $Re > 0$, the solution to this problem is given by:

$$Re = -\frac{1 + Ta^{0.5}}{2} + \sqrt{ \left(\frac{1 + Ta^{0.5}}{2} \right)^2 + Gr \cdot \frac{|\mathbf{b} \times \nabla T|}{(Ng)(\Delta T / h)} } \tag{13}$$

if $|\mathbf{b} \times \nabla T|$ is known. In the next step we approximate the vector ∇T by the characteristic temperature gradients in the axial and non-axial directions. The first one is given by $\Delta T/h$. The temperature differences in the y and z directions are assumed to be proportional to the product of interface deflection Δx_i and axial temperature gradient $\Delta T/h$, as indicated by Motakef.[11] With these approximations:

$$\nabla T \approx \frac{\Delta T}{h} \cdot \left[1, \frac{\Delta x_i}{r}, \frac{\Delta x_i}{r} \right] \tag{14}$$

Next, we approximate the resultant acceleration $\mathbf{b}$ (Equation 3) by its characteristic components in each direction. For this purpose, we consider two configurations of the cylinder of melt on the centrifuge arm. In setup A, the furnace is pivoted on the centrifuge arm as shown in Figure 1 (Equation 1 is valid). In setup B, the cylinder is fixed on the centrifuge arm so that the cylinder axis is always perpendicular to the rotation axis at any rotation rate ($\theta = 90°$ in Equation 3, and Equation 1 is not valid). For these two growth configurations - pivoted (setup A) and fixed mounted furnace (setup B) - the characteristic components of the resulting acceleration are different. Using the condition $R \gg h > r$, we find for configurations A and B:

$$
\mathbf{b} \approx \begin{pmatrix} -\omega^2 R (\sin\theta) - g(\cos\theta) \\ -\omega^2 h (\sin\theta)(\cos\theta) \\ \omega^2 r \end{pmatrix}
\qquad
\mathbf{b} \approx \begin{pmatrix} -\omega^2 \cdot R \\ -g \\ \omega^2 \cdot r \end{pmatrix}
\tag{15}
$$

Using Equations 14 and 15 and noting that $\Delta x_i \ll r$, we get for setup A: $\dfrac{|\mathbf{b} \times \nabla T|}{(\Delta T / h)} \approx$

$$
\sqrt{\left(r\omega^2 + (\Delta x_i / r)(R\omega^2 \sin\theta + g\cos\theta)\right)^2 + \left(g(h/R)\sin^2\theta - (\Delta x_i / r)(R\omega^2 \sin\theta + g\cos\theta)\right)^2}
$$

while for setup B: $\dfrac{|\mathbf{b} \times \nabla T|}{(\Delta T / h)} \approx \sqrt{\left(r\omega^2 + (\Delta x_i / r)R\omega^2\right)^2 + \left(g - (\Delta x_i / r)R\omega^2\right)^2}$

These results are put into Equation 11 in order to calculate the non-dimensional characteristic flow velocity Re from our order of magnitude analysis.

RESULTS

We have considered three different cases. In cases 1 and 2, molten gallium is considered with the same geometry and thermal boundary conditions. The "interface" is flat in both cases. The only difference between case 1 and case 2 is that in case 1 the containing cylinder is fixed (setup B) while in case 2 the ampoule can swing out (setup A). In case 3, the crucible with molten germanium is pivoted as in case 2, but the interface is curved. For case 3 we have considered different lengths of the centrifuge arm. An overview of the geometrical, thermal and centrifugal parameters is given in Table 1. The material properties of Ga and Ge are summarized in Table 2.

Table 1. The geometrical and centrifugal parameters and thermal boundary conditions used in the calculations.

case	setup	melt	r [cm]	h [cm]	Δx_i [mm]	ΔT [K]	R [m]
1	B	Ga	0.5	2.0	0.0	50	18.02
2	A	Ga	0.5	2.0	0.0	50	18.0 + 0.02 sinθ
3	A	Ge	1.0	4.0	0.1	10	0.5 (1 and 5.5) + 0.16 sinθ

In Figure 2 (case 1 left; case 2 right) and Figure 3 (case 3), the characteristic flow velocity v_{ref} calculated with Equations 11 and 16 is shown as a function of the rotation rate ω (solid lines). The symbols correspond to the maximum flow velocity obtained from our three-dimensional numerical simulations. The numerical results of Ramachandran[12] are also shown in Figure 2 for case 1. Ramachandran performed these numerical calculations with the same parameters as indicated in table 1 for case 1. In contrast to our simulations where the whole gravity field was considered, Ramachandran used only the characteristic value of the centrifugal force in the axial direction of the melt cylinder.

Table 2. Properties of molten Ga and Ge used in the calculations. The data for Ga were taken from Ramachandran[12] and for Ge from Camel and Favier.[8]

Property of melt	Ga	Ge
density ρ [g/cm^3]	6.1	5.5
thermal diffusivity κ [cm^2/s]	0.15	0.15
kinematic viscosity ν [10^{-3}cm^2/s]	3.14	1.35
expansion coefficient β [10^{-4}cm^2/s]	1.26	1.0

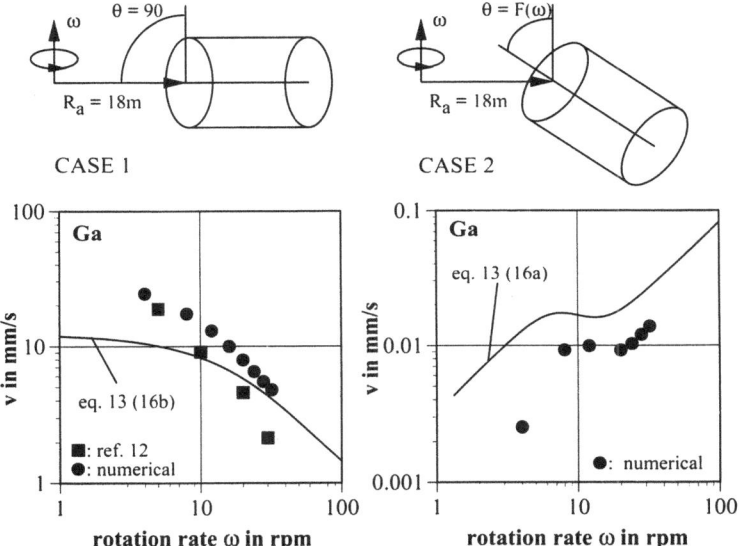

Figure 2. Dependence of the characteristic flow velocity on the rotation rate ω for case 1 (left) and case 2 (right). The solid lines are calculated with Equations 13 and 16. The symbols correspond to the maximum flow velocity resulting from our three-dimensional numerical calculations and Reference 12.

In all three cases, the numerical results for the different configurations agree with the approximations given by Equations 13 and 16. But the behaviour of the flows is completely different for the three cases. For case 1, the flow velocity decreases with increasing rotation rate, while for case 2 the convection intensity increases. The absolute value of the flow velocity for case 1 is several orders of magnitude higher than for case 2 at the same rotation rate.

Figure 3 shows that a minimum occurs in the vigor of convection for case 3. For small values of ω, the flow velocity decreases with increasing rotation rate ω. For this branch, the flow velocity is independent of the centrifuge radius. Then, at higher rotation rates the intensity of the flow increases again with increasing rotation rate. For this branch, the flow velocity increases with increasing length of the centrifuge arm.

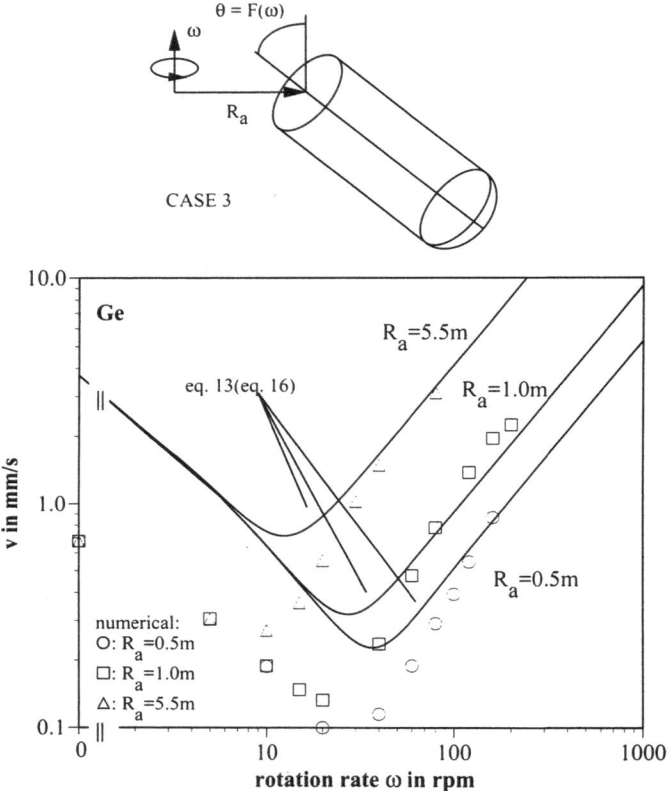

Figure 3. Dependence of the characteristic flow velocity on the rotation rate ω for case 3. The solid lines are calculated with Equations 13 and 16. The symbols correspond to the maximum flow velocity resulting from our three-dimensional numerical calculations.

The flow patterns in molten germanium (case 3) for R = 1m at rotation rates ω = 0, 40 and 80 rpm are shown in Figure 4. The Coriolis force causes a large convective cell near the curved interface. The rotation sense of the flow in this cell is equal to the rotation sense of the centrifuge. The dimension of this cell in the axial direction seems to decrease with increasing rotation rate. Beside this large cell, a very small convective cell exists near the curved bottom for ω = 40 rpm. The rotation sense of the flow in this cell is opposite to that of the centrifuge. This theoretically predicted flow pattern is confirmed by the flow visualization experiments of Skudarnov et al.[13] In these experiments, the flow pattern in water was visualized by a laser-cut technique in a test cell with similar geometry and thermal boundary conditions to those we used in our numerical calculations.

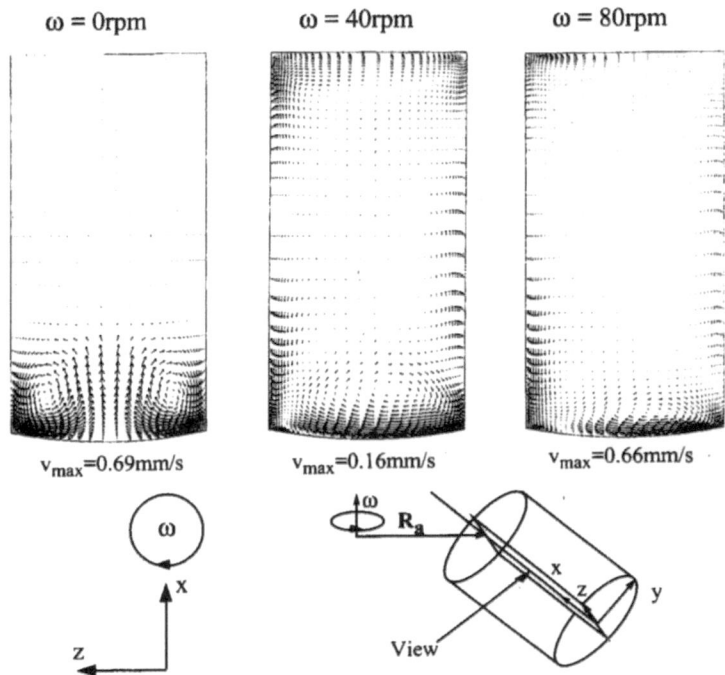

Figure 4. Numerically calculated flow patterns in the x-z plane in molten germanium (case 3, R = 1m) at rotation rates ω = 0, 40 and 80rpm.

DISCUSSION

The only driving force for convection is the buoyancy force, even in a centrifuge.[3,5,10] It has been shown that $\mathbf{b} \times \nabla T$ is the scale for the buoyant driving force.[3,10] We can split $\mathbf{b} \times \nabla T$ into two terms, $\mathbf{b}_{\parallel} \times \nabla T_{\perp}$ and $\mathbf{b}_{\perp} \times \nabla T_{\parallel}$, where the subscripts $\parallel$ and $\perp$ denote the components of the vectors in the axial and non-axial directions. Thus the driving force for buoyancy-driven convection is proportional to the sum of:

Effect a): the product of the axial acceleration component and the non-axial temperature gradient.

Effect b): the product of the non-axial acceleration component and the axial temperature gradient.

The buoyant driving force is zero for normal growth conditions (without rotation) only in a vertical configuration, when a linear stabilizing temperature gradient is applied along the crucible wall, and if all isotherms in the melt, including the interface, are perpendicular to the gravity vector. These conditions are valid for case 2 at ω = 0. Therefore, the flow velocity approaches zero as ω→0, as shown in Figure 2. For case 3, without rotation the curved isothermal interface ($\Delta x_i \neq 0$) indicates non-axial temperature gradients near the solidification front. These radial temperature gradients cause convection (effect a). For case 1 (without rotation), the earth's gravity is perpendicular to the axis of the melt cylinder. Therefore convection occurs always in this horizontal case, because of effect b, if an axial temperature gradient exists and even if the interface is flat.

The momentum resulting from the buoyancy force must be preserved according to the Navier-Stokes Equation (Equation 6). Without rotation, the buoyant driving force is balanced by either the diffusive force when the buoyancy force is small (Stokes regime) or by the inertia force when the buoyancy force is large (boundary layer regime).[8] It has been shown[3,10] that for the parameters that we used in our theoretical analysis, the Coriolis force in the momentum equation is the largest reacting force. The Coriolis force is much more important than the diffusive or inertial forces. Therefore, we assume that the buoyant driving force is in balance with the Coriolis force. Both forces depend on conditions in the centrifuge, especially on the rotation rate ω. For a first approximation, we assume that the Coriolis force is proportional to the product of flow velocity v_{ref} and rotation rate ω. The buoyancy force depends on the geometrical configuration of the melt cylinder on the centrifuge, on the centrifuge radius, and on the rotation rate. This dependence is complex and is discussed below.

For case 1, the flow must be driven by the non-axial acceleration component $\mathbf{b}_\perp$ (effect b), because the interface is assumed to be flat and thus radial temperature gradients are negligible near the interface. In this case, however, $\mathbf{b}_\perp$ is equal to earth's gravity for all rotation rates because the cylinder axis is held perpendicular to earth's gravity. The non-axial acceleration component caused by the non-uniform centrifugal field are much smaller than earth's gravity. Thus, the buoyant driving force is constant and independent of the rotation rate. To keep the constant driving force in balance with the Coriolis force ($\omega \cdot v$), the flow velocity must decrease with increasing rotation rate.

For case 2, the flow must also be driven by the non-axial acceleration component (effect b), as in case 1, because the interface is also assumed to be flat. But $\mathbf{b}_\perp$ is not earth's gravity as in case 1. For this pivoting melt cylinder, the characteristic value of the axial component of the resulting gravity is that at the centre of mass. The inhomogeneous centrifugal field always causes components of the resulting gravity in the radial direction in the melt cylinder, proportional to ω^2, except at the centre of mass. Therefore a good scale for $\mathbf{b}_\perp$ is the maximum difference of the centrifugal force in the melt volume. Thus, it follows from the balance between the driving force (ω^2) and the Coriolis force ($\omega \cdot v$) that the flow velocity must increase with increasing rotation rate.

From these considerations we see why, for example, the flow velocity in case 1 is approximately 1000 times higher than in case 2 at a rotation rate of 13 rpm. The reason is that the driving force for case 1 (proportional to 1 g) is also 1000 times higher than the driving force for case 2, which is proportional to $r \cdot \omega^2$ as a first approximation.

For case 3 (pivoted melt cylinder, curved interface) both mechanisms a or b have to be considered. But if we compare the scale of both mechanisms (see Equation 17), we see that for a typical parameter set ($r = 1$ cm, $\Delta x_i = 1$ mm, $\omega = 40$ rpm, $R = 1$ m) effect a ($\mathbf{b}_\parallel \times \nabla T_\perp$) should be much more important than effect b ($\mathbf{b}_\perp \times \nabla T_\parallel$):

$$\frac{\text{Effect} - a}{\text{Effect} - b} = \frac{\left| \mathbf{b}_{\parallel} \times \Delta T_\perp \right|}{\left| \mathbf{b}_\perp \times \Delta T_{\parallel} \right|} = \frac{(N \cdot g)(\Delta x_i / r)}{r \omega^2} > 10 \tag{17}$$

At high rotation rates ($R \omega^2 \sin\theta \gg g \cos\theta$), the influence of the buoyancy force due to earth's gravity can be neglected. Then the flow is mainly driven by the product of the centrifugal force acting in the axial direction of the melt cylinder and the non-axial temperature gradient, as manifested by the curved interface. Thus, for the centrifugal force ($R \cdot \omega^2$) to be in balance with the Coriolis force ($\omega \cdot v$), the flow velocity must be proportional to the product of centrifuge radius and rotation rate ($R \cdot \omega$), and therefore increase with increasing rotation rate for a constant arm length.

Our approach also describes the influence of the centrifuge arm length on the vigor of convection. At a rotation rate of 80 rpm, for example, the ratio of the maximum flow velocities obtained from our numerical calculation is approximately equal to the ratio of the centrifuge arm lengths.

At low rotation rates ($R \omega^2 \sin\theta \ll g \cos\theta$), the additional centrifugal acceleration is only a very small fraction of the resulting acceleration acting in the axial direction of the melt cylinder because the swinging angle θ of the furnace is relatively small (e.g., for $R = 1m$ and $\omega = 15$ rpm: $g \cos\theta \approx 0.93\ g$ and $R \omega^2 \sin\theta \approx 0.07\ g$). Thus, earth's acceleration g remains the only important axial acceleration. Under this condition, the buoyant driving force can be considered independent of the rotation rate. To keep the (constant) driving force in balance with the Coriolis force, the flow velocity must decrease with increasing rotation rate. Therefore, in this range of rotation rate, the behaviour of the flow velocity must be independent of the centrifuge radius R because both forces, buoyant and Coriolis, are independent of R.

The two effects above cause a minimum in convection intensity. The suppression of convection is up to one order of magnitude for our parameter set. This convective minimum was confirmed by our experimental results.[10,14] In these experiments, gallium-doped germanium crystals were grown on a centrifuge at different rotation rates. The axial concentration profiles in these crystals reflect the theoretically predicted behaviour for convection. We believe that the occurrence of this convective minimum is also responsible for the experimental results of Rodot and Regel.

CONCLUSION

The influence of centrifugation on buoyancy-driven convection is very complex. The resulting effect depends on the configuration of the melt cylinder on the centrifuge, on the centrifuge radius, and on the rotation rate. In all cases considered here, the behaviour of the flow can be described by the interaction of the buoyant driving force and the Coriolis force. The main result for directional solidification on centrifuges, is the existence of a range of parameters where convection is reduced in comparison to earth's gravity. At the convective minimum the suppression of flow velocity can be up to one order of magnitude for the growth parameters considered. Thus, a centrifuge can be used to study the influence of reduced convection as well as enhanced convection on the growth of crystals.

Acknowledgment

This work was supported by the Bundesministerium für Bildung und Forschung (BMBF) under the project management of the German space agency (DARA), contract no. 50WM9301.

REFERENCES

1. G. Müller, G. Neumann, and W. Weber, The growth of homogeneous semiconductor crystals in a centrifuge by the stabilizing influence of the Coriolis force, *J. Crystal Growth* 129:8 (1992).
2. H. Rodot, L.L. Regel, and A.M. Turtchaninov, Crystal growth of IV-VI semiconductors in a centrifuge, *J. Crystal Growth* 104:280 (1990).
3. V.A. Urpin, Convective flows during crystal growth in a centrifuge, *in*: "Materials Processing in High Gravity," L.L. Regel and W.R. Wilcox, eds., Plenum Press, New York (1994) 35.
4. W.A. Arnold and L.L. Regel, Thermal stability and the suppression of convection in a rotating fluid on earth, *ibid*, p. 17.

5. W.A. Arnold, W.R. Wilcox, F. Carlson, A. Chait, and L.L. Regel, Transport mode during crystal growth in a centrifuge, *J. Crystal Growth* 129:24 (1992).

6. H.J. Leister and M. Peric, Vectorized strongly implicit solving procedure for a seven-diagonal coefficient matrix, *Int. J. Num. Meth. Heat Fluid Flow* 4:159 (1994).

7. H.J. Leister, "Numerische Simulation dreidimensionaler zeitabhängiger Strömungen unter dem Einfluß von Auftriebs- und Trägheitskräften," Ph.D. Thesis, University Erlangen Nürnberg (1994).

8. D. Camel and J.J. Favier, Scaling analysis of convective solute transport and segregation in Bridgman crystal growth from the doped melt, *J. Physique* 47:1001 (1986).

9. I.N. Bronstein and K.A. Semendjajew, "Taschenbuch der Mathematik," Verlag Harri Deutsch, Thun und Frankfurt/Main (1987)

10. J. Friedrich, J. Baumgartl, H.J. Leister, and G. Müller, Experimental and theoretical analysis of convection and segregation in vertical Bridgman growth under high gravity on a centrifuge, *J. Crystal Growth*, in press.

11. S. Motakef, Interference of buoyancy-induced convection with segregation during directional solidification: scaling laws, *J. Crystal Growth* 102:197 (1990).

12. N. Ramachandran, J.P. Downey, P.A. Curreri, and J.C. Jones, Numerical modeling of crystal growth on a centrifuge for unstable natural convection, *J. Crystal Growth* 136:655 (1993).

13. P. Skudarnov, L.L. Regel, and W.R. Wilcox, *In - situ* observation of convection on the centrifuge, *in present volume*.

14. J. Friedrich and G. Müller, Segregation in crystal growth under high gravity on a centrifuge: a comparison between experimental and theoretical results, *in present volume*.

SEGREGATION IN CRYSTAL GROWTH
UNDER HIGH GRAVITY ON A CENTRIFUGE:
A COMPARISON BETWEEN EXPERIMENTAL
AND THEORETICAL RESULTS

J. Friedrich and G. Müller

Institut für Werkstoffwissenschaften LS 6
Universität Erlangen - Nürnberg
Erlangen, Germany

ABSTRACT

Germanium crystals doped with gallium were grown by the gradient freeze technique on a centrifuge at different rotation rates and with different lengths of the centrifuge arm. The resulting dopant distributions of these crystals indicate that convection was reduced during growth at some certain rotation rates, resulting in an effective segregation coefficient closer to one than for normal growth conditions. The experimental results agree with the theoretical prediction of a modified segregation model using a special boundary layer (Ekman layer) that occurs on a centrifuge. The results with respect to the uniformity of the crystals grown under high gravity on a centrifuge are compared with results obtained under microgravity.

INTRODUCTION

In previous papers,[1,2] we theoretically studied the influence of centrifugation on buoyancy-driven convection during crystal growth. We predicted that the vigor of convection can be reduced even under high gravity on a centrifuge. It was shown that for our idealized growth configuration, the flow velocities in molten germanium should be reduced by up to one order of magnitude over a certain range of the rotation rate of the centrifuge.

In the field of crystal growth, the influence of convection on species transport is normally discussed in terms of an effective segregation coefficient K_{eff}, which is the ratio between solute concentration in the solid and that in the melt. With the following equation, we can define a dimensionless segregation parameter Δ that is a function of the growth velocity V_G, diffusion coefficient D, and convection:[2,3]

$$K_{eff} = \frac{K_0}{1 - (1 - K_0) \cdot \Delta} \qquad (1)$$

where K_0 is the equilibrium segregation coefficient (at $V_G = 0$).

Figure 1 shows two limiting cases expressed in terms of K_{eff}. With strong convection, the axial dopant distribution in the crystal is non-uniform. Such segregation profiles are characterized by $K_{eff} = K_0$, which corresponds to $\Delta = 0$. With negligible convection, the dopant distribution in the middle part of the crystal is uniform. In this case, $K_{eff} = 1$, which corresponds to $\Delta = 1$. Such growth conditions have been obtained, for example, in microgravity and with a strong magnetic field imposed on the melt during solidification.

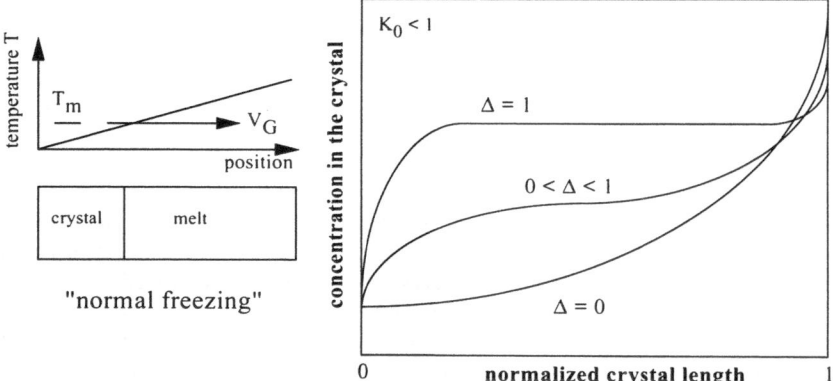

Figure 1. Influence of convection on the axial dopant distribution. The segregation profile is non-uniform and characterized by $\Delta = 0$ when the convection is very strong. Under growth conditions with weak convection, the dopant distribution is homogeneous in the middle part of the crystal, corresponding to $\Delta = 1$.

The aim of the present work was to determine whether one can achieve a more uniform solute distribution during crystal growth on a centrifuge compared to normal growth conditions. (We would expect that the dopant distribution in crystals grown at such rotation rates where we predict a minimum[1,2] of the convection to correspond to a value of K_{eff} closer to 1.) Furthermore we investigated if it is possible to grow crystals under high gravity dopant distributions similar to those obtained under microgravity conditions.

We chose the well-known system gallium-doped germanium for our growth experiments. This material has often been used in microgravity experiments, and has become a reference system for the study of the influence of convection on segregation. All material parameters needed for calculations are well known, e.g. the diffusion coefficient D of Ga in molten Ge at the melting temperature equals $1.9 \times 10^4 \text{cm}^2/\text{s}$, and the equilibrium segregation coefficient K_0 is 0.087.[3] This is a great advantage in comparison to materials like Ag-doped PbTe, which was used in the experiments of Rodot and Regel,[4] where these properties are unknown.

EXPERIMENTAL METHODS

The setup for our growth experiments was already described in detail.[5] The centrifuge consists of a horizontal double arm (maximum 2.6m) rotating around a vertical axis. The rotation speed ω can be up to 250 rpm.

The growth experiments were carried out in a furnace designed to resist up to 50 g (g = earth's gravity). The furnace with a water-cooled aluminum case has six separate resistance heated zones. The heating wires is coiled around a ceramic tube in a thread. A maximum temperature of about 1200°C is available in the furnace. The voltages from the Pt - PtRh10% thermocouples placed in each heating zone are amplified on the centrifuge to obtain a good signal to noise ratio. These amplified voltages and the heating power are transferred via slip rings to the control unit of the furnace outside the centrifuge. The furnace is mounted in a way that it can swing out along its vertical axis in the direction of the resulting acceleration, as in the configuration we treated theoretically.[1-2] The distance between the bearing point of the furnace and the rotation axis of the centrifuge R_a can be varied between 60cm and 1m.

A sealed silica ampoule with a graphite crucible containing a <100> oriented, undoped germanium seed, the polycrystalline material, and the dopant gallium (c_0 about 5×10^{18} atoms/cm^3) was placed in the furnace. The top of the seed was positioned approximately at the centre of mass of the furnace, with the distance between the bearing point of the furnace and the centre of mass being about 17 cm. During growth, an axial stabilizing temperature gradient $\Delta T/h$ of about 10 K/cm was produced in the melt. The crystals were grown by the "Vertical" Gradient Freeze technique. The growth rate V_G was approximately 5.6 μm/s.

EXPERIMENTAL RESULTS AND QUALITATIVE ANALYSIS

We grew six Ge:Ga crystals (radius r about 10 mm, length h of 40 - 60 mm) under the same thermal conditions using a centrifuge radius R_a of 1 m and rotation rates ω of 0, 20, 30, 40, 50 and 65 rpm, which was kept constant during a growth run. One additional experiment was performed at a rotation rate ω of 50 rpm and a centrifuge radius R_a of 64 cm. Five crystals were totally single crystalline. The other two crystals were only polycrystalline at the very end of the ingots. The position of the seeding solid-liquid interface was approximately the same in all growth runs. These initial interfaces had in all cases a concave shape towards the melt. The maximum bending of the initial phase boundary Δx_i was about 0.8 mm for all crystals. The axial dopant distribution of the crystals was determined from conductivity measurements using the four point probe method by the relations given by Favier et al.[6]

Figure 2 shows the normalized, radially averaged, dopant concentration versus the solidified fraction for crystals grown at a radius R_a of 1 m and at different rotation rates. The measured profiles were analyzed according to the theory of Favier[7] to determine the corresponding segregation parameter Δ. In this model, the one-dimensional, time-dependent continuity equation for solute in the liquid is solved. Following Favier,[7] the segregation profile can be calculated independent of the *a priori* unknown segregation parameter Δ only. The other parameters in this model are material properties, especially the diffusion coefficient of the solute in the melt D, the equilibrium segregation coefficient K_0, and the growth rate V_G. The solid lines in Figure 2 are calculated with this model for different values of Δ.

It can be seen in Figure 2 that the measured segregation profiles of the crystals grown

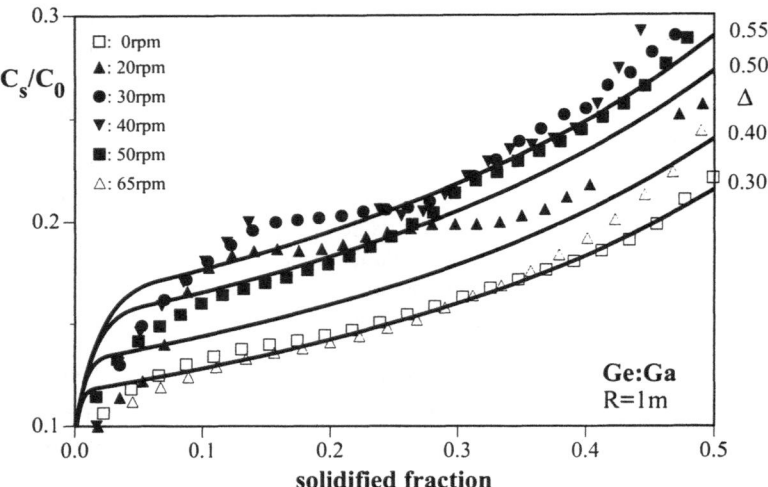

Figure 2. Gallium concentration C_S normalized by the initial concentration C_0 versus fraction solidified for Ge crystals grown during centrifugation at the rotation rates shown, with the growth conditions otherwise identical. The curves are the segregation profiles calculated with the model of Favier[7] for different values of the segregation parameter Δ.

at rotation rates of 0 rpm and 65 rpm are nearly identical. These two curves can be described by a value of the segregation parameter Δ of about 0.3. The dopant distributions of these two crystals differ distinctly from those grown at rotation rates of 20, 30, 40 and 50 rpm. These curves can be characterized by values of Δ greater than 0.5. This result confirms that the convection in the melt was reduced for our experimental growth parameters in the range of rotation rate between 20 rpm and 50 rpm. Therefore, our experimental results correspond qualitatively to the theoretical prediction given in our previous papers[1,2] for very similar growth conditions.

Figure 3 shows concentration profiles of crystals grown at the same rotation rate and different lengths of the centrifuge arm, together with profiles calculated with the model of Favier[7] for two different values of Δ. The measured distribution of the crystal grown at an R_a of 0.64 m can be described by a value of Δ of about 0.7, while the crystal grown at the longer arm length (R_a = 1 m) is characterized by a value of Δ of about 0.5. According to our theoretical prediction,[1] we expect this result because the flow is more suppressed when the centrifuge radius is shorter. All of these results confirm that convection can be reduced at certain rotation rates during growth on a centrifuge.

QUANTITATIVE DISCUSSION: SEGREGATION MODEL FOR CRYSTAL GROWTH ON A CENTRIFUGE

The results obtained so far can be used only for a qualitative analysis of the influence of centrifugation on the solute transport and segregation. Now we want to combine the models

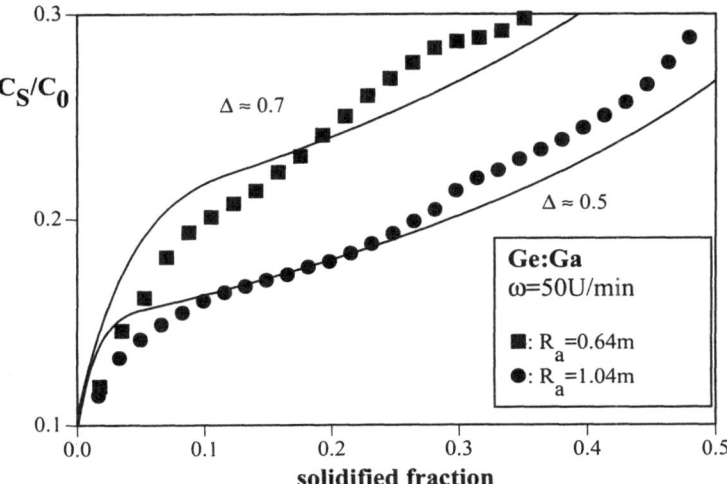

Figure 3. Gallium concentration C_S normalized by the initial concentration C_0 versus fraction solidified for crystals grown at two different centrifuge arm lengths and using the same growth conditions. The solid lines are concentration profiles calculated with the model of Favier[7] for two different values of Δ.

on convection given in the literature[1,2] with already existing segregation models, to obtain a quantitative description of mass transport in directional solidification under high gravity. For the analysis of solute segregation we use the one-dimensional segregation model of Ostrogorsky and Müller.[9] The effective segregation coefficient K_{eff} and the segregation parameter Δ_{OM} are computed by using the integral boundary layer analysis taking into account convection in the melt adjacent to the growth interface. According to the model of Ostrogorsky and Müller,[9] the relation between K_{eff} and the relevant parameters is:

$$K_{eff} = \frac{1 + \xi}{1 + \xi/K_0} = \frac{K_0}{1 - (1 - K_0) \cdot (\xi + 1)^{-1}} = \frac{K_0}{1 - (1 - K_0) \cdot \Delta_{OM}} \tag{2}$$

where

$$\xi = a \cdot \frac{v(x_c) \cdot x_c}{V_G \cdot L}$$

x_c is the diffusion boundary layer thickness, $v(x_c)$ is the flow velocity parallel to the interface at the edge of the boundary layer, and a is a constant (1/6 for a linear approximation and 1/7.2 for a cubic approximation to the velocity profile), and L is a characteristic length of the system, assumed to be equal to the ampoule radius r. After introducing the normalized solute boundary layer thickness $x^\circ_c = x_c/r$, the non - dimensional flow velocity $v^\circ(x^\circ_c) = v(x_c)\,r/v$, the Peclet number $Pe = V_G\,r/D$ and the Schmidt number $Sc = v/D$, we can write Δ_{OM} as:

$$\Delta_{OM} = \left(1 + a \cdot \frac{v(x^\circ_c) \cdot Sc \cdot x^\circ_c}{Pe}\right)^{-1} \tag{3}$$

Following Camel and Favier,[3] we assume that the velocity $v^o(x^o_c)$ at the edge of the boundary layer is related to the characteristic non-dimensional flow velocity Re in the melt. The Reynolds number Re can be calculated independent of the growth parameters using the scaling laws given in the literature for microgravity,[2,3] for normal earth's gravity,[2,3] and for high gravity conditions.[1,2] This velocity $v^o(x^o_c)$ is equal to the characteristic flow velocity Re if the solute layer thickness x^o_c is greater than the momentum boundary layer thickness $x^o_v = x_v/r$, i.e.:[2,3]

$$\text{If } x^o_c > x^o_v, \text{ then } v^o(x^o_c) = Re \qquad (4a)$$

On the other hand, if $x^o_c \leq x^o_v$ the ratio of $v^o(x^o_c)$ to the flow velocity $v^o(x^o_v) = Re$ at the edge of the momentum layer x^o_v should be proportional to the ratio of the boundary layer thicknesses x^o_c/x^o_v because of the no slip condition at the interface.[2,3,9]

$$\text{If } x^o_c \leq x^o_v, \text{ then } v^o(x^o_c) = Re \cdot x^o_c/x^o_v \qquad (4b)$$

Taking into account Equations 4a and 4b, we get the following relationships for Δ_{OM} from Equation 3. For $x^o_c \leq x^o_v$:

$$\Delta_{OM} = \left(1 + a \cdot \frac{Re \cdot Sc \cdot x^{o2}_c}{Pe \cdot x^o_v} \right)^{-1} \qquad (3a)$$

while for $x^o_c > x^o_v$:

$$\Delta_{OM} = \left(1 + a \cdot \frac{Re \cdot Sc \cdot x^o_c}{Pe} \right)^{-1} \qquad (3b)$$

Thus, to calculate Δ_{OM} the film thicknesses, x^o_v and x^o_c, must be estimated.

Formulation for momentum boundary layer thickness x^o_v without rotation ($\omega = 0$)

For normal Bridgman configurations without rotation ($\omega = 0$), the momentum boundary layer thickness x^o_v can be estimated by the scaling laws of Camel and Favier:[3]

$$\text{For } Gr < 1^{-6}_{s0}: \quad x^o_v = l_{s0} \qquad (5a)$$

$$\text{For } Gr \geq 1^{-6}_{s0}: \quad x^o_v = Re^{-0.5} \qquad (5b)$$

where l_{s0} is a factor (0.16 here) that takes into account the interaction between the crucible wall and the flow in the Stokes flow regime,[3] and Gr is the Grashof number as defined in the table of nomenclature. Jung and Müller[10] showed that these scaling laws agree very well with the momentum boundary layer thickness x^o_v determined from numerical calculations for an idealized vertical Bridgman configuration. Figure 4 compares the numerical results of Jung with the scaling law of Favier.

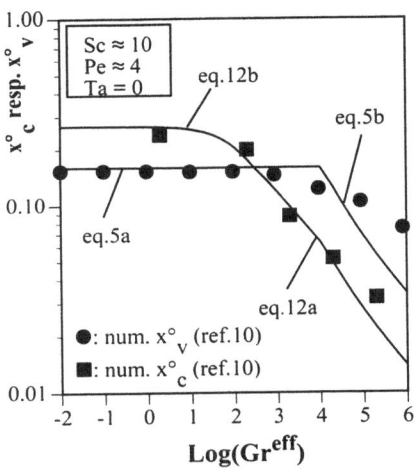

Figure 4. Momentum (x°_v) and solute (x°_c) boundary layer thicknesses versus $Gr^{eff} = Gr \cdot \Delta x_i/r$. The symbols are from numerical simulations.[10] The curves were calculated with Equation 5 for x°_v and with Equation 12 for x°_c (with Ta = 0).

Formulation for momentum boundary layer thickness x°_v with rotation ($\omega \neq 0$)

On a centrifuge, we have to consider the so-called Ekman or Taylor boundary layers near the solid-liquid interface because of the strong influence of the Coriolis force.[2,11] The thickness of the Ekman layer depends on the orientation of the wall to the rotation axis.[12] This is similar to the case of magnetic fields, where the thickness of the Hartmann layer depends on the orientation of the wall to the direction of the magnetic field.[13] Now we apply the results of Hide[12] for a momentum boundary layer in a rotating fluid annulus, as shown in Figure 5, to the boundary layer at a solidification front.

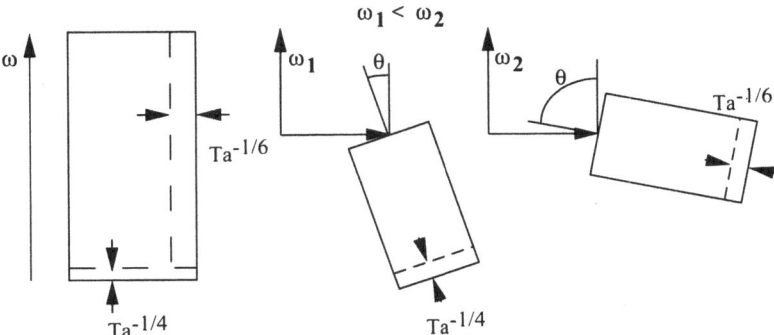

Figure 5. Scaling laws for the momentum boundary layer thickness in a rotating system. Left: rotating annulus.[12] Middle: Centrifuge at low rotation rate (solidification front approximately perpendicular to rotation axis). Right: Centrifuge at high rotation rate (solidification front approximately parallel to axis).

According to Hide,[12] at low rotation rates (depending on the centrifuge arm length R), the boundary layer x°_v is approximately perpendicular to the rotation axis of the centrifuge and, therefore, proportional to $Ta^{-1/4}$. Here Ta is the Taylor number, as defined in the table of nomenclature. At high rotation rates, the furnace axis is nearly perpendicular to the rotation axis. Therefore the interface is parallel to the rotation axis, as shown in Figure 5. In this case the momentum layer thickness x°_v is proportional to $Ta^{-1/6}$.[12] Taking this into account, we use Equation 6 to calculate the dependence of x°_v on the rotation rate of the centrifuge:

$$\text{For } Ta \gg 1: \quad x^\circ_v = Ta^{-m} \quad \text{where } 1/6 \le m \le 1/4 \tag{6}$$

To check Equation 6, we determined the dependence of the maximum flow velocity in horizontal cross sections on the distance to the solid-liquid interface using three-dimensional numerical simulations of the buoyant convection for case 3 in Reference 1 (R = 1 m). The distance from the interface to the first maximum of the flow velocity is assumed to be equal to the momentum boundary layer thickness x°_v. As shown in Figure 6, the numerical results agree with the scaling law given in Equation 6 with m = 1/5.

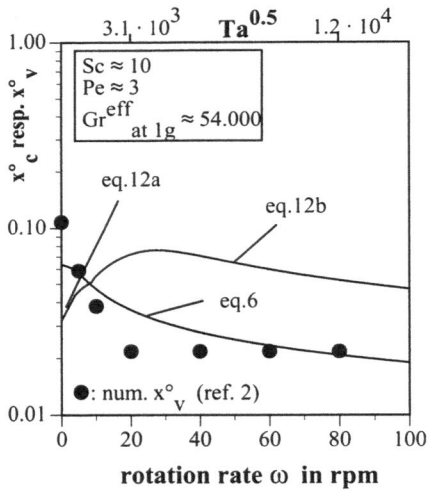

Figure 6. Momentum (x°_v) and solute (x°_c) boundary layer thicknesses versus rotation rate ω for case 3 of Reference 1, with R = 1 m. The line for x°_v was computed using Equation 6, while that for x°_c was via Equation 12. The symbols are for x°_v determined at the solid - liquid interface from numerical calculations.[2]

Formulation for solute layer thickness x°_c

The model of Garandet[14] was used to estimate the solute boundary layer thickness x°_c. The starting point is the stationary one-dimensional mass transport equation:

$$[V_G + u(x)]\frac{\partial C}{\partial x} + D\frac{\partial^2 C}{\partial x^2} = 0 \tag{7}$$

where $u(x)$ is the convective flow velocity normal to the interface and C the concentration of the solute at distance x from the interface. It is assumed that inside the solute boundary layer the concentration C varies from $C_{x=0}$ at the interface ($x = 0$) to C_∞ at the edge of the boundary layer ($x = x_c$). Outside of x_c the dopant distribution should be uniform and equal to C_∞. With the dimensionless variables $x^* = x/x_c$, $C^\circ = (C - C_\infty)/(C_{x=0} - C_\infty)$, and $u^\circ = u \cdot r/\nu$, Equation 7 can be rewritten taking into account the definitions for x°_c, Pe and Sc:

$$\left[Pe + u^\circ(x^*)Sc\right]x^\circ_c \frac{\partial C^\circ}{\partial x^*} + D\frac{\partial^2 C^\circ}{\partial x^{*2}} = 0 \tag{8}$$

Following Garandet,[14] we assume that at the edge of the boundary layer ($x^* = 1$) the derivatives in Equation 8 are on the order of one. From this assumption we get the defining equation for x°_c:

$$1 = Pe\, x^\circ_c + u^\circ(x^\circ_c)\, Sc\, x^\circ_c \tag{9}$$

Camel and Favier[3] gave the following formulation for the flow velocity $u^\circ(x^\circ)$ for Bridgman configurations as well as for rotating growth systems, especially for the Czochralski configurations. We assume that this formulation is also valid for gradient freeze growth on a centrifuge.

$$\text{For } x \le 1: \quad u^\circ(x^\circ_c) = v(x^\circ)\, x^\circ \tag{10}$$

Now we insert Equation 10 into Equation 9 and take into account Equations 4a and 4b for $v^\circ(x^\circ_c)$. For $x^\circ_c \le x^\circ_v$:

$$1 = Pe\, x^\circ_c + \frac{Re\, Sc}{x^\circ_v}\, x^{\circ 3}_c \tag{11a}$$

while for $x^\circ_c > x^\circ_v$:

$$1 = Pe\, x^\circ_c + Re\, Sc\, x^{\circ 2}_c \tag{11b}$$

Equations 11a 11b can be solved analytically.[15] It follows with the requirement $x^\circ_c > 0$, that for $x^\circ_c \le x^\circ_v$:

$$x^\circ_c = \frac{\left(0.5\left[x^\circ_v/(Re\, Sc)\right](1+\sqrt{1+\frac{4}{27}\left[x^\circ_v/(Re\, Sc)\right]Pe^3}\,)\right)^{2/3} - \frac{1}{3}\left[x^\circ_v/(Re\, Sc)\right]Pe}{\left(0.5\left[x^\circ_v/(Re\, Sc)\right](1+\sqrt{1+\frac{4}{27}\left[x^\circ_v/(Re\, Sc)\right]Pe^3}\,)\right)^{1/3}} \tag{12a}$$

while for $x^\circ_c > x^\circ_v$:

$$x^\circ_c = \frac{-Pe + \sqrt{Pe^2 + 4\, Re\, Sc}}{2\, Re\, Sc} \tag{12b}$$

Figure 4a shows the dependence of the solute boundary layer thickness x°_c (calculated with Equation 12 considering the scaling laws for Re and Equation 4 for x°_v) on the effective Grashof number, $Gr^{eff} = Gr \cdot \Delta x_i/r$ for $\omega = 0$. The estimation with Equation 12 is in good agreement with the numerical results of Jung.[10] At low Gr, the solute boundary layer thickness x_c is constant and equal to the characteristic diffusion length D/V_G. At high Gr, x°_c decreases because the convective mixing in the melt increases.

With rotation ($\omega \neq 0$), no numerical data are available to check the estimate for x°_c given in Equation 12 taking into account the scaling laws for Re and Equation 6 for x°_v. In Figure 6, the dependence of x°_c on the rotation rate ω of the centrifuge is shown for molten Ge:Ga with Gr^{eff} at 1 g equal to 54,000 and Pe = 3. Note that the solute layer thickness x°_c decreases with increasing flow velocity, as expected (see case 3 in Reference 1).

Now, we compare the analytical model with our experimental results. For that purpose we put Equation 4 for $v^{\circ}(x^{\circ}_c)$, Equation 6 for x°_v, and Equation 12 for x°_c in the definition equation of the segregation parameter Δ_{OM} (Equation 3). The geometry (pivoted furnace, melt - radius, interface deflection) and the thermal conditions (axial temperature gradient) of the performed growth experiments on the centrifuge are very similar to the data used for case 3 in Reference 1.

Figure 7 shows the theoretical dependence of the segregation parameter Δ_{OM} on rotation rate ω for our Ge:Ga growth experiments. The theoretical curves were calculated for two different arm lengths (R = 0.6 m and 1 m). The values of the segregation parameter Δ for all crystals determined from the measured dopant distributions are also shown in Figure 7. The theoretical curves are in a fairly good agreement with the experimental data. There is in both cases, experiments and theory, an increase of Δ with ω caused by a reduction of convection on the centrifuge. The rotation rate ω where Δ is a maximum corresponds to that rotation rate ω where the minimum of the flow velocity occurs for case 3 (see Reference 1). The difference between experimental results and theoretical predictions at higher rotation rates is probably due to the very simple approach taken to estimate the flow velocity near the solidification front.

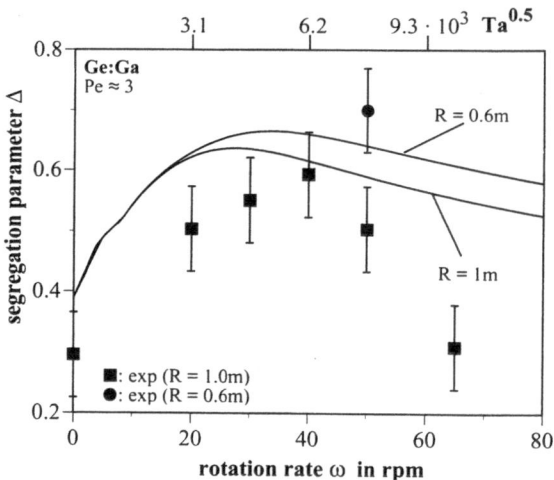

Figure 7. Segregation parameter Δ versus the rotation rate ω. The symbols are for Δ evaluated from crystals grown at different rotation rates ω. The curves are from our segregation model.

From Figure 8, it can be seen that a reduction of acceleration to 10^{-1} g to 10^{-2} g would be necessary to achieve the same value of Δ, about 0.7, which we observed for the same growth parameters in one of our experiments on the centrifuge. Figure 8 also shows the dependence of Δ on the Grashof number, which is directly proportional to the resulting gravity, for the growth of Ge:Ga crystals under normal and low gravity conditions.

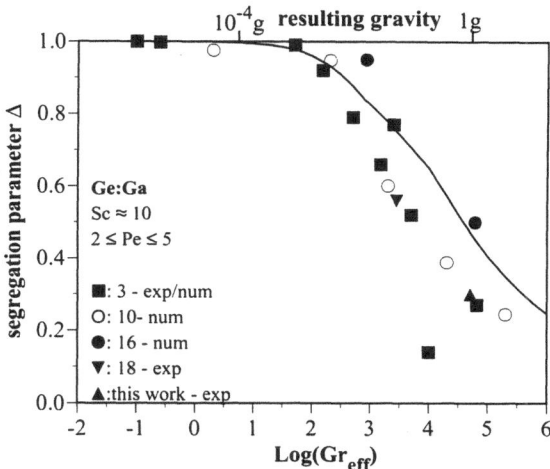

Figure 8. Dependence of Δ on Gr_{eff} (proportional to the resulting acceleration, shown as a second ordinate for our growth parameters) for Ge:Ga growth experiments. The solid line is estimated using our segregation model. The symbols are experimental and theoretical results from the literature.[3,10,16,17]

The experiments and the numerical results from the literature are plotted as symbols in Figure 8, and are characterized by values of Pe between 2 and 5. Thus, these results can be compared with our experiments. The solid line was calculated with our segregation model using the scaling laws for Re given by Favier.[3] It can be seen from Figure 8 that under microgravity a uniform dopant distribution characterized by $\Delta = 1$ can be achieved. A resulting acceleration of about 10^{-3} g to 10^{-4} g would be necessary for this, using the same thermal boundary conditions as in our experiments.

Maximum Flow Velocity for a Uniform Dopant Distribution ($K_{eff} \geq 0.9$)

We did not achieve a uniform dopant distribution in our growth experiments with Ge:Ga, in contrast to Rodot and Regel.[4] The reason for this is shown in Figure 9, where we have plotted the maximum dimensionless flow velocity $Re_{0.9}$ that will permit a uniform dopant distribution. The curves are plotted for two different values of the ratio Pe^2/Sc versus the equilibrium segregation coefficient K_0 as solid lines. We assume for determining $Re_{0.9}$ that the segregation profile is uniform when K_{eff} is greater than 0.9. The critical velocity $Re_{0.9}$ can then be calculated from Equations 2 and 3b using D/V_G as the diffusion boundary layer thickness x_c like Ostrogorsky.[9] The result for $Re_{0.9}$ is:

$$Re_{0.9} = 0.6 \, \frac{Pe^2}{Sc} \left(\frac{K_0}{0.9 - K_0} \right) \text{ for } K_0 < 0.9 \tag{13}$$

The critical flow velocity $Re_{0.9}$ increases with increasing K_0. This means that in systems with a large value of K_0 a higher convection intensity in the melt permits achievement of a homogeneous dopant distribution than in systems with a small value of K_0. The critical flow velocity $Re_{0.9}$ for our Ge:Ga experiments ($K_0 = 0.087$) on our centrifuge (characterized by $Pe^2/Sc \approx 4$) is more than one order of magnitude lower than the typical flow velocity in the melt calculated with our scaling laws.[1,2] This characteristic flow velocity is shown as symbol 1 in Figure 9. Thus, we could not grow crystals with a homogeneous Ga concentration because the reduction of convection in our growth experiments on the centrifuge was not sufficiently high. A more effective suppression of convection is possible in space. Therefore, Favier[6] could observe a uniform dopant concentration in the Ge:Ga crystal grown in microgravity because in this experiment, also characterized by $Pe^2/Sc \approx 4$, the typical flow velocity shown as symbol 2 was more than one order of magnitude lower than the critical flow velocity.

It follows from Figure 9 what conditions have to be chosen to achieve a homogeneous dopant concentration under high gravity on a centrifuge. Material systems with a high value of K_0 must be selected. Such materials are, for example, Sn:Bi ($K_0 \approx 0.35$)[18] or InSb:Te ($K_0 \approx 0.5$)[19]. We would expect that an axially uniform dopant concentration should be possible on a centrifuge at certain rotation rates and at very short arm lengths with these materials, while without centrifugation K_{eff} would be smaller than 0.9. For these materials, the suppression of convection by the interaction between the buoyancy force and the Coriolis force should be sufficient to dampen the typical flow velocities in the melts (shown as symbols 3 and 4 in Figure 9) below the critical flow velocity when conditions are similar to those we used in our experiments. Such experiments would be characterized by a value of the parameter Pe^2/Sc of about 10.

Theoretical analysis of the experiments of Rodot and Regel[4] with Ag-doped PbTe are not shown in Figure 9 because the diffusion coefficient D as well as the equilibrium segregation coefficient K_0 are unknown. Therefore we could not calculate K_{eff}. In a previous paper, we showed that our theoretical prediction about the homogeneity of the PbTe:Ag crystals grown on the centrifuge by Rodot and Regel agree qualitatively with the experimental results.[2] In this calculation we used the growth parameters indicated in the papers of Regel[4,8] and assumed a typical value of $10^{-5} cm^2/s$ for the diffusion coefficient D. The predicted curves agree qualitatively with the experimental results.[2] The theoretical values do not reach the conditions of the convection-free regime ($\Delta = 1$) as claimed by the authors for the growth experiment at a rotation rate ω of about 16 rpm. But it was shown by our model, in agreement with the experiments, that with increasing deviation from the optimum rotation rate of 16 rpm the axial dopant distribution should become more non-uniform because Δ decreases.[2] This is consistent with the data of Rodot and Regel, who performed their experiments only at rotation rates higher than 16 rpm and not at lower values.[4]

CONCLUSIONS

We have shown experimentally and theoretically that a reduction of the vigor of convection is possible on a centrifuge under some certain growth conditions. The suppression of convection is caused by the interaction between the buoyancy force and the Coriolis force. For the parameters considered in our investigations, the damping of the flow velocity is up to one order of magnitude on a centrifuge in comparison to the case without centrifugation. A stronger suppression of convection is possible under microgravity. Therefore, we expect an improvement of the homogeneity of species concentration in crystals grown on a centrifuge can only for systems with $K_0 > 0.1$.

The use of the centrifuge as a tool to study transport processes under reduced convective conditions has two advantages: First, the costs for experiments on a centrifuge are very low in contrast to space experiments. Second, the investigations under reduced convective conditions with a centrifuge are not limited to electrically conductive melts in contrast to the case when magnetic fields are used for suppression of convection.

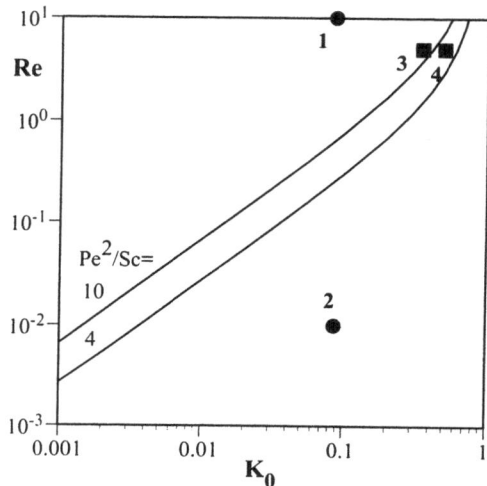

Figure 9. Maximum allowed flow velocity, $Re_{0.9}$, to achieve $K_{eff} \geq 0.9$, versus the equilibrium segregation coefficient K_0. Calculated with Equation 13 for different values of the parameter Pe^2/Sc (solid lines). The symbols are for flow velocities estimated with our scaling laws[1,2] for experiments performed under low and high gravity. (1) centrifuge (Ge:Ga): $Re \approx 10$; $Pe^2/Sc \approx 4$; (2) microgravity[6] (Ge:Ga): $Re \approx 0.01$; $Pe^2/Sc \approx 4$; (3) centrifuge (Sn:Bi): $Re \approx 5$; $Pe^2/Sc \approx 10$; (4) centrifuge (InSb:Te): $Re \approx 5$; $Pe^2/Sc \approx 10$;

Acknowledgment

This work was supported by the Bundesministerium für Bildung und Forschung (BMBF) under the project management of the German space agency (DARA), contract number 50WM9301.

REFERENCES

1. J. Friedrich and G. Müller, Convection in crystal growth under high gravity on a centrifuge, *in present volume.*
2. J. Friedrich, J. Baumgartl, H.J. Leister, and G. Müller, Experimental and theoretical analysis of convection and segregation in vertical Bridgman growth under high gravity on a centrifuge, *J. Crystal Growth*, in press.
3. D. Camel and J.J. Favier, Scaling analysis of convective solute transport and segregation in Bridgman crystal growth from the doped melt, *J. Physique* 47:1001 (1986).
4. H. Rodot, L.L. Regel, and A.M. Turtchaninov, Crystal growth of IV-VI semiconductors in a centrifuge, *J. Crystal Growth* 104:280 (1990).
5. G. Müller, G. Neumann, and W. Weber, The growth of homogeneous semiconductor crystals in a centrifuge by the stabilizing influence of the Coriolis force, *J. Crystal Growth* 129:8 (1992).

6. J.J. Favier, J. deGoer, R. LeMaguet, Analyse de la segregation du gallium dans des barreux de germanium solidifies unidirectionnallement en fusee sonde (missions TEXUS IV et TEXUS VI), C.E.A. Internal Report, Grenoble (1985).

7. J.J. Favier, Macrosegregation I: unified analysis during non-steady state solidification, *Acta Metallurgica* 29:197 (1981).

8. W.A. Arnold, W.R. Wilcox, F. Carlson, A. Chait, and L.L. Regel, Transport mode during crystal growth in a centrifuge, *J. Crystal Growth* 129:24 (1992).

9. A.G. Ostrogorsky and G. Müller, A model of effective segregation coefficient, accounting for convection in the solute layer at the growth interface, *J. Crystal Growth* 131:587 (1992).

10. T. Jung and G. Müller, Effective segregation coefficients: a comparison of axial solute distributions predicted by analytical boundary layer models and numerical calculation, *J. Crystal Growth*, in press.

11. M.A. Fikri, G. Labrosse, and M. Betrouni, The melt phase hydrodynamics for the "stabilized" Bridgman procedure applied under centrifugation; preliminary analysis and numerical results, *J. Crystal Growth* 119:41 (1992).

12. R. Hide, Theory of axisymmetric thermal convection in a rotating fluid annulus, *Phys. Fluids* 10:56 (1967).

13. N. Ma and J.S. Walker, Liquid-metal buoyant convection in a vertical cylinder with a strong magnetic field with a nonaxisymmetric temperature, *Phys. Fluids* 7:2061 (1995).

14. J.P. Garandet, T. Duffar, and J.J. Favier, On the scaling analysis of the solute boundary layer in idealized growth configurations, *J. Crystal Growth* 106:437 (1990).

15. I.N. Bronstein, and K.A. Semendjajew, "Taschenbuch der Mathematik," Verlag Harri Deutsch, Thun und Frankfurt/Main (1987).

16. C.J. Chang and R.A. Brown, Radial segregation induced by natural convection and melt/solid interface shape in vertical Bridgman growth, *J. Crystal Growth* 63:343 (1983).

17. D.H. Matthiesen, M.J. Wargo, S. Motakef, D.J. Carlson, J.S. Nakos, and A.F. Witt, Dopant segregation during vertical Bridgman-Stockbarger growth with melt stabilization by strong axial magnetic fields, *J. Crystal Growth* 85:557 (1987).

18. A.G. Ostrogorsky, F. Mosel, and M. Schmidt, Diffusion-controlled distribution of solute in Sn-1%Bi solidified by the Submerged Heater Method, *J. Crystal Growth* 110:950 (1991).

19. A.G. Ostrogorksy, H.J. Sell, S. Scharl, and G. Müller, Convection and segregation during growth of Ge and InSb crystals by the Submerged Heater Method, *J. Crystal Growth* 128:207 (1993).

TABLE OF NOMENCLATURE

b	vector of the resultant acceleration, m/s^2
b	thermal expansion coefficient, K^{-1}
C	concentration of the solute in the liquid, cm^{-3}
C_S	concentration of the solute in the crystal, cm^{-3}
$C_{x=0}$	concentration of the solute in the melt at the interface, cm^{-3}
C_∞	concentration of the solute in the melt at the edge of the solute layer, cm^{-3}
C_0	initial dopant concentration in the liquid, cm^{-3}
D	diffusion coefficient of the solute in the melt, m^2/s
f	fraction solidified
g	vector of the acceleration due to gravity, m/s^2
g	$9.81 m/s^2$ (acceleration due to earth's gravity)
h	melt height, m
K_0	equilibrium segregation coefficient
K_{eff}	effective segregation coefficient
l_{s0}	dimensionless parameter in the Stokes regime used to calculate x_v with $\omega = 0$
$N \cdot g$	absolute value of the resultant acceleration in multiples of g
m	exponent in Equation 28 ($1/6 \leq m \leq 1/4$)
p	dynamic pressure, N/m^2

r	location vector, with components x, y and z, m
r	melt radius, m
R_a	distance between the rotation axis and the bearing point of the furnace, m
R	vector from the rotation axis and the origin of the local coordinate system, m
R	absolute value of R, m
T	temperature, K
u	flow velocity normal to the interface, m/s
v	flow velocity, m/s
v	flow velocity parallel to the interface, m/s
v_{ref}	characteristic flow velocity, m/s
V_G	growth rate, m/s
x_c	solute boundary layer thickness, m
x_v	momentum boundary layer thickness, m
x_S	distance between the bearing point of the furnace and the center of mass, m
Δ	segregation parameter
Δx_i	maximum deviation of the interface from planarity, m
κ	thermal diffusivity, m^2/s
ν	kinematic viscosity, m^2/s
ω	vector of the angular velocity, rad/s
ω	absolute value of **w**
ρ	density of the melt, kg/m^3
θ	swinging angle of the furnace, rad
$^\circ$	corresponding dimensionless value
Gr	Grashof number, $\beta(\Delta T/h)(Ng)\, r^4 / \nu^2$
Gr^{eff}	$Gr \cdot \Delta x_i / r$
Pe	Peclet number, $V_G\, r / D$
Re	Reynolds number, $v_{ref}\, r / \nu$
Sc	Schmidt number, ν / D
Ta	Taylor number, $4\, r^4\, \omega^2 / \nu^2$

ANALYSIS OF THERMAL CONVECTION IN MOLTEN TIN UNDER CENTRIFUGAL CONDITIONS

L. Bergelin and A. Chevy

Physiques des Milieux Condensés, C.N.R.S., U.R.A. 782
Université Pierre et Marie Curie
Paris, France

ABSTRACT

The hydrodynamic behavior of molten tin in an ampoule was studied under centrifugal conditions with the temperature decreasing with height. When the resultant acceleration was initially increased, the system remained unsteady. Instabilities occurred in the form of bursts of hot fluid with dimensions between 10 and 65 mm. During this phase, the inertial force balances the centrifugal force. Above an acceleration threshold, the system becomes steady with organized convective rolls and enhanced heat transfer. The Coriolis force leads to the emergence of an azimuthal component of the fluid velocity. The other velocity components do not change, or change only slightly, for a resultant acceleration up to 1.2 g.

INTRODUCTION

These last twenty years have seen a remarkable growth of electronics in everyday life. As a result, the electronics industry needs very good quality crystals, especially of silicon, to manufacture integrated circuits. Silicon single crystals are grown commercially by the Czochralski technique. This technique consists of extracting a crystal from a melt that is heated from below, that is, under destabilizing thermal conditions. This growth technique has many technological advantages, but it also has a major disadvantage: it causes turbulent free convection that results in temperature fluctuations at the liquid-solid interface. These fluctuations cause crystal striations, and these striations affect the homogeneity of the crystal slice used for integrated circuit substrates. Temperature fluctuations and the resulting striations can be stopped by use of a magnetic field or by centrifugation.[1,2]

The crystallographic quality and electronic properties of Czochralski-grown crystals critically depend on the fluid flow pattern in the melt during growth. Thus, it appears to be very important to study the fluid flow pattern under centrifugal conditions.

In this paper, experimental results and theoretical analysis will be reported that allow us to begin to understand the convection mechanism in an ampoule under centrifugal conditions.

EXPERIMENTAL

The centrifuge used for the experiments (Figure 1) is at Nantes in France, at the "Laboratoire Central des Ponts et Chaussées." This centrifuge has a vertical central axis (1) that allows rotation of two asymmetric arms: arm (2) carrying basket (3), and arm (4) with balancing counterweight (5). The basket is attached to the arm by a pin, so that it can swivel and allow the basket's platform to align itself, at every moment during rotation, normal to the resultant of the net acceleration (gravitational acceleration and centrifugal acceleration). The distance between the centrifuge's axis and the basket's platform varies from 3.96 m to 5.50 m due to the orientation of the basket.

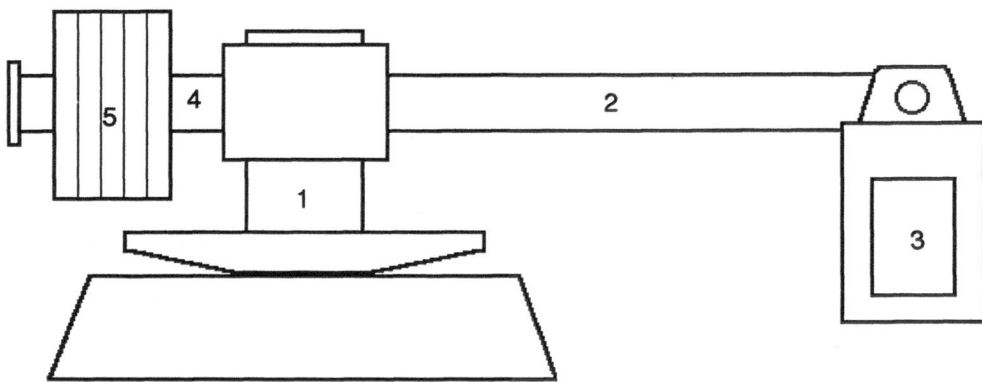

Figure 1. Schematic diagram of the Nantes centrifuge.

Figure 2 shows the furnace on the centrifuge basket (1). This furnace (2) was installed such that its axis (3) coincided with the resultant acceleration direction. It was insulated against heat losses with two caps of quartz wool (4) to close each end. To provide a constant thermal environment during rotation, the furnace was placed in an enclosure (5) maintained at a constant temperature by circulating water. The furnace was divided into eight independent temperature zones (6). These zones allowed us to establish a computer-controlled temperature gradient in the furnace. The temperature-control was connected to a PC3000 Eurotherm computer. However, despite the temperature-control instructions sent to the furnace computer, it was not always possible to enforce the desired gradients, when the resultant acceleration was high, due to the thermal characteristics of the furnace. Four thermocouples (7) were placed at strategic spots in or near the furnace to control temperature and detect possible problems.

Figure 3 shows a quartz ampoule (1) containing a tin melt (2) and placed in the furnace in such a way that the ampoule's axis (3) coincides with the resultant acceleration direction. Both the ampoule diameter and the melt height were varied. Four thermocouples (4) were held in the ampoule at a fixed height and in direct contact with the melt.

During a typical experiment, an ampoule was placed in the furnace, after which a destabilizing temperature gradient was applied. The resultant acceleration was increased step-wise. Each step duration was about 10 minutes. Temperatures were measured every 10 seconds or every 4-5 seconds. When the acceleration reached 10 or 13 times earth's gravity g, the centrifuge speed was reduced either step-wise or continuously.

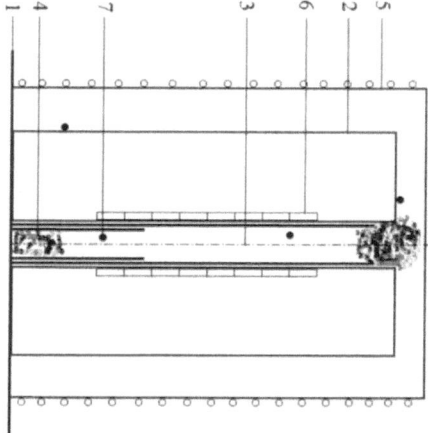

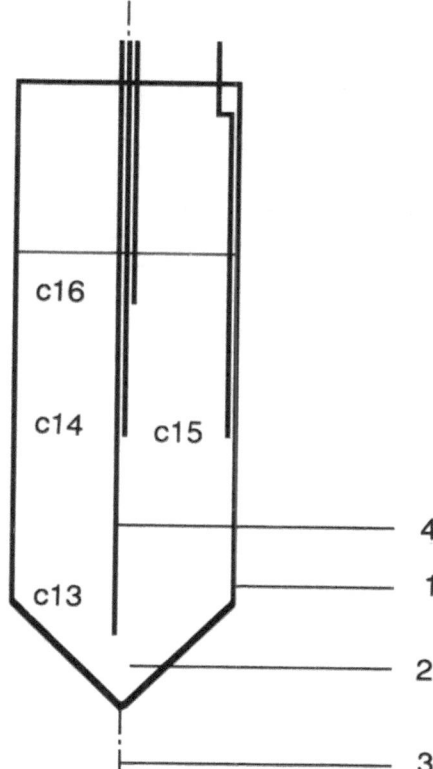

Figure 2. Schematic diagram of the furnace. **Figure 3.** Schematic diagram of an ampoule.

RESULTS

As shown in Figure 4, when the acceleration was increased, a decrease in the amplitude of the temperature fluctuations was noted. This decrease was either sudden or gradual; the temperatures eventually stabilized. This stabilization occurred before the accleration reached 2 g. So, stabilization was observed at a relatively-low value of the resultant acceleration. When the acceleration was increased further, the melt temperatures increased and the temperature differences between thermocouples tended to diminish.

Table 1. Experimental data

Ampoule	Tin height (mm)	Ampoule diameter (mm)	Geometric ratio	Tin weight (gram)	distance between c14-c15 (mm)	distance between c14-c13 (mm)	distance between c13-c16 (mm)
A	65	28.5	2.3	312.7	14	30	62
B	140	28.5	4.9	615.6	14	65	120
C	149	22.0	6.8	331.0	11	65	120

Temperature fluctuations reappeared for one of the ampoules (C in Table 1) when the resultant acceleration was high (at 5, 10 and 13g). But these fluctuations were more regular and appeared as rhythmic beatings. The resultant acceleration was not allowed to exceed 13g, because we believed that this was the most the furnace and electronic equipment could support.

When the centrifuge was slowed, temperature fluctuations appeared again at the same threshold resultant acceleration. Therefore, this phenomenon was reversible.[3,4]

During the unsteady phase, a thermal layer develops in the lower part of the ampoule.[6] This layer grows and when it is too thick, hot fluid rises in the ampoule, producing instability in the form of a burst of hot fluid. In a burst of hot fluid, temperatures are greater than the surrounding fluid. When the hot fluid arrives at a thermocouple, the temperature there increases. When the hot fluid moves off the thermocouple, the temperature decreases.

Initially, Lissajou diagrams were plotted to reveal any phase correlations between the temperature fluctuations recorded at different points in the melt. In a Lissajou diagram,[5] the

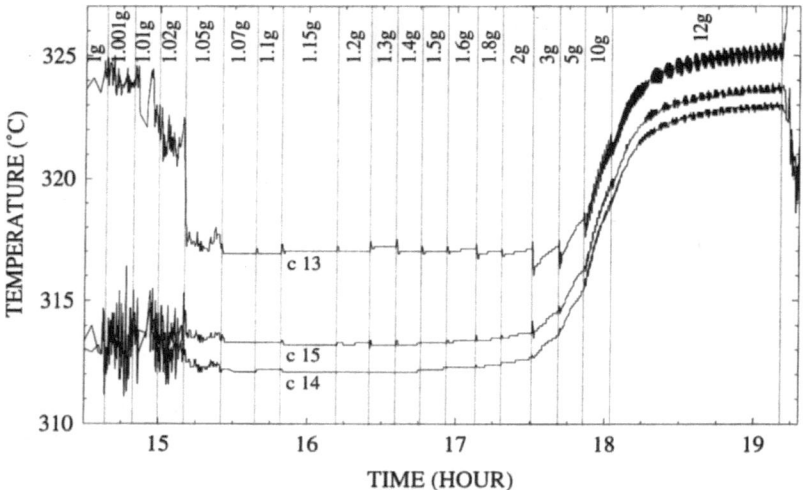

Figure 4. Temperatures measured inside a destabilizing tin melt (ampoule C in Table 1). The vertical dashed lines indicate changes in rotation speed.

temperature of one thermocouple is plotted against the corresponding temperature of another thermocouple. A perfect correlation would be represented by a plot that forms a single straight line with a positive slope for in-phase or with a negative slope for out-of- phase behavior. A distribution of points close to a straight line implies that the oscillations of the thermocouples remain closely in phase or out of phase. When measurements at two thermocouples are in phase for several minutes, this suggests that they are in contact with the same instabilities at the same time. On the other hand, when the thermocouples are out of phase, or when there is no dominant phase relationship, this suggests that they are not crossed by the same instabilities at the same time.

The Lissajou diagrams between thermocouple pairs c13 and c14, c13 and c16, c13 and c15, c14 and c16, c14 and c15, and c15 and c16 are given in Figure 5.

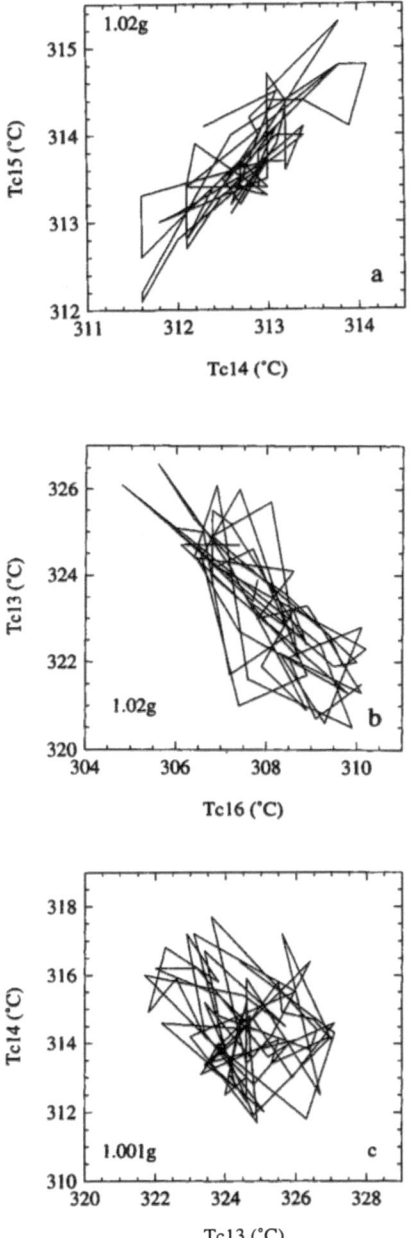

Figure 5. Lissajou diagrams. Temperature readings for pairs of thermocouples at same times. a: Ampoule C, thermocouples c14 & c15.   b: Ampoule B, TC's c13 & c16.   c: Ampoule B, c14 & c13.

Thermocouple c14 in ampoule A was defective, and so was ignored. As c15 and c14, in ampoules B and C, were almost in perfect phase (Figure 5a), and as they were about 10 mm apart, the bursts of hot fluid must have been thicker than 10 mm.

For other combinations, sometimes the thermocouple readings were almost 180° out of phase (Figure 5b). Sometimes there was no dominant phase relationship (Figure 5c). These thermocouples were 65 mm apart (Table 1). By considering all combinations of thermocouples, we conclude that the thermal instabilities must have had dimensions between about 10 and 65 mm.

The influence, on the Lissajou diagrams, of the acceleration and of the furnace temperature gradient is very difficult to detect. Nevertheless, it does seem that a high g produced the best phase relationship, whereas a high furnace temperature gradient did not lead to the best phase relationship. So, an increase of rotation rate coalesced the instabilities, while an increase of the furnace temperature gradient destabilized the system.

Figure 6 shows the acceleration at which the unsteady-steady transition occured versus the temperature gradient in the furnace. The centrifuge's speed was increased stepwise, and each step appears on the plot as a bar, whose magnitude represents the change in the resultant acceleration, i.e. the bar is not an error bar. For an ampoule, the curve seems to be almost linear, with a positive slope, which implies that a higher furnace temperature gradient destabilizes the system and a higher rotation rate is needed to stabilize the temperature. A correlation between the aspect ratio (liquid height divided by diameter) and this slope is observed. In fact, when the geometric ratio was small, the slope was high. This indicates that a long thin ampoule is required to stabilize the temperature at a low acceleration. It is still generally accepted that high aspect ratios (thin ampoules) tend to stabilize a flow due to the fluid being close to a fixed boundary.

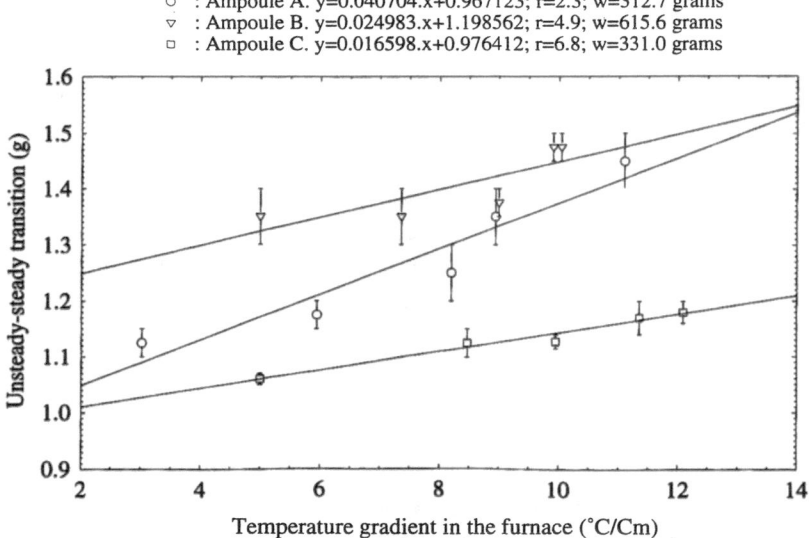

Figure 6. Threshhold acceleration for the transition between steady and unsteady temperature in the molten tin versus temperature gradient measured in the furnace.

The intercept in Figure 6 appears to increase with the mass of molten tin. Perhaps, a simple relationship exists between the intercept and the amount of tin, but we need more data points to confirm this.

In order to understand the fluid dynamics, we first consider the Rayleigh number,

$$Ra = \frac{V_{th}^2}{(\frac{\upsilon}{H})(\frac{\kappa}{H})}$$, where the coefficients υ and κ are the momentum and thermal energy

diffusivities, respectively, H is the height of the melt, $V_{th} = (\beta g_0 \Delta T H)^{\frac{1}{2}}$ is the free-fall velocity in the buoyancy field, g_0 is the total resultant acceleration, and ΔT is the temperature difference between the two furnace zones at the bottom and the top of the molten tin. As the usual first approximation, we take $\rho(T) = \rho_0(1 - \beta(T - T_0))$, where $\beta = -\frac{1}{\rho_0}(\frac{\partial \rho}{\partial T})_{T_0}$ and ρ_0 corresponds to some average density in the melt.

The experimental axial and radial gradients are plotted against Rayleigh number in Figures 7a and 7b, respectively. The axial gradient curve has the form of a hyperbola.

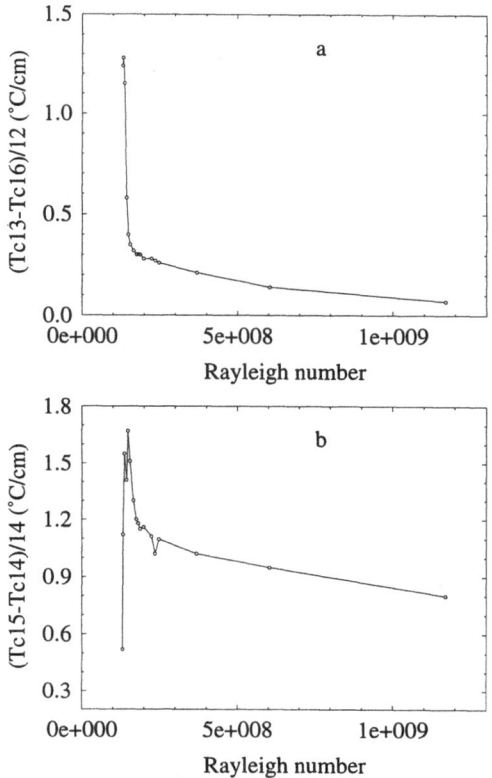

Figure 7. Temperature gradients inside molten tin under thermally unstable conditions versus Rayleigh number. Ampoule B. (a) axial temperature gradient. (b) radial temperature gradient.

So, when the Rayleigh number was low (the acceleration was close to earth's gravity), although the melt was very turbulent, the instabilities did not result in good heat transfer. A small increase of the centrifuge's speed enhanced the heat transfer, implying that the scale of the convective motion became larger, even while it was still unsteady. When the rotation rate was high and the convection became steady, the axial gradient tended towards zero, implying that the steady thermoconvective rolls resulted in very good heat transfer within the ampoule.

Initially at low Rayleigh numbers, the radial gradient increased (see Figure 7b); the thermal instabilities did not promote effective radial heat transfer. After the abrupt reduction in the amplitude of the temperature fluctuations (Figure 4), the radial gradient curve has the same form of the axial gradient curve, which implies that the heat transfer increased with increasing rotation rate. Müller et al.[1] noticed a change of sign in the radial temperature gradient when the Rayleigh number was increased. This change corresponded to a change in the direction of the convective roll. In our experiments, no such change was observed.

In our experiments, the time dependence of temperature T(t) was determined by recording temperatures at sequential time intervals $t_k = p(\Delta t)$, where p=0,1,...,N (N is the total number of data points). The time series $T(t_k)$ was recorded with a computer and its power spectral density $P(\omega)$ (the modulus squared of the Fourier transform) was calculated using the Cooley-Turkey fast Fourier transform algorithm[7]. Knowledge of the power spectra makes it possible to distinguish between periodic, quasi-periodic and chaotic regimes. Figure 8 shows the power spectral density versus frequency, which is given by $f_k = \dfrac{k}{N\Delta t}$, where k=0,1,..., $\dfrac{N}{2}$. A particular frequency appears for ampoule A (Figure 8). This frequency is about 0.04 Hz, corresponding to a time interval of about 25 seconds.

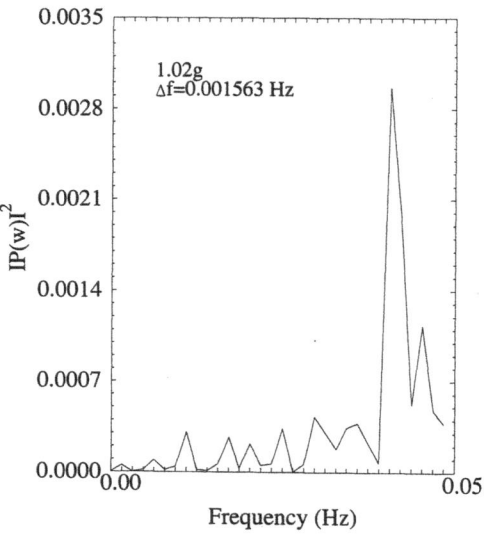

Figure 8. Power spectra for temperature measurements in Ampoule A at a resulting acceleration of 1.02 g and a temperature gradient in the furnace of 5°C.

ANALYSIS

Consider the axisymetric system shown in Figure 9, where the centrifuge rotation axis coincides with the ampoule's axis[8]. This simplifies the equations. Taking into account only the temperature dependence of the fluid density and starting with the Boussinesq conditions (assumed to be fulfilled), the Navier-Stokes equation is given by:

$$\frac{D\vec{u}}{Dt} = \frac{\partial \vec{u}}{\partial t} + \vec{u}.\nabla\vec{u} = -\frac{1}{\rho_0}\vec{\nabla}P + \frac{\mu}{\rho_0}\nabla^2\vec{u} - \beta\Delta T(\vec{g}_0 + \Omega^2\vec{r}) - 2\vec{\Omega}\wedge\vec{u} \qquad (1)$$

where $\vec{u}$ is the velocity vector in the rotating reference system. As internal scales, we have V, π and τ for fluid velocity, pressure and time, respectively. As external scales, $H\Omega$, Ω^{-1}, H and ΔT represent the velocity, time, length and temperature, respectively. In this way, the dimensionless Navier-Stokes equation can be written as:

$$\frac{\partial \vec{u}}{\partial t} + \vec{u}.\nabla\vec{u} = -\vec{\nabla}P + \nu.\frac{1}{V.H}\nabla^2\vec{u} + \beta\Delta T\frac{H}{V^2}Tg_0\vec{z} - \beta\Delta T\Omega^2\frac{H^2}{V^2}\vec{r} - 2\frac{\Omega H}{V}\vec{\Omega}\wedge\vec{u} \qquad (2)$$

where the orders of magnitude of the thermal buoyancy, Coriolis, inertia and centrifugal forces are $\dfrac{\beta\Delta THg_0}{V^2}$, $\dfrac{2\Omega H}{V}$, 1 and $\dfrac{\Omega^2H^2\beta\Delta T}{V^2}$, respectively. If the inertia force is balanced by the centrifugal force, then the dimensionless centrifugal and inertial forces are equal, i.e., $\dfrac{\Omega^2H^2\beta\Delta T}{V^2}=1$. The internal time scale $\tau=\dfrac{H}{V}$ can be calculated from the experimental data: $\Omega = 6.49$ revolutions per minute (for 1.02 g, $\Omega = 0.68$ rad/s), H = 0.065 m, $\beta = 7.57.10^{-5}$ K^{-1}, and $\Delta T \sim 5°$C/cm. For these experimental parameters, the internal time scale τ is about 30 seconds. This time and the periodicity obtained from the spectral power density are nearly the same, indicating a possible balance of inertial and centrifugal forces. With these chosen scales, the orders of magnitude of the thermal buoyancy and Coriolis forces are $\dfrac{g_0}{\Omega^2H}=\dfrac{1}{Fr}$ (Fr is the Froude number) and $\dfrac{2\Omega H}{V}=\dfrac{2}{\sqrt{\beta\Delta T}}=\dfrac{2}{Ro.Fr^{\frac{1}{2}}}$ (Ro is the thermal Rossby number), respectively.

Thus, the dimensionless Navier-Stokes equation becomes:

$$\frac{\partial \vec{u}}{\partial t} + \vec{u}.\nabla\vec{u} = -\vec{\nabla}P + (\frac{Pr}{Ra})^{\frac{1}{2}}\nabla^2\vec{u} + \frac{1}{Fr}T\vec{z} + \vec{r}T - \frac{2}{Ro.Fr^{\frac{1}{2}}}\vec{\Omega}\wedge\vec{u} \qquad (3)$$

To obtain some orders of magnitude estimates, we use the data for the ampoule A experiment. With $\Delta T \sim 5°$C/cm , $\Omega = 6.49$ revolutions per minute (1.02 g, $\Omega = 0.68$ rad/s), $T_0 \approx 400°$C, $\rho_0 = 6.88$, $\beta = 7.57.10^{-5}$ K^{-1}, $\nu = 2.57.10^{-3}$ cm^2/s, $\kappa = 0.200$ cm^2/s, and H = 0.065 m, one gets

$$V = 0.21 \text{ cm/s}, \quad Pr = \frac{\nu}{\kappa} = 0.013, \quad Ra = \frac{V^2}{(\frac{\nu}{H})(\frac{\kappa}{H})}=3.6.10^3, \quad (\frac{Pr}{Ra})^{\frac{1}{2}}=1.9.10^{-3}, \quad Fr=\frac{\Omega^2H}{g_0}=0.003,$$

$\dfrac{1}{Fr}=333.3$, and $Ro.Fr^{\frac{1}{2}}=\sqrt{\beta\Delta T}=0.048$, $\dfrac{2}{Ro.Fr^{\frac{1}{2}}}\approx42$. Thus, in this experiment the Coriolis force was much greater than the inertial force. Therefore, the thermal buoyancy force must have balanced the Coriolis force, leading to:

$$-\vec{\nabla}P+\frac{1}{Fr}T\vec{z}-\frac{2}{Ro}\vec{\Omega}\wedge\vec{u}=\vec{0} \tag{4}$$

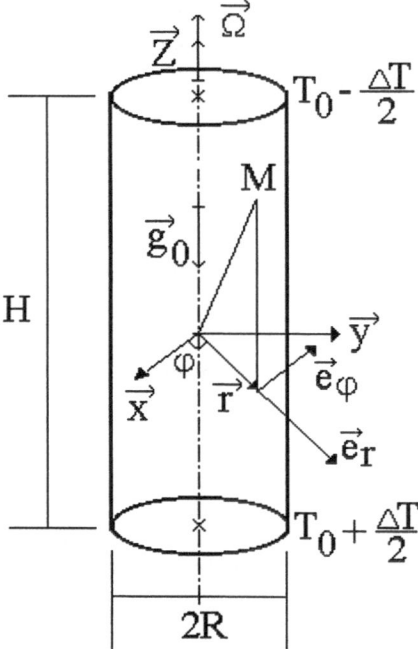

Figure 9. Axisymetric coordinate system used for analysis.

The vectors of equation (4) are projected onto the corresponding directions of the cylindrical reference system, where $\vec{u}$ has the components u, v and w:

$$\frac{\partial P}{\partial z}=\frac{T}{Fr} \text{ along the z axis, } (-\vec{\nabla}P)_{horizontal}=\frac{2}{Ro.Fr^{\frac{1}{2}}}\vec{\Omega}\wedge\vec{u} . \tag{5}$$

Taking the curl of equation (4) and noting that the system is axisymetric we find:

$$\frac{\partial u}{\partial z}=0, \quad \frac{\partial v}{\partial z}=-\frac{Ro}{2\Omega Fr^{\frac{1}{2}}}\frac{\partial T}{\partial r}, \quad \frac{\partial w}{\partial z}=0. \tag{6}$$

In addition, the continuity equation is $\nabla.\vec{u}=0$, which implies $\dfrac{\partial u}{\partial r}=-\dfrac{u}{r}$.

The balance of the inertia and centrifugal forces is described by:

$$\frac{\partial \vec{u}}{\partial t} + \vec{u}.\nabla \vec{u} = r\vec{u} \tag{7}$$

In a steady regime, taking into account only the temperature dependency of the fluid density and starting with the Boussinesq conditions (assumed to be fulfilled), the equations for the momentum and thermal energy balances are:

$$\frac{D\vec{u}}{Dt} = -\frac{1}{\varrho_0}\vec{\nabla}P + \frac{\mu}{\varrho_0}\nabla^2\vec{u} - \beta\Delta T(\vec{g}_0 + \Omega^2\vec{r}) - 2\vec{\Omega}\wedge\vec{u} = \vec{u}.\nabla\vec{u} \tag{8}$$

$$\nabla.\vec{u} = 0 \tag{9}$$

$$(\vec{u}.\vec{\nabla})T = \kappa\nabla^2 T \tag{10}$$

The external velocity, time, length and temperature scales are, respectively, $H\Omega$, Ω^{-1}, H, ΔT. As internal scales, we have V_{th}, π, τ for fluid velocity, pressure and time, respectively. Here $V_{th} = \sqrt{\beta\Delta T g_0 H}$ is the free-fall velocity in the buoyancy field, $\pi = \rho_0 V_{th}^2$ and $\tau = \frac{H}{V_{th}}$.

Therefore in a steady regime, the dimensionless equations are:

$$\frac{D\vec{u}}{Dt} = \vec{u}.\nabla\vec{u} = -\vec{\nabla}P - T(-\vec{z} + Fr.\vec{r}) - \frac{2}{Ro}\vec{\Omega}\wedge\vec{u} + (\frac{Pr}{Ra})^{\frac{1}{2}}\nabla^2\vec{u}, \tag{11}$$

$$(\vec{u}.\vec{\nabla})T = \frac{1}{Pe}\nabla^2 T, \tag{12}$$

$$\nabla.\vec{u} = 0, \tag{13}$$

where $Fr = \frac{H\Omega^2}{g_0}$ = Froude number, $Ro = \frac{V_{th}}{\Omega H}$ is the thermal Rossby number, $Pr = \frac{\nu}{\kappa}$ is the Prandtl number and $Ra = \frac{V_{th}^2.H^2}{\nu\kappa}$ is the Rayleigh number.

The ampoule is heated through the wall, imposing the following boundary conditions: $T + \alpha\frac{\partial T}{\partial r} = T_{ext}(z)$ at r=R and $T_{ext}(z = \pm\frac{1}{2}h) = T_0 \pm \frac{1}{2}h$, where α describes the heat transfer from the furnace to the liquid. It can be related to a Biot number through $Bi = \frac{R}{\beta}$. This heating creates a positive radial temperature gradient towards the ampoule's axis.

If the centrifugal force is neglected with respect to the Coriolis force, the Navier-Stokes equation becomes:

$$\vec{u}.\nabla\vec{u} = -\vec{\nabla}P + T\vec{z} - \frac{2}{Ro}\vec{\Omega}\wedge\vec{u} + (\frac{Pr}{Ra})^{\frac{1}{2}}\nabla^2\vec{u} \tag{14}$$

To obtain some order of magnitude estimates, we use the data from the ampoule A experiment. With $\Delta T \approx 5°C/cm$, $\Omega=11.3$ revolutions per minute ($g_0=1.2$ g, $\Omega=1.2$ rad/s), $T \approx 400°C$, $\rho_0=6.88$, $\beta=7.57.10^{-5}$ K^{-1}, $\nu=2.57.10^{-3}$ cm^2/s, $\kappa=0.200$ cm^2/s, and H=0.065 m, we get Pr=0.013, Ro=0.48, $\dfrac{2}{Ro}=4.1$ and $(\dfrac{Pr}{Ra})^{\frac{1}{2}} \approx 10^{-4}$. Because 10^{-4} is very small compared to 4.1, the viscous forces are negligible with respect to the Coriolis and thermal buoyancy forces.

If there is no rotation and no centrifugal force, the Navier-Stokes equation becomes:

$$\vec{u}.\nabla\vec{u}=-\vec{\nabla}P+T\vec{z}+(\frac{Pr}{Ra})^{\frac{1}{2}}\nabla^2\vec{u} \qquad (15)$$

We project the vectors of this equation onto the corresponding directions of the cylindrical reference system, where $\vec{u}$ has the components u, v and w:

$$\text{(a) } (\vec{u}.\vec{\nabla})u=-\frac{\partial P}{\partial r}; \quad \text{(b) } (\vec{u}.\vec{\nabla})v=0; \quad \text{(c) } (\vec{u}.\vec{\nabla})w=-\frac{\partial P}{\partial z}+T \qquad (16)$$

The continuity equation and the thermal energy balance equation become:

$$\nabla.\vec{u}=0 \rightarrow \frac{1}{r}\frac{\partial(ru)}{\partial r}+\frac{\partial w}{\partial z}=0 \qquad (17)$$

$$u\frac{\partial T}{\partial r}+w\frac{\partial T}{\partial z}=\frac{1}{r}\frac{\partial}{\partial r}(r\frac{\partial T}{\partial r})+\frac{\partial^2 T}{\partial z^2} \qquad (18)$$

In dimensionless form, $(\vec{u}.\vec{\nabla})T=\frac{1}{Pe}\nabla^2 T$ with $\frac{1}{Pe}\approx 10^{-2}$. Equation 16 shows that w is determined by the thermal buoyancy force and the pressure (and itself determines u through equations 17 and 18), but there is no azimuth component of velocity. Outside the boundary layers, we have $(\vec{u}.\vec{\nabla})T\approx 0$. Thus, the velocity and temperature gradients are perpendicular to one another. The isotherms in the ampoule are deduced to be as shown in Figure 10. The horizontal velocity component is very small in the center of the ampoule, existing principally at the top and the bottom of the melt to complete the flow circulation (1 in Figure 10) (outside the boundary layers). This horizontal flow is decribed by equation 16a. The variation in the vertical component of the velocity along an ampoule's radius is shown schematically in Figure 11. The flow is downward in the central region of the ampoule and upward along the vertical walls, because of heating through the ampoule wall. This heating creates a positive radial temperature gradient toward the ampoule's axis.

Consider now the case where there is rotation, but no centrifugal force. The vectors of dimensionless equations for the momentum and thermal energy balances are projected onto the corresponding directions of the cylindrical reference system:

$$(\vec{u}.\vec{\nabla})u=-\frac{\partial P}{\partial r}+\frac{2}{Ro}v \qquad (19)$$

$$(\vec{u}.\vec{\nabla})v=-\frac{1}{\phi}\frac{\partial P}{\partial \phi}-\frac{2}{Ro}u=-\frac{2}{Ro}u \text{ because of axisymetry} \qquad (20)$$

$$(\vec{u}.\vec{\nabla})w=-\frac{\partial P}{\partial z}+T \qquad (21)$$

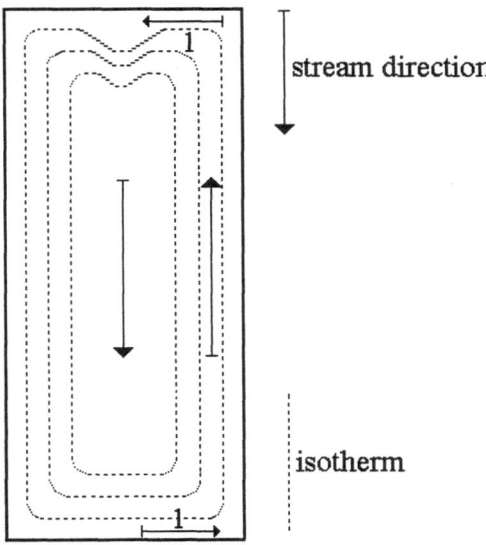

Figure 10. Schematic diagram of isotherms and directions of flow in the ampoule.

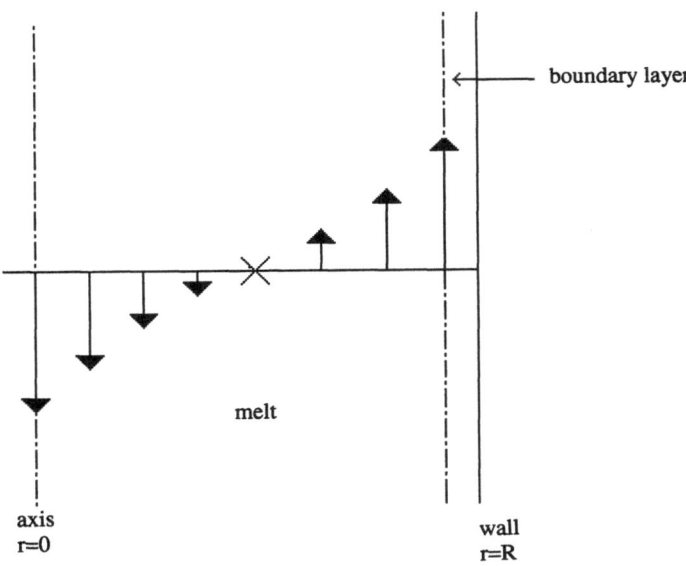

Figure 11. Radial variation of the vertical component of velocity in the melt.

$$\nabla.\vec{u}=0 \Rightarrow \frac{1}{r}\frac{\partial (ru)}{\partial r}+\frac{\partial w}{\partial z}=0 \qquad (22)$$

$$u\frac{\partial T}{\partial r}+w\frac{\partial T}{\partial z}=\frac{1}{r}\frac{\partial}{\partial r}(r\frac{\partial T}{\partial r})+\frac{\partial^2 T}{\partial z^2} \qquad (23)$$

Here, w is determined by the thermal buoyancy force and the pressure (equation 21). As before, w determines u through equations 22 and 23, and u determines v via equation 20. We find that u only changes slightly with regard to a non-rotating system, because Ro is close to 1. Therefore, because of the Coriolis force, an azimuthal component of the velocity appears. The velocity component along the z axis is independent of the presence or absence of rotation. The amplitude of the component u changes, but remains the same order of magnitude, because Ro is close to 1.

CONCLUSIONS

During the unsteady phase, bursts of hot fluid rise in the ampoule. Lissajou diagrams show that these instabilities have dimensions between 10 and 65 millimeters. Although the system is very turbulent, the instabilities do not result in good heat transfer. A study of the spectral power density implies a balance between inertial and centrifugal forces. A small increase of the centrifuge's rotation rate results in an enhanced heat transfer, and produces a marked reduction in the vigor of the instabilities (as illustrated by the decrease in the amplitude of the temperature fluctuations). This implies that the convective motion becomes organized, even if it is still unsteady. When the resultant acceleration exceeds a certain threshold, the system becomes steady. This transition can be correlated with the mass of the molten tin and the aspect ratio of the melt. Increasing the temperature gradient in the furnace destabilizes the system. A higher temperature gradient in the furnace requires a higher resultant acceleration for the transition to steady convection to occur. The temperature differences measured in the ampoule tend to diminish above the transition, implying that steady thermoconvective rolls produce very good heat transfer in the ampoule. It seems to be that with the Coriolis force, an azimuthal component of the fluid velocity appears. The velocity component along the z azis does not change with the Coriolis force, and the amplitude of the radial component of velocity keeps the same order of magnitude.

Investigations of the spectral power density will be continued and numerical experiments are required to understand exactly the balances between the various forces.

REFERENCES

1. W. Weber, G. Neumann, and G. Müller, Stabilizing Influence of the Coriolis Force During Melt Growth on a Centrifuge, *J. Crystal Growth* 100:145 (1990).
2. W. A. Arnold, L. L. Regel, and W.R. Wilcox, paper IAF-92-0913, 43rd Congress of the International Astronautical Federation, Washington DC (1992).
3. A. Chevy, P. Williams, M. Rodot, and G. Labrosse, Removal of convective instabilities in liquid metals by centrifugation, *in*: "Materials Processing in High Gravity," edited by L.L. Regel and W.R Wilcox, Plenum Press, New York, 1994.
4. P. Williams, A. Chevy, S. Bobèche, and M. Rodot, Stabilization of unsteady convective flows by centrifugation, *J. Phys. D: Appl. Phys.* 27:920 (1994).
5. D. J. Knuteson, A. L. Fripp, G. A. Woodell, W. J. Debnam, and R. Narayanan, Oscillation phase relations in a Bridgman system, *J. Crystal Growth* 109:127 (1991).
6. J. R. Carruthers, Origins of convective temperature oscillations in crystal Growth melts, *J. Crystal Growth* 32:13 (1976).
7. H. L. Swinney, Observations of order and chaos in nonlinear systems, *Physica* 7D:3 (1983).
8. M. A. Fikri and G. Labrosse, The melt phase hydrodynamics for the "stabilized" Bridgman procedure applied under centrifugation; preliminary analysis and numerical results, *J. Crystal Growth* 119:41 (1992).

THERMAL STABILITY DURING CENTRIFUGATION
FLOW VISUALIZATION EXPERIMENT:
NUMERICAL RESULTS

William A. Arnold[1] and Liya L. Regel[2]

[1]W. A. Arnold and Associates Engineering, Inc.
Akron, Ohio 44310

[2]International Center for Gravity Materials Science and Applications
Clarkson University
Potsdam, NY 13699-5814

ABSTRACT

By definition, in a thermally stable fluid there is no buoyancy-driven flow. The theory of thermal stability predicts that in a rotating fluid in microgravity there exists at least one family of thermal configurations where convective flow ceases. Thermal stability theory predicts that this family of thermal fields resulting in an absence of convection has circular isotherms centered about the axis of rotation. In earth-based centrifuges, one family of thermal fields that leads to a thermally stable configuration has isotherms that are paraboloids centered on the axis of rotation.

A fluid flow visualization cell was designed to test the predicted thermally stable temperature fields during centrifugation in ground-based experiments. The flow cell consists of a quartz glass ampule with aluminum end caps, filled with water. An axial density gradient is set up by imposing a thermal gradient between the two ends of the cell. Radial density gradients are controlled by the curvature of the aluminum end caps and by the use of heat sinks and sources. Numerical simulations were used to aid in the design of the flow visualization cell. The simulations of the ground-based fluid flow cell are presented here.

BACKGROUND

Rotating fluids are known to exhibit several unique and counter-intuitive qualities. Examples are geostrophic flows,[1] Taylor columns,[1] and the stabilizing effect of the Coriolis force on buoyancy driven flow.[2] Experimental evidence is also mounting indicating an apparent suppression of convective transport at a well defined acceleration level during solidification of semiconductor materials in a centrifuge.[3,4,5]

The last of the above listed phenomena is of special interest as it has direct applicability to the crystal growth industry. The apparent convective suppression leads to nearly uniform axial doping in the crystal. The acceleration for reduced convection depended on the arm length of the centrifuge used to directionally solidify Ag-doped PbTe crystals.

Figure 1 highlights some experimental features of crystal growth in a Bridgman furnace. The important feature is the unavoidable radial temperature gradients. In a constant acceleration field, such as earth's gravity, these gradients result in convection in the melt.[6] In the idealized case of flat isotherms, even a slight misalignment of the gravity vector with the ampule would cause natural convection in the melt. The presence of radial temperature gradients always causes convection. For a concave interface, this convection can be minimized at a particular rotation rate as explained by thermal stability theory.

Although progress is being made, much research still needs to be done in basic materials processing and fluid dynamics research in centrifuges. Basic research is needed to improve the understanding of the influence of gravity on materials processing and fluid flow processes. Fluid flow research in centrifuges on earth is very complicated, and has a lower acceleration bound (earth's gravity).

In a centrifuge, the acceleration field varies spatially. Experimental conditions can be tailored such the buoyancy-driven flow is effectively driven by a homogeneous acceleration field analogous to a gravitational acceleration field. Alternatively, the experimental conditions can be altered so that the buoyancy-driven flow responds to the inhomogeneity of the acceleration field.

In order to investigate earth-based centrifugal fluid flow phenomena, in particular thermal stability, a flow visualization system was needed. There are no known semiconductor or metal systems that are transparent in the visible spectrum. The cell presented here uses water, which has a relatively high Prandtl number as compared to semiconductors. However, thermal stability theory predicts that the thermally stable configurations resulting in a convectionless regime are independent of the fluid used.

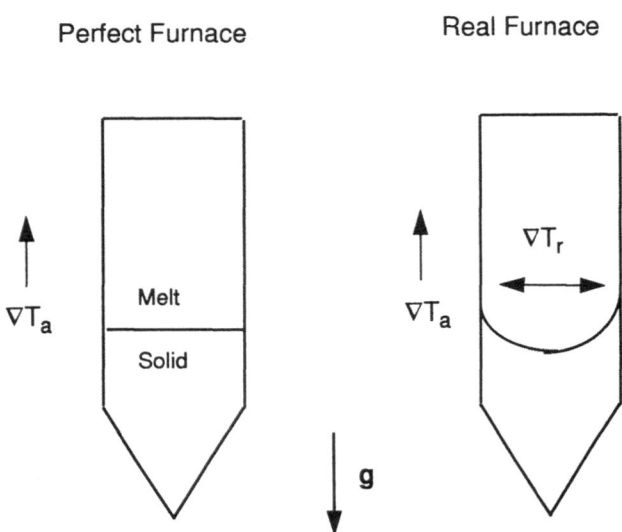

Figure 1. Bridgman crystal growth and temperature variations leading to convection on earth.

GOVERNING EQUATIONS

For the experiment presented here, the Boussinesq approximation should be valid for the convective terms and the Coriolis acceleration, since the temperature gradients are relatively low and because the coefficient of thermal expansion is small ($1 \gg \beta\Delta T$). The governing conservation equations, with the Boussinesq approximation applied to the convective terms and the Coriolis acceleration, in dimensional form are:[7,8]

$$\nabla \cdot \hat{u} = 0 \tag{1}$$

$$\rho_o\left(\frac{\partial}{\partial t}\hat{u} + \hat{u} \cdot \nabla\hat{u}\right) = -\nabla P + \nabla \cdot [\mu(\nabla\hat{u} + \nabla\hat{u}^T)]$$
$$- \rho_o\beta(T - T_{ref})[\hat{g} - \vec{\omega} \times \vec{\omega} \times \hat{R}] - \rho_o(2\vec{\omega} \times \hat{u}) \tag{2}$$

$$\rho_o C_P\left(\frac{\partial}{\partial t}T + \hat{u} \cdot \nabla T\right) = \nabla \cdot (k\nabla T) + q_s \tag{3}$$

Equations 1-3 can be nondimensionalized. The nondimensionallization produces several relevant nondimensional groups, some of which are referenced in this work. The relevant nondimensional groups are:

$Gr = a\rho^2\beta(T_H - T_C)_L L^3/\mu^2$

$Pr = \mu C_p/k$

$Ra = GrPr = \rho^2 C_p a\beta(T_H - T_C)_L L^3/\mu k$

$Ta = 4\rho^2\omega^2 L^4/\mu^2$

where the average value of the material properties is typically used. See the table of Nomenclature for definitions of symbols.

THERMAL STABILITY THEORY

The analysis of thermal stability[8,9,10] begins by examining the conservation of momentum equation in full form:

$$\rho\left(\frac{\partial}{\partial t}\hat{u} + \hat{u} \cdot \nabla\hat{u}\right) = -\nabla p + \nabla \cdot [\mu(\nabla\hat{u} + \nabla\hat{u}^T)] + \rho\hat{a} - \rho(2\vec{\omega} \times \hat{u}) \tag{4}$$

where $\hat{a} = \hat{g} - \vec{\omega} \times \vec{\omega} \times \hat{R}$ and ω is constant in time. In the absence of convection, equation 4 reduces to:

$$\nabla p = \rho\hat{a} \tag{5}$$

Taking the curl of equation 5 with constant gravitational acceleration and rotation rate yields:

$$0 = \hat{a} \times \nabla\rho \tag{6}$$

When equation 6 is applied to a fluid in a constant acceleration field, such as earth's gravitational field, thermal fields exist that are stable with respect to buoyancy. Such thermal fields have flat isotherms that are perpendicular to the gravitational field everywhere. That is, there is no horizontal temperature gradient. In addition, in order to be valid at all Rayleigh numbers the density must decrease with height (the hotter fluid must be above the cooler fluid with respect to the gravitational vector, assuming a positive coefficient of thermal expansion). A more precise way of stating this is that the acceleration field is parallel to the density gradient at all points in the fluid, or that:

$$\vec{g} \times \nabla\rho = 0 \text{ and } \vec{g} \bullet \nabla\rho > 0 \tag{7}$$

This analysis assumes that the coefficient of thermal expansion is positive. Extending this concept and using the Boussinesq approximation, it is seen that the acceleration gradient is antiparallel to the thermal gradient at each and every point in the fluid, or in general that:

$$\vec{g} \times \nabla\rho_o(1 - \beta\Delta T) = 0 \tag{8}$$

where $\Delta T = T - T_{ref}$. With the assumption that $\beta > 0$, even though β may be a function of temperature equation, 8 reduces to:

$$\vec{g} \times \nabla T = 0 \tag{9}$$

Likewise, the second part of equation 7 reduces to:

$$\vec{g} \bullet \nabla T < 0 \tag{10}$$

In a rotating fluid, the acceleration field is not homogeneous, i.e. the acceleration vector varies in magnitude and direction throughout the fluid. Application of the above analysis to equation 6 predicts that there still exists at least one family of thermal configurations in a rotating fluid where convective flow ceases. Here this state is called the thermally stable configuration. To illustrate this, figure 2 shows the centrifugal acceleration field in a centrifuge without the inclusion of a background gravitational acceleration. This scenario would be experienced by a rotating fluid in space (i.e., a centrifuge in space). The acceleration field always points radially out from the axis of rotation and the magnitude of the acceleration is proportional to the radial distance. In addition, this field is two-dimensional, i.e. there is no acceleration along the axis of rotation. Thermal stability theory predicts that one family of thermal fields that results in an absence of convection has circular isotherms centered about the axis of rotation. The fluid is cooler as one moves radially outward from the axis of rotation, so that the hotter fluid is "over" the colder fluid in relation to the acceleration vector. Here the acceleration field is perpendicular to the isotherms everywhere.

With the inclusion of a constant background acceleration, the thermally stable field is not readily recognizable. This scenario occurs in centrifuges on earth where the background acceleration is earth's gravitational field. One family of thermal fields that leads to a thermally stable configuration has isotherms that are paraboloids centered on the axis of rotation, as shown in figure 3. In figure 3, a cylindrical coordinate system is shown. The explanation for the paraboloidal isotherms begins by examining the acceleration field, which is:

$$\vec{a} = \vec{g} + \omega^2\vec{r} = -g\hat{z} + \omega^2 r\hat{r} \tag{11}$$

The unit directional of the acceleration vector is:

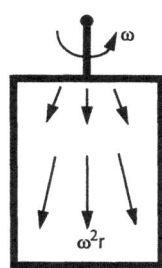

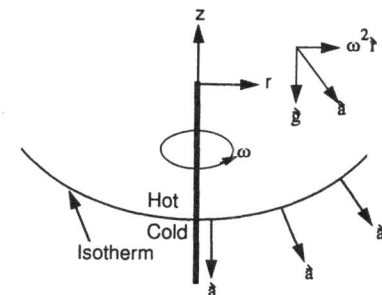

Figure 2. Acceleration field for an enclosed rotating fluid without a constant background acceleration field.

Figure 3. Thermally stable configuration for a rotating fluid with a background gravitational acceleration. (Cross-section through the axis of rotation shown).

$$\hat{e} = \frac{-g\hat{z} + \omega^2 r\hat{r}}{\sqrt{g^2 + \omega^4 r^2}} \qquad (12)$$

A family of paraboloids centered about the z axis can be represented by the equation:

$$bz = d + cr^2 \qquad (13)$$

where b, c and d are constants specific to each individual paraboloid. The outward unit normal to equation 13 is:

$$\hat{n} = \frac{-b\hat{z} + 2cr\hat{r}}{\sqrt{b^2 + 4c^2}} \qquad (14)$$

which is in the same form as equation 12. Thus, the acceleration field in a fluid held on a centrifuge on earth is paraboloidal.

THE GROUND-BASED FLOW VISUALIZATION CELL

The flow visualization cell is shown in figure 4. The cell was originally designed to simulate convective flows during directional solidification.[8,11] The cell has been modified here to test thermal stability when rotated about its axis. The cell consists of a 12 mm I.D. by 16 mm O.D. quartz tube with aluminum end caps. Water is shown as the fluid inside, but other usable fluids may include silicone oil, alcohol, methanol, etc. When water is used, small amounts of deuterated water, glycerin, alcohol or salt can be added for density matching the fluid to the tracer particles or dye. For instance, if monobasic aluminum stearate is used as the tracer particle material, salt can be used for density matching. Ivory hand soap can be used to leave a streak enabling visualization of the flow. The ampule is sealed by an O-ring in the groove in the aluminum end cap, into which the quartz ampule is compressed. Mechanical stress is taken off the quartz ampules by 4 stainless steel tubing stops. In the experiments, heat transfer out of the ampule could be reduced by the use of an outer quartz tube that serves as

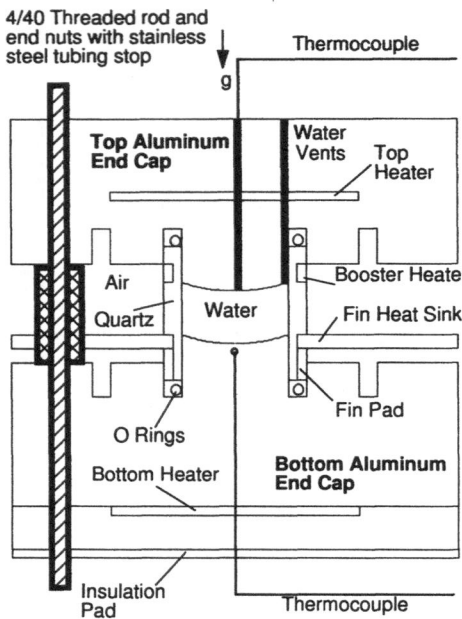

Figure 4. The ground-based flow visualization cell. Cross-section view shown. Outer quartz ampule not shown.

a wind break. The axial temperature gradient is set up by a heater embedded inside the top end cap. Radial temperature gradients are primarily induced by the curvature in the end caps. A heat source and a quartz fin (heat sink) on the inner quartz ampule are used to adjust the radial temperature gradient. In this cell, the end caps were inserts, thereby allowing the shape of the interface to be changed by machining additional inserts. A 12 inch long, type T, sheathed thermocouple was coiled and used as a heater. Water vents allow the cell to be filled without trapping air. In addition, the water vents can be used to insert thermocouples allowing temperature profiles inside the cell to be mapped.

As stated previously, the radial temperature gradients can be controlled with this cell design. From equation 13, the parabolic isotherms are determined by the constants b,c and d. For this analysis, it was important to design the experiment such that thermal stability would occur at a rotation rate less than 30 rad/s. If an interface depth, denoted Δx in figure 5, of 1 mm is used, the resulting equation for the isotherms with d = 0 is:

$$z = 0.278r^2 \tag{15}$$

The outward normal to this equation is:

$$\hat{n} = \frac{-\hat{z} + 0.556r\hat{r}}{1.144} \tag{16}$$

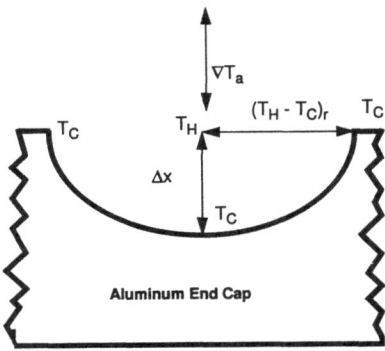

Figure 5. Estimation of the radial temperature gradient near the end cap.

Relating equation 16 to equation 12 allows the rotation rate for thermal stability to be calculated. The thermally stable rotation rate is $\omega = 23.35$ rad/s for these conditions.

The ideal interface would follow equation 15, i.e., paraboloid. However, it was much simpler to machine a spherical surface with a radius of 1.85 cm. The machined surface does not vary by more than 0.003 mm (0.0011 inch) from the ideal surface described by equation 15. In fact, a small amount of hand finishing would render the two surfaces indistinguishable.

From figure 5 it is seen that the temperature at the axis of the cell can be approximated by:

$$T_H \cong T_C + (\Delta T_a / l)\Delta x \tag{17}$$

which leads to an approximation for the radial temperature difference:

$$T_H - T_C \cong (\nabla T_a)\Delta x \tag{18}$$

NUMERICAL MODEL

In this section are results obtained from a fully nonlinear two-dimensional axisymmetric numerical model for the experimental cell described above. The model included cell geometry and temperature-dependent material properties. Only steady-state results are presented. The objectives were to test the feasibility of inducing a thermally stable temperature field, to inspect the impact on the flow of heat loss through the ampule wall, to compute the power required for this system, and to visualize the expected flow modes. The model was used with earth's gravity perfectly aligned with the ampule axis and the cell rotated about this axis.

The model used for this study is shown in figure 6. The numerical work presented here does not include the outer quartz cylinder. Each O-ring groove contains a rubber O-ring, air, and water. Because of the complicated geometry of the various materials filling the O-ring groove and the small volume these materials occupy, the O-ring groove area was approximated as a single material and given a volume-weighted average of the rubber, air and water material properties. Those areas are shown as black squares in figure 6. Similarly, the

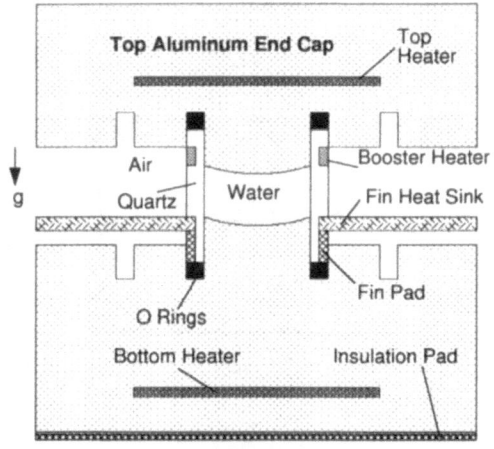

Figure 6. Modeled portion of the visualization cell.

heater was given a volume-weighted average of the material properties of the thermocouple and air. As with the actual cell, the interface used in the model was not a paraboloid, but rather had a constant radius of curvature of 1.85 cm. This imperfection was introduced into the model because it would occur in the experimental cell.

Because the desired temperature field is parabolic, the radial heat transfer must be nearly constant when the axial temperature gradient is constant. The gradient of the temperature field is:

$$\nabla T = \frac{\partial T}{\partial z}\hat{z} + \frac{\partial T}{\partial r}\hat{r} \tag{19}$$

In experiments, the imposed axial temperature gradient was to be on the order of 30 °C/cm. With this axial temperature gradient, then it would be expected that everywhere:

$$\frac{\partial T}{\partial z} \cong 30 \ ^{\circ}C/cm \tag{20}$$

In general, the temperature field can be expressed by:

$$T = [\nabla T_a]z - \left[\frac{4\nabla T_a}{d^2}\right]r^2 + T_o \tag{21}$$

Thus, the temperature field desired here is expressed by:

$$T = 30z - 8.34r^2 + T_o \tag{22}$$

where T_o is the temperature of the bottom cap. Hence, at the water-quartz boundary ($r = 0.6$ cm):

$$\frac{\partial T}{\partial r}\hat{r} = -10.0 \ ^\circ C/cm \tag{23}$$

which, with $k = 0.0015$ cal/cm•s•$^\circ$C, leads to the approximate boundary condition at the water-quartz interface:

$$-k\frac{\partial T}{\partial r}\hat{r} = -k\frac{\partial T}{\partial n}\hat{n} = q" = 0.015 \ cal/s•cm^2 \tag{24}$$

At the outer quartz boundary (r=0.8 cm):

$$q" = 0.0112 \ cal/s•cm^2 \tag{25}$$

Convective boundary conditions were applied to the outer ampule boundary and end caps. Assuming an emissivity of one, the linearized radiative heat transfer coefficient has a value of 1.4×10^{-4} cal/$^\circ$C•s•cm^2 at 22°C and 1.8×10^{-4} cal/$^\circ$C•s•cm^2 at 77°C. Convective heat transfer coefficients typically have a value of 0.5 to 1.0×10^{-4} cal/$^\circ$C•s•cm^2 when the Grashof number is small. Therefore, a constant heat transfer coefficient that approximated the combined radiative and convective effects, had a value of 2.4×10^{-4} cal/$^\circ$C•s•cm^2, and was used on all outer surfaces. (Physically, the heat transfer coefficient at the outer boundaries could be held nearly constant by encasing the experiment in a closed container while undergoing centrifugation.)

The bottom end cap can be held at room temperature by resting the cell on a large metal block. However, if the bottom end cap is held at room temperature, then a parabolic temperature profile would be difficult or nearly impossible to set up because the heat transfer from the ampule near the bottom cap would be near zero since the temperature difference between the ampule and the ambient air would be near zero. It is better to let the bottom cap achieve a temperature 40 to 50 $^\circ$C above ambient in order to make the radial heat transfer more constant. This was achieved by placing an insulating layer between the bottom face of the cell and a large metal block. Relating the heat transfer coefficient described above to equation 25 indicates that the average temperature of the cell should be about 46.7°C above ambient, or about 68.7°C. In the model, the temperature of the ambient air was taken as 22°C. Because of the low temperatures and complicated geometry involved, exact radiative heat transfer was not incorporated into the model.

NUMERICAL METHODS

The above set of equations in dimensional form was solved using a modified version of FIDAP 7.5, a finite element based code.[7] Non-slip boundary conditions were imposed on all solid walls. For the results presented, the cell was assumed to be at steady state. In previous studies, transient analyses were used for flow stability verification.[10] In all cases a fixed-grid approach (nodal points spatially fixed) was used.

The results presented hereafter were checked for convergence to within a specified absolute tolerance of 0.0001 for both the normalized velocity and the residual error norms. Spatial convergence was ascertained by comparing results obtained with different grid spacings. Typical simulations involved 22005 nodes using 4 node isoparametric quadrilateral elements for the 2-D simulations. The thermophysical properties of the various materials are listed in table 1.

NUMERICAL RESULTS

The power to the cell was adjusted to produce an axial temperature gradient of about 30°C/cm. The numerical model predicted that 11.22 watts of power would be required to attain a 30°C/cm gradient. With 3.61 watts of power applied to the upper main heater, 0.15 watts applied to the booster heater and 7.46 watts applied to the bottom heater, a 30°C/cm axial temperature gradient was set up. Here, the temperature of the top cap was 77°C and of the bottom cap was 59°C. The average temperature of the cell was 68°C. The resulting thermal field in the absence of convection is shown in figure 7. Note that the end caps are isothermal to within about 1°C. The isotherms in the water next to an aluminum end cap follow the curvature of the cap. This result was expected due to the high thermal conductivity of the aluminum in comparison with water. Figure 8 shows the radial temperature profile at the center of the cell. The radial temperature difference at the center of the cell was approximately 2.4°C. The average of the calculated maximum radial temperature differences agrees reasonably well with that predicted by use of equation 18, i.e., 3.0°C. The effect of using curved end caps to induce radial temperature gradients is seen. Figure 9 shows the axial temperature profile in the water at the centerline of the cell. Note that the profile is nearly linear, supporting the approximation made in equation 20.

The calculated velocity vectors in water in the cell at rotation rates 0, 23.35, and 25 rad/s are shown in figure 10 with the absence of Coriolis effects (We demonstrated in earlier numerical work that the Coriolis force alters flow patterns but does not alter the rotation rate

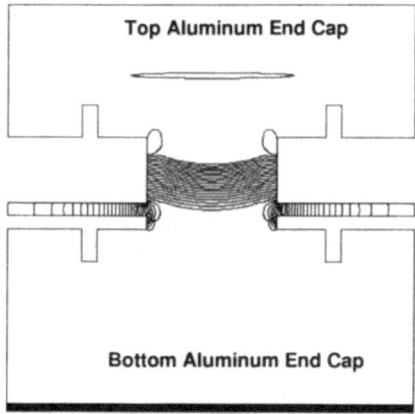

Figure 7. Calculated isotherms inside the test cell in the absence of convection (1.0°C between isotherms).

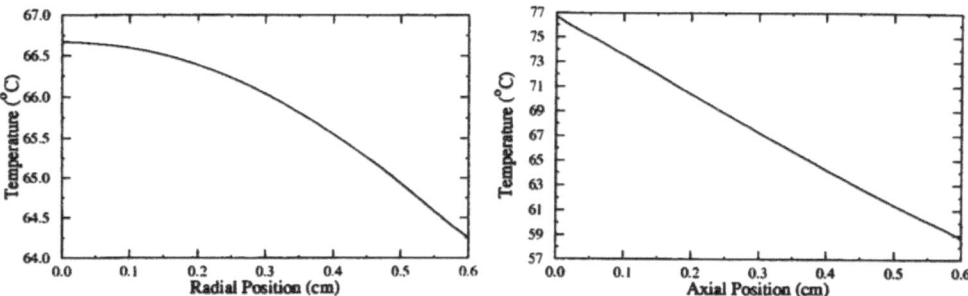

Figure 8. Calculated radial temperature profile at the midpoint of the cell in the absence of convection.

Figure 9. Calculated axial temperature profile in the water at the centerline of the cell in the absence of convection.

producing a minimum of convection.[10]) At $\omega = 0$, the maximum fluid velocity is 242 μm/s and, in general, flows up the center and down along the ampule wall. At $\omega = 23.35$ rad/s, the flow has nearly stopped with a maximum velocity of 38 μm/s, which is nearly an order of magnitude below the value at $\omega = 0$. At $\omega > 23.35$ rad/s, the flow is, in general, in a direction opposite to that at $\omega < 23.35$ rad/s, with the fluid flowing up the ampule wall and down the center. The maximum velocity as a function of rotation rate is shown in figure 11. As expected, a minimum of convection occurred at the rotation rate were the cell was predicted to be nearly thermally stable. However, the rotation rate at this point was $\omega = 23.75$ rad/s, which is slightly higher than that calculated by equations 16 and 12, i.e., $\omega = 23.35$ rad/s. This can be attributed partly to the spherical vs. paraboloidal interface. Experimentally, as well as here numerically, the flow would never be identically zero. Some level of residual convection occurs because the isotherms are never perfectly parabolic and because other misalignments, inhomogeneities and hardware imperfections will always be present.

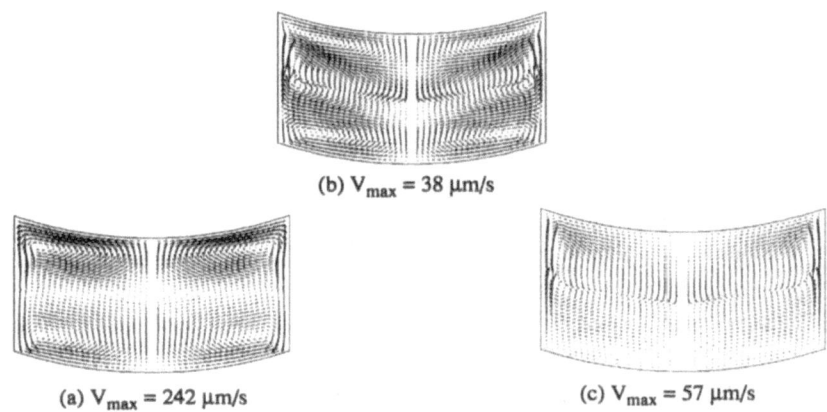

(b) $V_{max} = 38$ μm/s

(a) $V_{max} = 242$ μm/s

(c) $V_{max} = 57$ μm/s

Figure 10. Computed velocity vectors inside the cell in the absence of the Coriolis effect. (a) $\omega = 0$, (b) $\omega = 23.35$ rad/s and (c) $\omega = 25$ rad/s

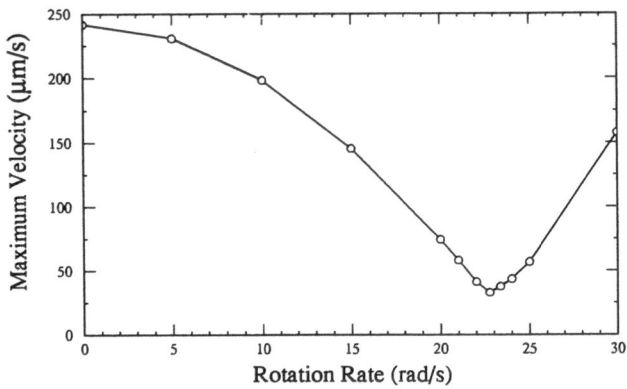

Figure 11. Calculated maximum fluid velocity in the cell as a function of rotation rate, neglecting Coriolis effect.

CONCLUSIONS

A flow visualization cell for investigating thermal stability in a centrifuge was designed using numerical simulations. It was shown that a thermally stable temperature field can be obtained in a simple flow visualization cell. Buoyancy-driven flow was nearly eliminated at the predicted rotation rate. The cell should operate with low power.

FUTURE DIRECTIONS

This numerical research lays the groundwork to build a flow visualization cell to experimentally test thermally stability in a centrifuge on earth.

ACKNOWLEDGMENTS

This work was made possible through a space act agreement between the Computational Materials Laboratory at NASA Lewis Research Center and W. A. Arnold and Associates Engineering, Inc.

NOMENCLATURE

Dimensional Quantities

a	Net acceleration; vector sum of gravitational and centrifugal accelerations, cm/s^2
C_p	Specific heat, $cal/g \cdot {}^{\circ}C$
ê	Unit directional
g	Gravitational acceleration, cm/s^2
h	The distance opposite to gravity from any chosen reference plane, cm
k	Thermal conductivity, $cal/cm \cdot s \cdot {}^{\circ}C$
l	Length of the fluid column, cm

L	Characteristic length, cm
$\hat{n}$	Unit normal
p	Pressure, $g/cm \cdot s^2$
P	Combined effect of local pressure and static centrifugal and gravitational forces ($P = p + \rho_o gh - \rho_o \omega^2 r^2/2$), $g/cm \cdot s^2$
q"	Heat flux, $cal/s \cdot cm^2$
q_s	Heat generation per unit volume, $cal/s \cdot cm^3$
r, θ, z	Cylindrical spatial coordinates, cm, rad, cm
$\vec{R}$	Position vector, cm
r	Radial distance perpendicular to the rotation axis, cm
T	Temperature, ^{o}C
t	Time, s
$\vec{u}$	Vector velocity, cm/s
V	Velocity in the fluid, cm/s
Δx	Depth of the machined interface, cm

Greek

β	Thermal expansion coefficient, $^oC^{-1}$
μ	Dynamic viscosity, $g/cm \cdot s$
ρ	Density, g/cm^3
ω	Rotation rate, rad/s

Subscripts

axial	Axial value
C	Cold
H	Hot
L	Refers to a length
max	Maximum value
o	Initial reference value
r, θ, z	Cylindrical spatial coordinates
radial	Radial value
ref	Reference value

Nondimensional Quantities

Gr	Grashof number, $\rho^2 a \beta (T_H - T_C)_L L^3/\mu^2$
Pr	Prandtl number, $\mu C_p/k$
Ra	Rayleigh number, $\rho^2 C_p a \beta (T_H - T_C)_L L^3/\mu k$
Ta	Taylor number, $4\rho^2 \omega^2 L^4/\mu^2$

Symbols and Diacritical Marks

$\wedge$	Unit Vector
$\rightarrow$	Vector quantity

TABLE 1. MATERIAL PROPERTIES

Water

Density = 1.0 g/cm^3
Specific Heat = 1.0 cal/g•C

Thermal Conductivity:

T (C)	k (cal/cm•s•C)
0	0.00135
20	0.00144
40	0.00151
60	0.00156
80	0.00160
100	0.00163

Viscosity:

T (C)	μ (g/cm•s)
0	0.0179
20	0.00982
40	0.00620
60	0.00471
80	0.00352
100	0.00297

Volumetric Expansion
Coefficient:

T (C)	β (1/C)
20	2.1 x 10^{-4}
40	3.8 x 10^{-4}
60	5.2 x 10^{-4}
80	6.4 x 10^{-4}
100	7.5 x 10^{-4}

Quartz

Density = 2.2 g/cm^3
Specific Heat = 0.15 cal/g•C
Thermal Conductivity = 3.4 x 10^{-4} cal/cm•s•C

Aluminum

Density = 2.7 g/cm^3
Specific Heat = 0.21 cal/g•C
Thermal Conductivity = 0.5 cal/cm•s•C

O-Rings

Density = 1.15 g/cm^3
Specific Heat = 0.5 cal/g•C
Thermal Conductivity = 3.5 x 10^{-4} cal/cm•s•C

Air

Density = 1.2 x 10^{-3} g/cm^3
Specific Heat = 0.24 cal/g•C
Thermal Conductivity = 6.3 x 10^{-5} cal/cm•s•C
Thermal Expansion Coefficient = 3.34 x 10^{-3} C^{-1}

REFERENCES

1. D.J. Tritton, "Physical Fluid Dynamics," 2nd edition, Oxford University Press, N.Y. (1988)
2. W. Weber, G. Neumann and G. Muller, Stabilizing Influence of the Coriolis Force During Melt Growth on a Centrifuge, *J. Crystal Growth*,100:145 (1990).
3. H. Rodot, L.L. Regel, and A.M. Turtchaninov, Crystal Growth of IV-VI Semiconductors in a Centrifuge, *J. Crystal Growth*, 104:280 (1990).

4. L.L. Regel, "Kosmicheskoye Materialovedeniye," Part 2, Volume 29 of the series Issledovaniye Kosmicheskovo Prostranstva, VINITI, Moscow (1987); translated into English as "Materials Science in Space: Theory, Experiments, and Technology," Plenum Press (1990).

5. L.L. Regel, "Kosmicheskoye Materialovedeniye", Part 3, Volume 34 and Part 4, Volume 39 of the series Issledovaniye Kosmicheskovo Prostranstva, VINITI, Moscow (1991).

6. G. T. Neugebauer and W. R. Wilcox, Experimental Observation of the Influence of Furnace Temperature Profile on Convection and Segregation in the Vertical Bridgman Crystal Growth Technique, *Acta Astronautica*, 25: 357 (1991).

7. M. Engelman, FIDAP Theoretical Manual (1987), Fluid Dynamics International, Inc., 500 Davis Street, Suite 600, Evanston, Illinois 60201

8. W. A. Arnold, Ph.D. Thesis, Department of Electrical and Computer Engineering, Clarkson University, Potsdam, NY (1993).

9. W. A. Arnold, L. L. Regel, W. R. Wilcox, Thermal Stability During Rotation in Space: A Scaling and Numerical Analysis, *Acta Astronautica*, 30:357 (1993).

10. W. A. Arnold and L. L. Regel, Thermal Stability and the Suppression of Convection in a Rotating Fluid on Earth, *in:* Materials Processing in High Gravity, Plenum Press (1994).

11. W. A. Arnold, W. R. Wilcox, L. L. Regel, and B. J. Dunbar, "Centrifuge in Space Flow Visualization Experiment," AIAA paper #93-0467, 31st AIAA Conference (1993).

FLOW VISUALIZATION STUDY OF CONVECTION
IN A CENTRIFUGE

Peter V. Skudarnov, Liya L. Regel, and William R. Wilcox

International Center for Gravity Materials Science and Applications
Clarkson University
Potsdam, NY 13699-5814

ABSTRACT

This research was dedicated to experimental observation of convection in a fluid subjected to a stabilizing thermal gradient during centrifugation. A laser light cut technique was employed to visualize convection in a test cell consisting of a transparent cylinder enclosed with two metal disks at the top and bottom. A concave shape was machined onto the bottom disk in order to simulate an interface typical for gradient freeze solidification of semiconductors. A constant vertical temperature gradient, similar to that used in gradient freeze crystal growth, was maintained in the cell. Convection was observed in water at several net acceleration levels, from 1 to 4g. Without rotation (1 g), buoyancy caused the usual two-dimensional axisymmetric flow pattern. Centrifugation introduced the Coriolis force and caused the flow near the "interface" to be predominantly rotational about the axis, as predicted by theory. With an uncooled plastic bottom, the flow there split to form two rotating cells in the cross sectional plane. No flow transition was found over the range of acceleration investigated, and the flow velocities were not strongly dependent on g.

INTRODUCTION

Regel and Rodot[1] showed experimentally that in directional solidification of a semiconductor in a centrifuge, it is possible to obtain a uniform doping concentration along the resulting ingot at a certain rotation rate. Theoretically, uniform doping is expected only when convection in the melt during solidification is negligible compared to the freezing rate. As pointed out by Arnold and Regel[2] and by Urpin,[3] this condition is expected when the acceleration vector is nearly normal to the solidifying interface (for very low doping levels). When this happens, the buoyancy driving force for convection in the neighborhood of the interface is a minimum.

Figure 1 shows a typical arrangement of equipment for centrifugal crystal growth. A furnace is placed inside a swing bucket so that when the centrifuge is rotating, the longitudinal axis of the furnace is aligned with the net acceleration vector. The first experiments in this configuration were performed by Regel et al.[4,5] The C-18 centrifuge at the Gagarin Space Training Center (Star City) was used for these experiments. This centrifuge has an 18 meter arm and can provide a net acceleration of up to 50 g (g = 9.81 m/s^2). Similar experiments were then conducted by Rodot et al.,[1,6] both in the C-18 centrifuge and in a 5.5 meter arm centrifuge at the Laboratoire Central des Ponts et Chaussees in France. Ag-doped PbTe and $Pb_{0.83}Sn_{0.17}Te$ crystals were grown in gradient freeze furnaces. These furnaces had a stabilizing thermal gradient, i.e. the temperature increased with height. Crystals grown at a net acceleration level of 5.5 g in the 18 m centrifuge and at 2 g in the 5.5 m centrifuge appeared to have an axially uniform composition. A constant dopant concentration is expected when there is a near absence of convection in the melt. For this reason, it was suggested that a convectionless flow regime existed in the melt at a certain acceleration level that depended on the centrifuge arm length. Later, it was pointed out that chemically homogeneous crystals were obtained at approximately the same rotation rate in both of the centrifuges.

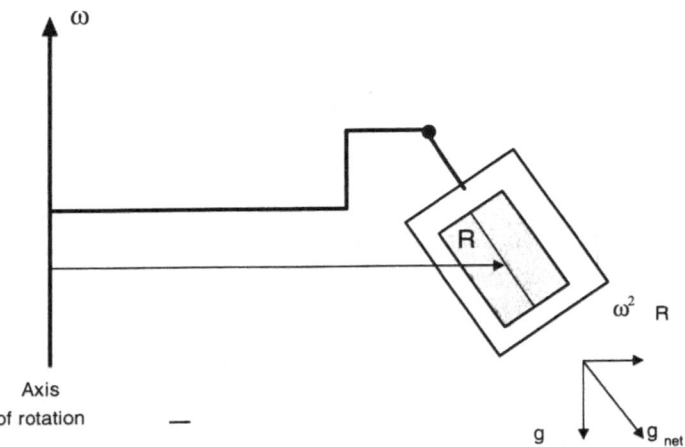

Figure 1. Schematic diagram of typical arrangement for gradient freeze growth on a centrifuge. The shaded area represents the furnace, which is attached to the centrifuge arm by a hinge so that the resultant acceleration g_{net} is aligned with the axis of the furnace tube.

Arnold et al.[2,7,8] considered convection in the gradient freeze experiments reported in references 1,4-6. They explained the reduction of convection at a critical g via the concept of thermal stability, as follows. The driving force for buoyancy-driven convection vanishes when the net acceleration vector is parallel to the density gradient throughout the fluid, with density decreasing with height. For gradient freeze growth in a centrifuge, it is important to recognize two facts. First, in a single component system the density varies only with temperature and the isotherms are normal to the density gradient. At equilibrium, the solid-liquid interface is an isotherm -- the melting point. The second fact, which is less apparent, is that the acceleration vector in a growth ampoule cannot be aligned everywhere with the ampoule's axis. Unlike growth in earth's gravity, the acceleration varies both axially and radially within the ampoule (because the distance to the centrifuge's center of rotation varies). It turns out that, in a centrifuge, the acceleration can be almost normal to a slightly concave freezing interface, which is the usual shape in gradient freeze growth. They also showed that the acceleration vector cannot be normal to isotherms throughout the entire melt, and so some convection must always be present. However, numerical simulation of convection in the PbTe experiments gave a sharp minimum in convection near the freezing interface at a particular rotation rate. This critical rotation rate would depend strongly on the exact shape of the freezing interface, on the arm length of the centrifuge, and to a lesser extent on the details of the thermal field throughout the melt.

In their numerical simulation, Arnold et al.[2,7,8] found that, for a given centrifugal acceleration field, introduction of the Coriolis force dramatically alters the flow field, but does not cause large changes in the axial and radial velocity components. Without the Coriolis force, the flow is predicted to be axisymmetric (as is well-known for Bridgman growth). With the Coriolis force, the flow is predicted to become three-dimensional, with circulation normal to the axis. The object of the present project was to observe the convection experimentally in a simulation of the gradient freeze configuration.

Although Arnold performed his modeling for the PbTe gradient freeze experiments, he pointed out[8] that the flow phenomena experienced during this type of crystal growth are not restricted to directional solidification. This provides an additional basis for conducting flow visualization experiments with a transparent working liquid instead of an opaque semiconductor melt.

The thermally unstable configuration, when temperature in a furnace decreases with height, was studied extensively by Muller et al.[e.g. 9-11] Experiments and numerical modeling showed that the unsteady convection regime expected in this configuration can be stabilized by centrifugation. A single steady convection roll, having the same rotation sense as the centrifuge, forms in the melt. Due to the action of the Coriolis force, the resulting flow is a nearly two-dimensional roll cell, mainly perpendicular to the axis of rotation of a centrifuge.

The only prior flow visualization experiments performed in a centrifuge appear to be those of Williams et al.[12] A column of water was heated from below, i.e. thermally unstable. A single convection roll formed, rotating with the same sense as the centrifuge. It was also noted that fluid motion was limited to the tangential planes. The flow patterns were explained by invoking the Coriolis effect. Results of this flow visualization are in very good agreement with the numerical modeling and crystal growth experiments described in reference 9.

EXPERIMENTAL PROCEDURE

In the present work, we studied a thermally stable configuration similar to the gradient freeze experiments of Regel and Rodot.[1,4-6] A cell filled with water was heated from above and cooled from below. This cell was placed in a swing bucket on the HIRB centrifuge,[13] so that the net acceleration vector coincided with the cell longitudinal axis (see figure 1). Observations were made both in tangential and cross sectional planes. The centrifuge is equipped with a data acquisition system and a triaxial accelerometer. Acceleration and temperature data are collected on a 486 computer while the centrifuge is spinning.

The test cell design is shown schematically in Figure 2. It consisted of a 22 mm inner diameter acrylic tube, with aluminum top and bottom end caps. The height of the tube was 45mm (aspect ratio 2:1). A concave shape was machined onto the bottom cap to simulate the concave crystal-melt interface typical for the gradient freeze technique. The top end cap was also curved. According to the numerical simulations of Arnold,[8] the surface at the hot end does not affect the flow near the cold end. The depth of curvature of both end caps was 2mm. An axial temperature gradient was maintained in the cell by a resistance heater at the top and a

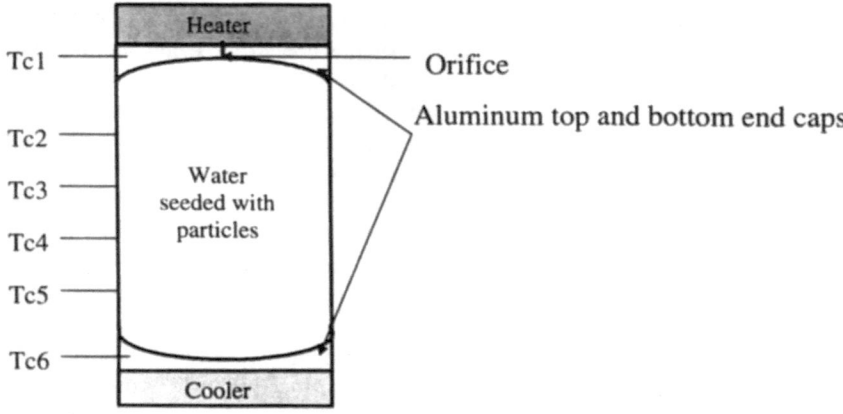

Figure 2. Schematic diagram of the test cell used in our flow visualization experiments.

thermoelectric cooler at the bottom. The temperature in the cell was monitored with two K-type thermocouples installed into the top and bottom end caps. Additional thermocouples were installed in the cell wall for temperature profile measurements. Polystyrene spheres (Duke Scientific Corporation's Polymer Microspheres, Polystyrene DVB, size range 2-120 μm, density 1.05 g/cm^3) and liquid soap (Colgate-Palmolive Company's Softsoap$^®$) were added to the water in the cell for visualization purposes. To reduce the amount of air in the water, it was boiled before pouring into the cell. Dry polystyrene spheres and liquid soap were mixed with preboiled water, then the cell was filled with this mixture through a small orifice in the top end cap. This hole also allowed extra water to escape when it expands due to heating. For observation of the flow pattern in cross sectional planes of the cell, the aluminum bottom end cap was replaced with a transparent cap made out of Plexiglas. This Plexiglas cap had the same curvature depth, 2 mm, as the aluminum one.

A laser light-cut technique was used for visualization of the convection. A sheet of laser light was produced by a glass rod placed in front of a He-Ne laser beam. Either the meridian or a cross sectional plane of the cell was illuminated with this sheet of light. When the water in the cell was heated, the liquid soap left streaks that were easy to observe. The convective pattern was recorded with a CCD camera connected to a video recorder. Since velocities in the thermally stable configuration were very low, we had to use a time-lapse video recorder in order to see the flow. This recorder allowed us to play back a videotape at up to 480 times faster than recorded. Usually 36-48 times accelerated playback was enough to see convection in our system. A 45° mirror was placed under the test cell for cross-sectional observations.

The test cell and video camera were mounted inside the centrifuge swing bucket. Thermocouple signals, video signal, and power for the heating and cooling elements were connected via slip rings to the control room. The standard experimental procedure consisted of spinning up the centrifuge, switching on the power to the heating and cooling elements, and recording the flow pattern after steady state was achieved in the cell. Usually it took 1.5 hours for the system to reach the steady state.

RESULTS

Temperature profile

The temperature profile in the cell was measured under normal gravity conditions, without centrifugation. Figure 3 shows the temperature along the longitudinal axis of the cell, measured with a moving thermocouple. In addition, the temperature along the wall was measured with the four thermocouples installed through small holes in the wall. The tips of these thermocouples were immersed in the water near the inner wall of the cell. All thermocouples used in these experiments were Omega K-type, stainless steel sheathed, 0.010 inch diameter. The temperature difference between the centerline and the wall was about 3 °C. As seen from Figure 3, the profile in the bottom part of the cell was very close to linear, which is typical for the gradient freeze crystal growth technique.

Flow visualization

Side view. An axial temperature gradient of about 15 °C/cm was maintained in the cell in all of the experiments described below. Figure 4 is a photograph of the streaks of liquid soap at 1 g, without centrifugation. Bright spots in the bulk of the cell are tracer polystyrene particles. A sketch of the flow pattern is shown in Figure 5. The primary bottom toroidal convection roll could clearly be seen in the time-lapse video recording. In addition, two secondary toroidal cells were present. One was adjacent to the primary cell, while the other was near the cell wall. This latter cell was much weaker than the other two; that is why it is almost indistinguishable in the photograph.

The bottom roll was generated by the radial temperature gradient imposed by the curved bottom end cap. The fluid rose near the centerline of the cell and descended near the cell walls. In the cross sectional view created by the light sheet, two convective cells could be seen. The right cell rotated clockwise and the left one counter-clockwise. Observations in several meridian planes showed the axisymmetry of this flow regime. Convective velocities in the primary bottom roll were estimated by tracking tracer particles. Measured flow velocities were about 25 μm/sec. Our normal gravity results are in good agreement with the numerical modeling and experiments[8] and in qualitative accordance with numerical simulations.[14]

During centrifugation, the tangential plane of the cell was illuminated with a laser light sheet, as shown in Figure 6. The tangential plane was parallel to the tangent of the circular motion of the cell during centrifugation, and passed through the longitudinal axis of the cell.

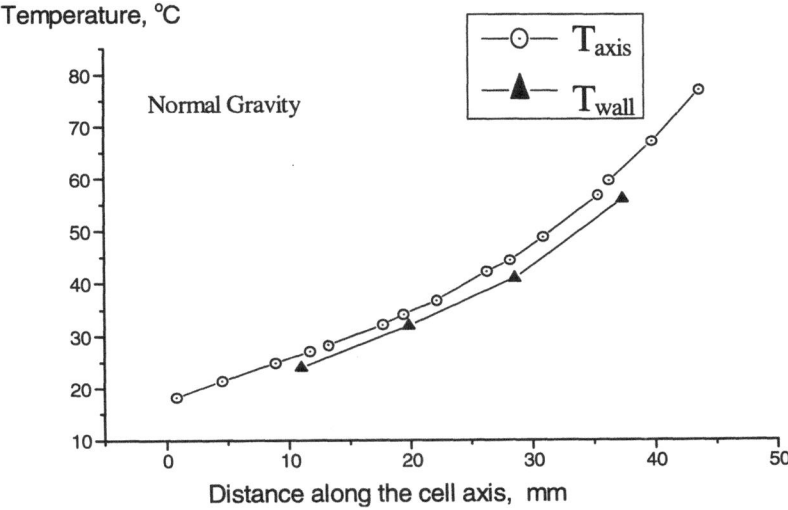

Figure 3. Temperature profile in the test cell, along the axis and at the wall.

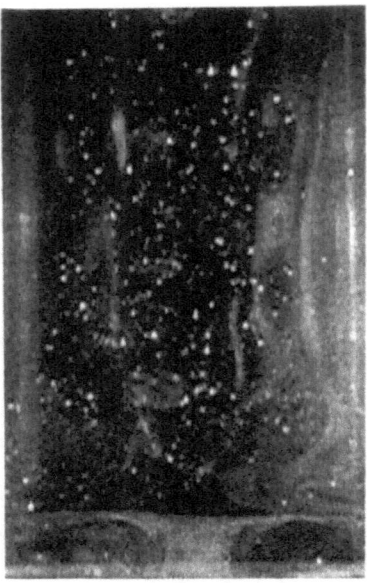

Figure 4. Side view, without centrifugation.

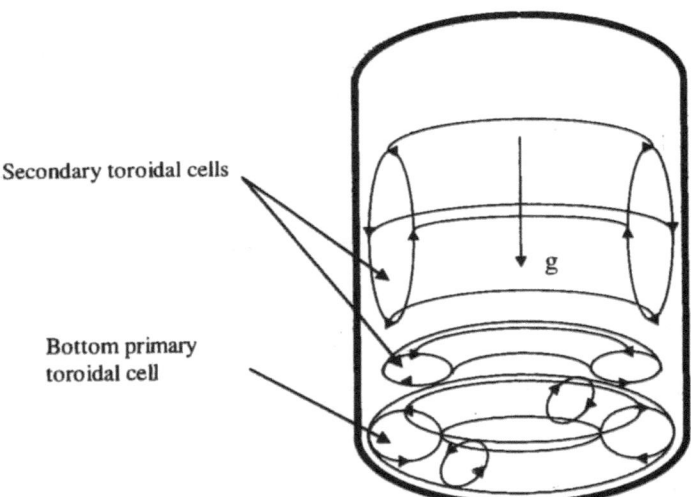

Figure 5. Schematic diagram of flow pattern at 1 g.

Figure 7 consists of photographs taken at different acceleration levels. Figure 8 is a sketch of the flow pattern. The photographs were taken from the monitor while the videotape was played back 36 times faster than real time.

The exposure time was about 6.5 s, which is equivalent to 4 min of experiment time. All photographs were taken 1 hr 35 min after the heating and cooling elements were turned on. One large convection roll formed near the bottom of the test cell. In the middle of the cell,

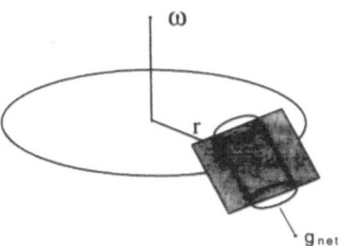

Figure 6. Tangential plane of the test cell illuminated with a light sheet.

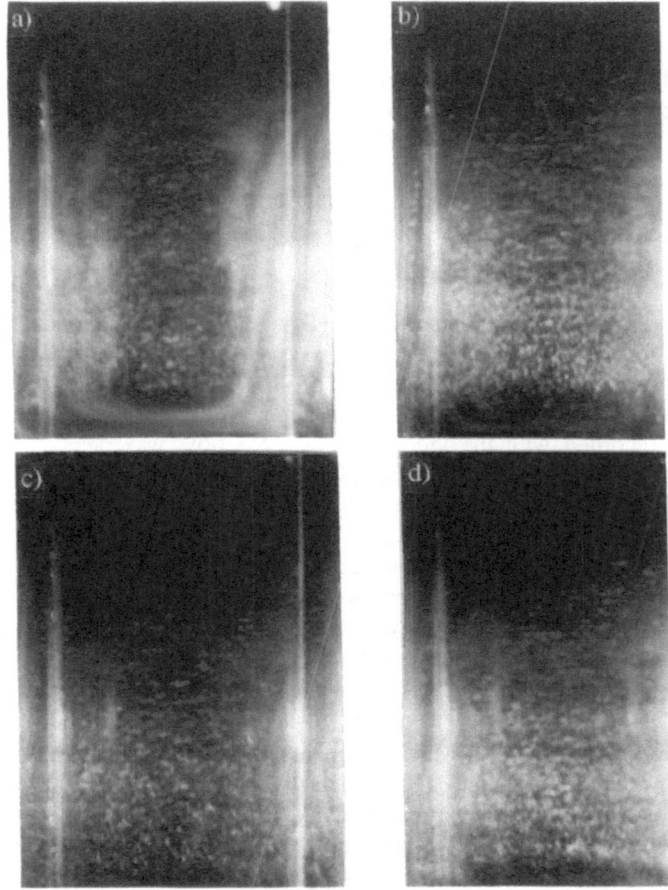

Figure 7. Streak photographs taken from videotape recordings of the illuminated tangential plane (see Figure 6) at different net accelerations with a cooled, curved bottom cap. (a) 1.5g; (b) 2 g; (c) 2.5g; (d) 3g

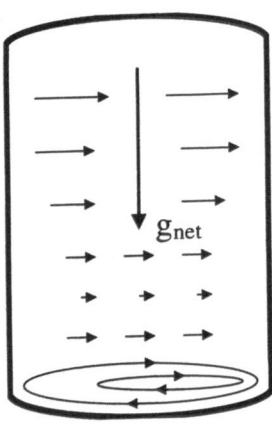

Figure 8. Flow pattern in the tangential plane of the cell during centrifugation, with a cooled, concave aluminum end cap on the bottom.

the particles had a velocity component directed perpendicular to the cell axis, from left to right, as depicted by the arrows in Figure 8. This velocity component was probably a consequence of the circular flow around the cell axis, with the same rotation sense as the centrifuge.

Convective velocities in the bottom roll were estimated by tracking individual tracer particles. The velocities were on the order of 65 μm/s, and increased only slightly with increasing acceleration.

Bottom end view. Cross-sectional planes were perpendicular to the longitudinal axis of the cell. For observation of convection patterns in cross sections, a transparent bottom end cap of plastic was used. There was no active cooling applied to the bottom in these experiments. The axial temperature gradient was about 10°C/cm. The bottom view, from a 45° mirror, was recorded with the video camera. No flow was detected in the cross-sectional plane at normal gravity, because all particles involved in the toroidal primary bottom convective roll had velocities directed almost perpendicular to this cross-sectional plane.

Figure 9 is a schematic view of the convective pattern during centrifugation is shown in Figure 9. Photographs taken at different accelerations, when the light sheet was positioned 5 mm above the bottom end cap, are shown in Figure 10. The other set of photographs, shown in Figure 11, was taken at the same net acceleration level of 1.5 g, but with different positions of the light sheet above the bottom end cap. All photographs were taken from the monitor while the videotape was played back 36 times faster than recorded.

The exposure time was about 6.5 s, which is equivalent to 4 min of experiment time. The photographs were taken 1 hr 55 min after the top heating element was turned on. Two separate counter rotating cells were present in the illuminated cross section. The fluid flowed from left to right near the tangential plane of the cell, split near the wall, went from right to left near the walls, and finally rejoined in the tangential plane. Based on our two-dimensional views in different planes, we can infer the three-dimensional structure of this flow.

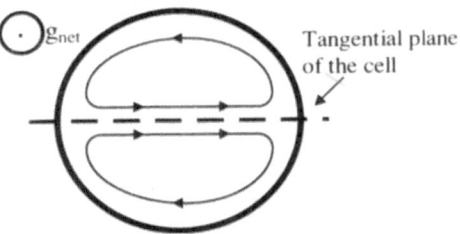

Figure 9. Flow pattern in a cross-section near an uncooled, plastic end cap on the bottom. The net acceleration vector is directed toward the viewer.

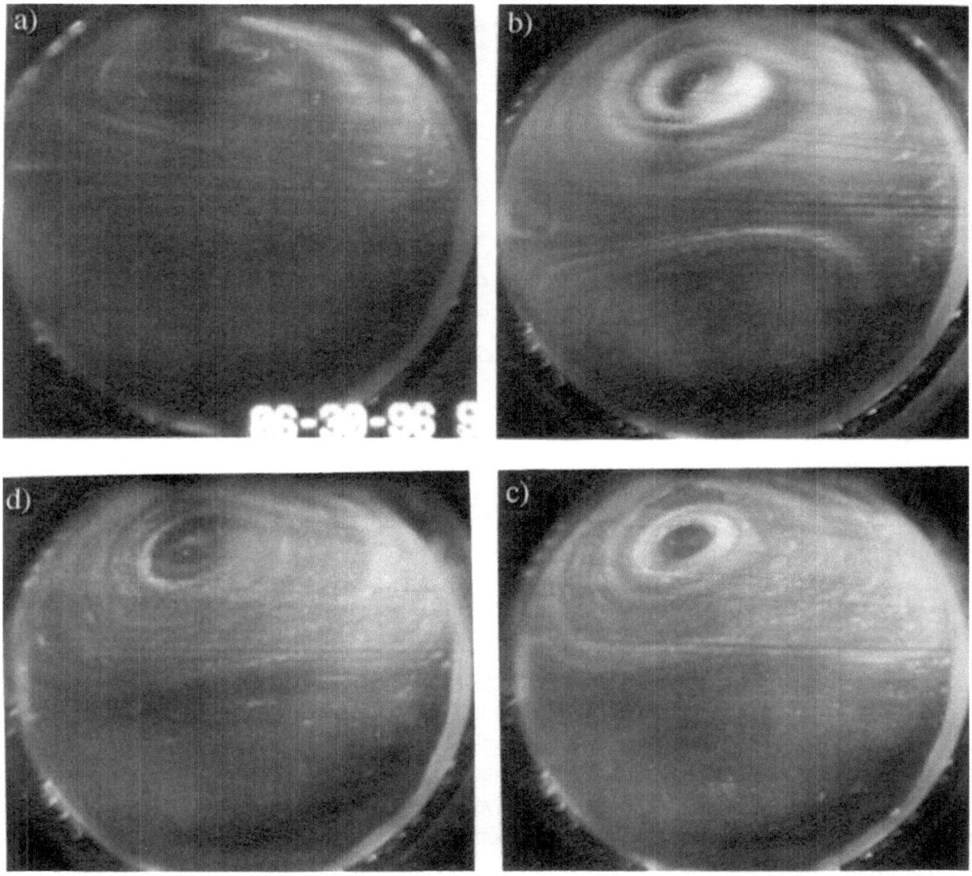

Figure 10. Streak photographs of a cross-section 5 mm above an uncooled, plastic end cap at the bottom of the cell at: (a) 1.5 g; (b) 2 g; (c) 3 g;. (d) 4 g

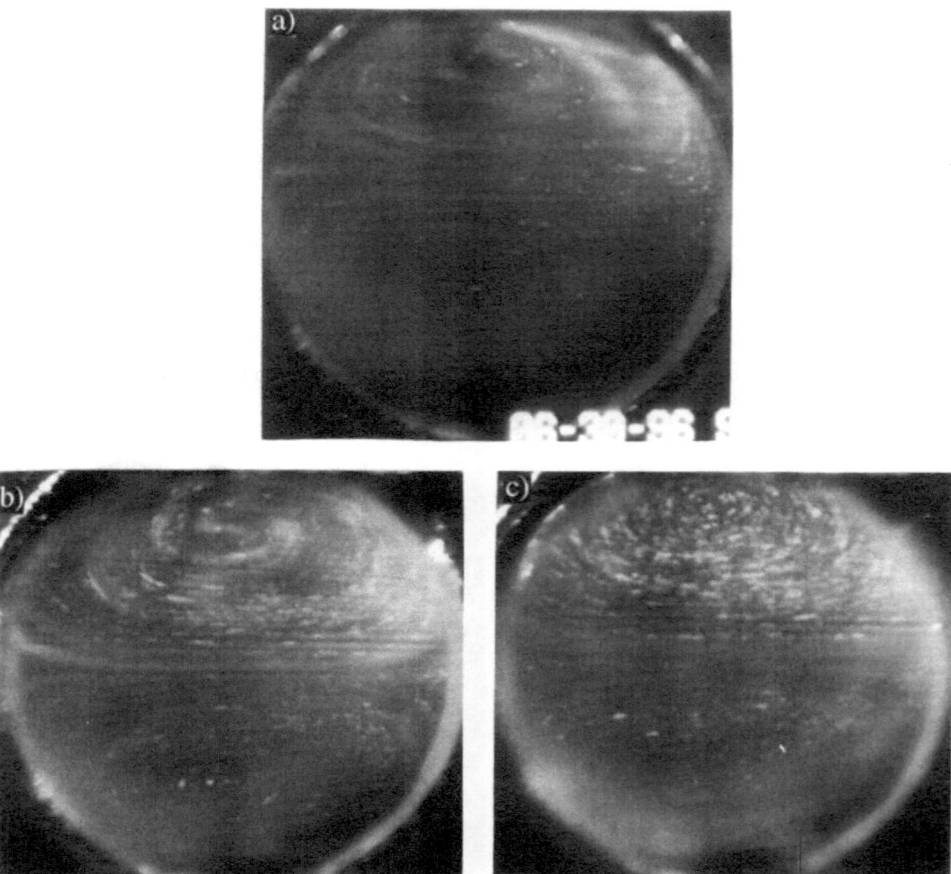

Figure 11. Streak photographs of a cross-section at 1.5 g with different positions of the light sheet above the bottom end cap: (a) 5mm; (b) 15mm; (c) 25mm

The flow near the bottom end cap consisted of two cells, separated from one another by the tangential plane, as shown in Figure 12.

DISCUSSION

The rotational flows observed near the bottom were a consequence of the action of the Coriolis force on buoyancy-driven convection, which otherwise would manifest itself as vertical cells with no circumferential component. It would be interesting to know the influence of this rotation on radial and axial segregation during solidification.

With a cooled aluminum bottom, the convection pattern corresponded to that predicted theoretically for flow that would be axisymmetric in the absence of centrifugation.[2,8] The Coriolis force caused the water near the bottom to rotate in a single cell. This rotation was superimposed on the usual flow, which rises in the center and falls at the wall. On the other hand, with a plastic bottom, two rotating cells were observed near the bottom. This behavior

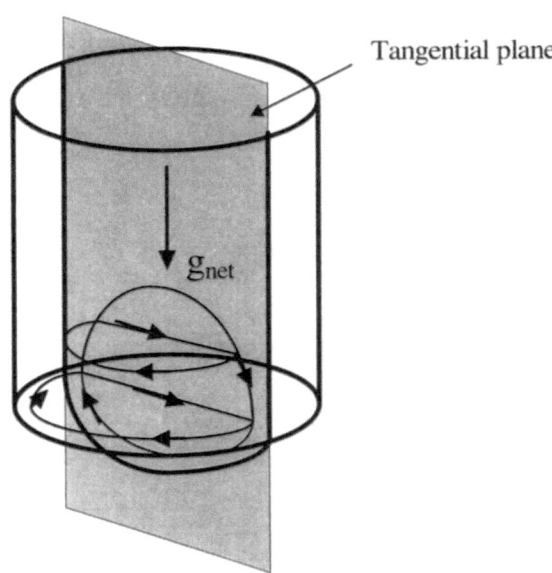

Figure 12. Schematic diagram of flow pattern near the bottom of an uncooled, plastic end cap during centrifugation. Only one of the two cells is shown.

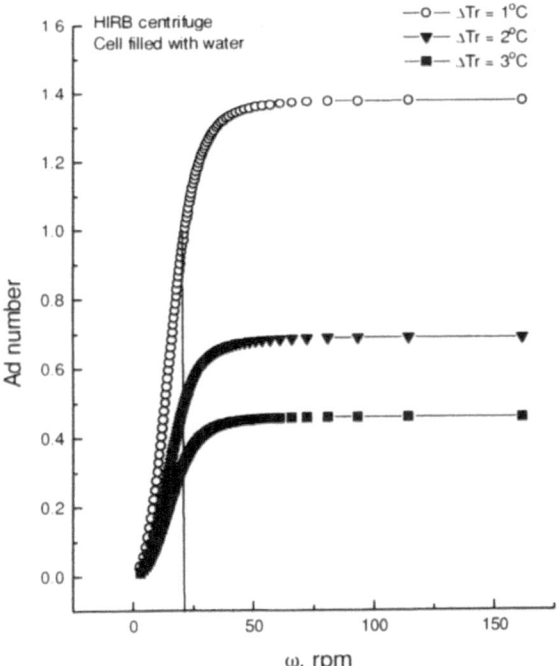

Figure 13. Ad versus rotation rate of the HIRB centrifuge for different values of the radial temperature difference.

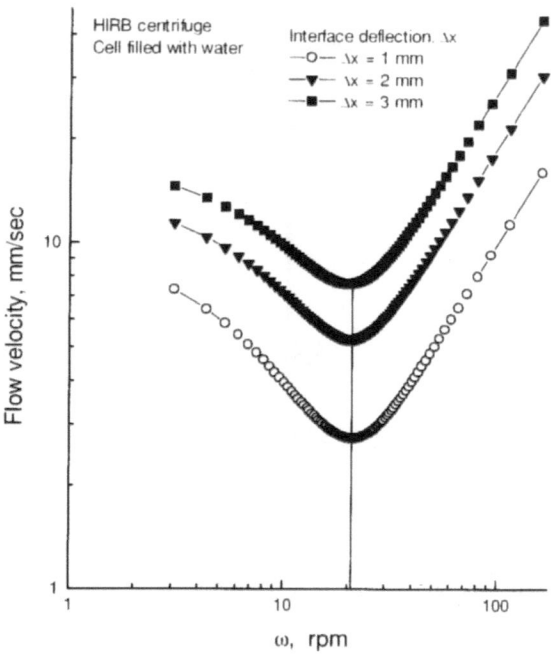

Figure 14. Predicted maximum flow velocity versus rotation rate of the HIRB centrifuge.[17,18]

corresponded to that predicted for the influence of Coriolis force on flow rising at one wall and falling at the other.[15] A primary lesson gained from our experiments is that, with centrifugation, the usual side view is insufficient to reveal convective patterns. One must also look from an end in order to see the circumferential flow components introduced by the Coriolis force.

In the range of net acceleration from 1.5 to 4 g, the general flow pattern stayed the same. In other words there was no flow transition. This is in agreement with theoretical predictions,[7,8] where the non-dimensional Ad number was introduced to identify flow transition conditions: $Ad = \omega^2 h \, [T_H - T_C]_a / g_{avg} \, [T_H - T_C]_r$, where ω is the centrifuge rotation rate, h is the height of the melt column, g_{avg} is the average net acceleration in the melt column, and $[T_H - T_C]_a$ and $[T_H - T_C]_r$ are the radial and axial temperature differences, respectively, in the melt column. This number determines whether the gradient of acceleration or the average net acceleration dominates the buoyancy-driven flow. When Ad > 1, the gradient of acceleration produces the dominant flow mode. On the other hand, when Ad < 1, the natural buoyancy flow mode prevails. The transition in flow mode is predicted to occur at Ad = 1. In our experiments, the maximum value of Ad was on the order of 0.4, and so natural buoyancy always controlled the flow, and no flow transition would be expected. Figure 13 shows a plot of Ad versus rotation rate of the centrifuge for different values of the axial temperature difference. It is seen that for 1°C axial temperature difference, Ad reaches a value of unity at about 22 rpm, which corresponds to 1.5 g net acceleration for the HIRB centrifuge.

Using the order of magnitude analysis constructed by Friedrich *et al.*,[16,17] we obtained the plot of maximum flow velocity versus rotation rate shown in Figure 14. This plot suggests that a minimum in the flow velocities should occur at a rotation rate of about 22 rpm.

Acknowledgment

This research was supported by the National Science Foundation under grant number DMR-9414304.

REFERENCES

1. H. Rodot, L.L. Regel, and A.M. Turtchaninov, Crystal growth of IV-VI semiconductors in a centrifuge, *J. Crystal Growth* 104:280 (1990).

2. W.A. Arnold and L.L. Regel, Thermal stability and the suppression of convection in a rotating fluid on earth, *in:* "Materials Processing in High Gravity," L.L. Regel and W.R. Wilcox, eds., Plenum Press, New York (1994).

3. V.A. Urpin, Convective flows during crystal growth in a centrifuge, *ibid.*

4. L.L. Regel, G.V. Sarafanov, and A.M. Turtchaninov, State of the art and development prospects for research on crystal growth and metal solidification under conditions of elevated gravity, *Inst. Kosm. Issled. Akad. Nauk SSSR,* Preprint No. 907 (1984). Cited in L.L. Regel, "Materials Processing in Space," volume 1, Consultants Bureau, New York (1990).

5. L.L. Regel, G.V. Sarafanov, I.V. Videnskii, A.M. Turtchaninov, and H. Rodot, Making lead telluride crystals doped with silver under conditions of elevated gravity, *Inst. Kosm. Issled. Akad. Nauk SSSR,* Preprint No. 908 (1984). Cited in L.L. Regel, "Materials Processing in Space," volume 1, Consultants Bureau, New York (1990).

6. H. Rodot, L.L. Regel, G.V. Sarafanov, M. Hamidi, I.V. Videnskii, and A.M. Turtchaninov, Cristaux de tellurure de plomb elabores en centrifugeuse, *J. Crystal Growth* 79:77 (1986).

7. W.A. Arnold, W.R. Wilcox, F. Carlson, A. Chait, and L.L. Regel, Transport modes during crystal growth in a centrifuge, *J. Crystal Growth* 119:24 (1992).

8. W.A. Arnold, "Numerical Modeling of Directional Solidification in a Centrifuge," Ph.D. Dissertation, Clarkson University, Potsdam, NY (1993).

9. W. Weber, G. Neumann, and G. Müller, Stabilizing influence of the Coriolis force during melt growth on a centrifuge, *J. Crystal Growth* 100:145 (1990).

10. G. Müller, E. Shmidt, and P. Kyr, Investigation of convection in melts and crystal growth under large inertial accelerations, *J. Crystal Growth* 49:387 (1980).

11. G. Müller, G. Neumann, and W. Weber, *J. Crystal Growth* 70:78 (1984).

12. P. Williams, A. Chevy, S. Bobeche, and M. Rodot, *J. Phys. D: Appl. Phys.* 27:920 (1994).

13. R. Derebail, W.A. Arnold, G.J. Rosen, W.R. Wilcox, and L.L. Regel, HIRB - the centrifuge facility at Clarkson, *in:* "Materials Processing in High Gravity," L.L. Regel and W.R. Wilcox, eds. Plenum Press, New York (1994).

14. D.H. Kim and R.A. Brown, *J. Crystal Growth* 109:66 (1991).

15. Sh. Mavlonov, The phenomena of crystallization in centrifugal force fields and the dynamo effect, *J. Crystal Growth* 119:167 (1992).

16. J. Friedrich and G. Muller, Convection in crystal growth under high gravity on a centrifuge, *in present volume.*

17. J. Friedrich, J. Baumgartl, H.-J. Leister, and G. Muller, *J. Crystal Growth* 167:45 (1996).

DETERMINATION OF SOLID/MELT INTERFACE SHAPE AND GROWTH RATE DURING GRADIENT FREEZE SOLIDIFICATION ON A CENTRIFUGE USING CURRENT INTERFACE DEMARCATION

Illa Moskowitz, Liya L. Regel, and William R. Wilcox

International Center for Gravity Materials Science and Applications
Clarkson University, Potsdam, NY 13699-5814

INTRODUCTION

Since the experiments conducted by Regel and Rodot[1] on the Star City and Nantes centrifuges, several attempts have been made to explain their surprising results. A few numerical models of Bridgman growth on a centrifuge have been developed, the most notable being those of Arnold[2] and of Friedrich et al.[3] Arnold[2] developed a non-linear, three-dimensional model for convection in the melt during crystal growth by the "thermally stable" gradient freeze technique on a centrifuge. The parameters used were those of the Ag-doped PbTe experiments on the Star City centrifuge in Reference 1. The effects of the average resultant acceleration, the acceleration gradient and the Coriolis force were studied separately and together. Arnold predicted that thermal stability (no convection) would occur when the acceleration is orthogonal to the isotherms and the density increases in the direction of the acceleration. From numerical simulations, he found that the rotation rate for uniform impurity doping is strongly influenced by the interface shape and the growth rate. A higher concave curvature of the interface corresponds to a larger radial temperature gradient, and requires a higher rotation rate to induce thermal stability. A lower growth velocity leads to a longer diffusive length and causes the impurity segregation to be more sensitive to convection near the interface. From computer simulation, Arnold also found that the temperature gradient along the axis of the charge in the PbTe experiment was only ~20% of that measured on the steel cartridge holding the ampoule and so the freezing rate was about 5 times larger than previously thought.

The model of Friedrich et al.[3] is also three dimensional and non-linear. In this model, the material properties of Ge were used. From scaling analysis, the authors found that the diffusive and convective terms in the momentum equation are negligible compared to the buoyancy and Coriolis terms. They concluded that the buoyancy force is balanced by the Coriolis force. For a concave interface with the ampoule aligned with the net acceleration vector, they predict that the flow velocity at first decreases with increasing rotation rate, reaches a minimum, and then increases at higher rotation rates. A numerical simulation, carried out using the data from the experiments in Reference 1, showed that uniform axial doping should not occur at the rotation rate observed in the experiments. However, the actual interface shape during the experiments was unknown and the growth rate used in the calculations was not mentioned.

Until now, no one has determined the interface shape and the growth rate during directional solidification on a centrifuge. Without this knowledge, comparison of theory and experiment is a risky business, and the fundamental physics remains uncertain. In the study described here, Current Interface Demarcation (CID) is being utilized to determine this vital information. We are growing Te-doped InSb using the "thermally stable" gradient freeze technique on the HIRB centrifuge at the International Center for Gravity Materials Science and Applications at Clarkson University. A series of current pulses is applied to the charge at known time intervals throughout the growth process. Each current pulse causes a sudden increase and decrease in freezing rate, forming a dopant striation in the resulting solid. Upon proper etching, these striations are visible under an optical microscope, and from them the growth rate and interface curvature can be measured versus distance down the ingot.

EXPERIMENTAL SET-UP

The HIRB centrifuge has an arm length of 1.25 m and is capable of ~90 rpm (13.5 g). A single zone Bridgman furnace is mounted onto a swing-bucket at the end of the arm. The swing-bucket is attached to the arm by a hinge, which allows it to swing out during rotation maintain alignment of the net acceleration vector with the axis of the ampoule. The temperature in the furnace is remotely controlled by computer.

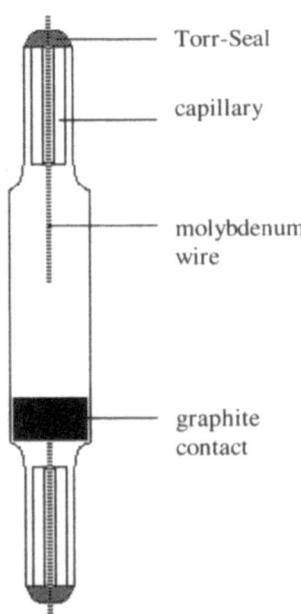

Figure 1. Quartz ampoule for CID growth on the centrifuge

For the experiments, 12 mm ID quartz ampoules are being used, as shown in Figure 1. Molybdenum wire is used to carry the current pulses to the charge. At the bottom of the charge a graphite electrical contact is used. At the top of the charge, the wire is inserted directly into the melt. The ends of the ampoule are sealed with an epoxy, which can withstand up to 200°C. In order to prevent the epoxy from being heated beyond this limit, the ends of the ampoule extend about 4 cm from the furnace at either end into ceramic caps which hold the ampoule in place. The ampoule is wrapped in quartz fabric and placed in a quartz liner, which is inserted into the furnace and held in place by ceramic holders. A diagram of the furnace set-up is given in Figure 2. The current pulses are provided by a power supply which is capable of producing 20A/28VDC and is designed to generate three different frequencies of square current pulses that are superposed on one another. The pulse width and time spacing in each of the frequencies can be selected by the user.

After growth, the samples are cut using a wire saw impregnated with diamond particles. The pieces are mounted in epoxy and polished mechanically using 2 grades of sandpaper and 3 grades of alumina suspension. In order to remove the larger polishing scratches, the samples are etched in a HNO_3: HF: HAc: Br_2 solution for a few seconds. After this, an etching solution of HF: HAc: $KMnO_4$ is used in order to reveal the striations. It is possible to see the striations with the naked eye and to characterize them using an optical microscope.

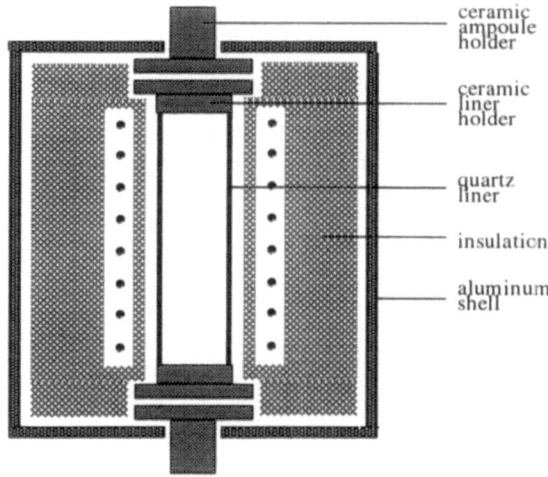

Figure 2. Furnace set-up for crystal growth with CID on the centrifuge.

RESULTS

This project is still in progress. A photomicrograph of an experiment at 1g is shown in Figure 3. For this experiment, a 9 mm inside diameter ampoule was used. The InSb was doped with Te at a concentration of $\sim 5 \times 10^{19}$ atom/cm^3. Current pulses of 10 ADC magnitude and 1 s duration were applied at intervals of 14 min. From the distances between neighboring striations, we estimate a growth rate of ~3 mm/h. From the cooling rate of the furnace and the measured temperature gradient one calculates a growth rate of 3.68 mm/h. The curvature of the interface has not yet been determined, but appears to be small.

Figure 3. Photo-micrograph of 9 mm diameter sample grown at 1 g with current interface demarcation. The growth direction was from right to left.

Acknowledgment

This research is supported by the National Science Foundation under grant DMR-9414304.

REFERENCES

1. H. Rodot, L.L. Regel, and A.M. Turtchaninov, Crystal growth of IV-VI semiconductors in a centrifuge, *J. Crystal Growth* 104:280 (1990).
2. W.A. Arnold, "Numerical Modeling of Directional Solidification in a Centrifuge," Ph.D. Thesis, Clarkson University (1993).
3. J. Friedrich, J. Baumgartl, H.J. Leister, and G. Müller, Experimental and theoretical analysis of convection in vertical Bridgman growth under high gravity on a centrifuge," *J Crystal Growth* 167:45 (1996).

IN SITU OBSERVATION OF DIRECTIONAL
SOLIDIFICATION IN HIGH GRAVITY

Y. Inatomi,[1] O. Kitajima,[2] W. Huang,[3] and K. Kuribayashi[1]

[1]The Institute of Space and Astronautical Science
3-1-1 Yoshinodai, Sagamihara, Kanagawa 229, Japan
[2]Tokyo Institute of Polytechnics
1583 Iiyama, Atsugi, Kanagawa 243-02, Japan
[3]Northwestern Polytechnical University
Xian 710072, P.R. China

ABSTRACT

In situ observation of directional solidification of transparent, faceting organic materials was carried out in a centrifuge at accelerations from 2 to 10 g. The solidification rate, temperature and concentration distributions, and flow pattern in front of the solid-liquid interface were simultaneously measured and observed using a microscopic interferometer. Temperature oscillations occurred in this high Prandtl number melt from 2 to 10g. Non-uniform temperature and concentration distributions in front of the interface were considered as the driving forces for interface breakdown at high acceleration.

INTRODUCTION

Recent work on the production of high-quality semiconductor devices, *e.g.*, thin-film silicon single crystals produced by zone-melting recrystallization[1,2] and bulk single crystals of compound semiconductors produced by the Czochralski technique,[3] have often reported the breakdown from a planar solid-liquid interface to a faceted cellular array. Since this morphological breakdown of the solid-liquid interface causes segregation of dopants and defects, there has been much interest in understanding pattern formation in faceted cellular array growth.[1,4,5] Many theoretical growth models, *e.g.*, two-dimensional nucleation,[6,7] persistent dislocation[8] and accumulated strain energy,[9] have been proposed.

Shangguan *et al.* performed *in situ* observations of unidirectional solidification of transparent, faceting organic compounds by a microscope, and did numerical work on the pattern formation.[5,10,11] They attributed the stable morphology of a cellular interface moving at a constant growth rate imposed by heater movement to solute pile-up in front of the bottom of the interface. The solute concentration at the bottom increased to keep constant the kinetic undercooling.[4-7]

Higashino, Inatomi and Kuribayashi[12] found recalescence regions, whose widths were about 200 μm, in front of the growing interface of salol by means of *in situ* observation using an interferometer (see Figure 1). In addition, they observed that the morphology of the cellular array was constant, although the salol had been purified by zone-refining. Faceting materials generally show large entropy changes upon fusion compared with metals. Therefore, it may be predicted that the latent heat of fusion released by growth changes the temperature field. This change is controlled by growth rate, thermal diffusivities of crystal, liquid and glass-cell wall, latent heat of fusion, configuration of the solid-liquid interface, and structure of the growth cell. However, there has been little discussion of the recalescence phenomenon in previous studies, because there were some difficulty in the precise measurement of heat transport. The conventional technique of employing only a thermocouple is not highly suitable for measurement of a two-dimensional temperature distribution without disturbing the growth environment.

In the present study, we performed temperature measurement and *in situ* observation of unidirectional solidification of faceted transparent organic compounds by two-wavelength interferometry on a centrifuge, in order to investigate the influence of acceleration upon mass and heat transport in the melt.

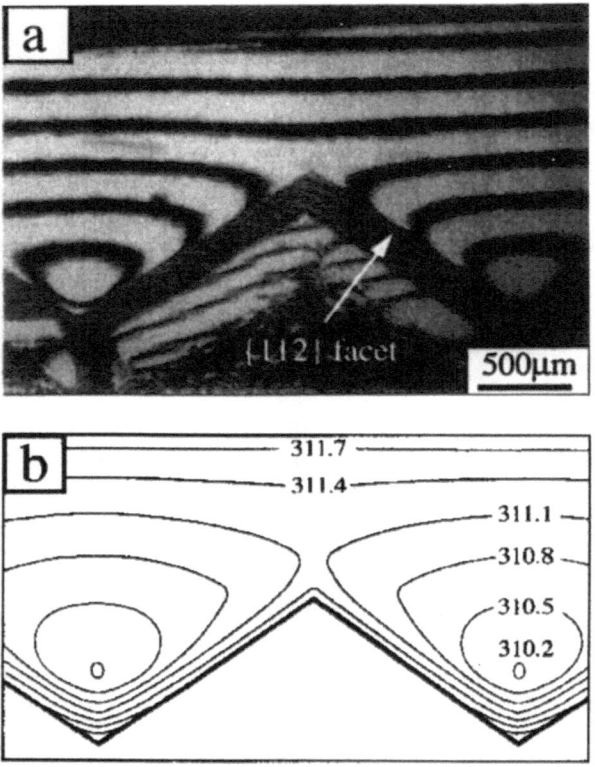

Figure 1. Temperature field in molten salol during solidification.[12]
(a) interference fringe pattern; (b) isotherms (in °C).

EXPERIMENTAL

Centrifuge

A table-type centrifuge system was designed and constructed for materials processing experiments. Figures 2 and 3 show a photo and a schematic drawing of the system. The radius and thickness of the aluminum table were 1.2 m and 30 mm, respectively. The rotation rate of the table was controlled with an inverter from 0 to ± 150 rpm. The center of inertia for the specimen cell was located at the point 0.855 m from the center of the centrifuge; therefore the maximum resultant acceleration at that point was 21.5 g (g = 9.81m/s^2). All components were bolted to the table. A maximum load of 100 kg on the table was certified using dummy weights. The positions of the components on the table were optimized for distance from the center of the centrifuge.

Figure 2. Photograph of the table-type centrifuge at the Institute of Space and Aeronautical Science (ISAS).

This experimental setup consisted of five functional parts: an optical system, a specimen cell, a data acquisition and processing system, a power supply for the specimen cell, and a telemetering system. Eighteen-channel slip rings were used for primary power supplies of the components and command-sending interfaces with computers, because the influence of the random noise, which may occur at the slip rings, on the signal should be suppressed as much as possible. Three components were included in the data acquisition and processing system; 1) a six-channel data collector for output of thermocouples or analog data, 2) a handheld FFT analyzer connected to a servo-type accelerometer through an amplifier, 3) four programmable temperature controllers. The accelerometer, whose measurable frequency range was 0 to 100 Hz, had been developed for an airplane and a rocket. Two components were included in the telemetering system; 1) three RS-232C interface circuits for computers, 2) a video-signal wireless transmitter.

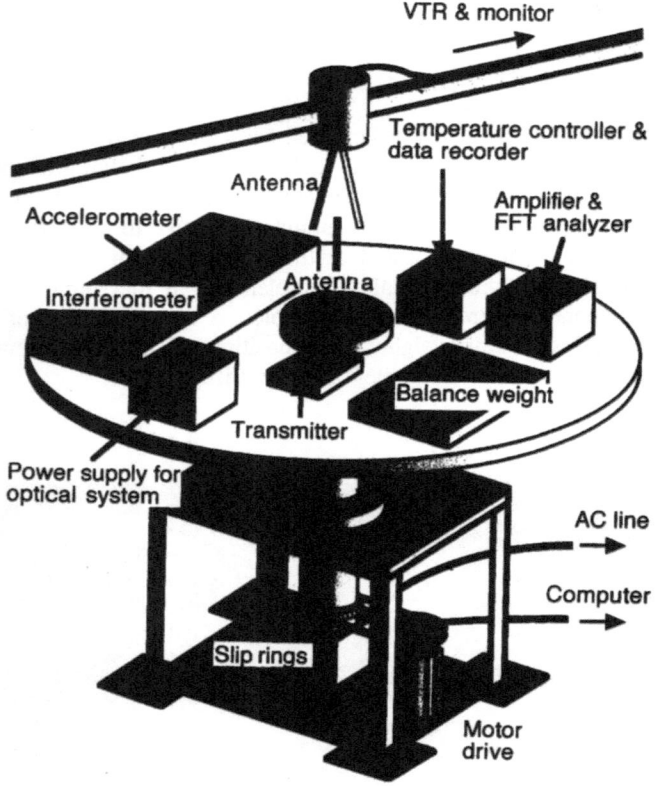

Figure 3. Schematic diagram of the centrifuge facility.

Optical system

Measurement of temperature and solute concentration profiles in melt growth with transparent systems, *e.g.*, succinonitrile-acetone mixtures[13] and salol-*t*-butyl alcohol alloys,[14] can be carried out with multi-wavelength interferometry. The refractive index of the melt depends on temperature, concentration, and the wavelength of the light source used. For small changes in the concentration and the temperature of the liquid, ΔC and ΔT, the linear relationship,

$$\Delta n_i \approx \left(\frac{\partial n_i}{\partial C}\right)_T \Delta C + \left(\frac{\partial n_i}{\partial T}\right)_C \Delta T \tag{1}$$

may be assumed with $(\partial n_i/\partial C)_T$ and $(\partial n_i/\partial T)_C$ constant, where n is the refractive index of the specimen. Therefore, the spatial distribution of the refractive index for two different wave lengths can be transformed into ΔC and ΔT relative to the values at a certain point in the liquid by solving Equation 1. By defining the values at a junction of a thermocouple set in the liquid as the origins of C and T, the concentration and the temperature distributions were finally obtained by measurement of the temperature and the change of refractive index near the junction.

Optical facilities used in a high acceleration environment must be stable mechanically. The optical system used in the experiment was a 2-prism-type microscopic interferometer (Figure 4), which had been newly designed and manufactured by Olympus Optical Company, Ltd. for a sounding rocket experiment. The functioning of this interferometer is based on multi-wavelength interferometry.

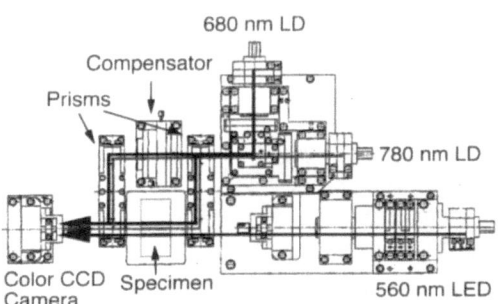

Figure 4. Schematic diagram of the two-prism interferometer used in the experiments.

As light sources in the present optical system, 5 mW laser diodes (LD) with 680 nm and 780 nm wavelengths and a high luminous light emitting diode (LED) with 560 nm, were used for observing the interference fringes and bright field images, respectively. Bright field images and interference fringe could be observed easily by switching the light sources. The former was used to measure solidification rate and morphological changes, and the latter was used to visualize temperature and concentration profiles in the melt. The objective lenses, with magnifications of 1, 2, 5 and 10, were adjusted to be both parfocal from visual to near-infrared wavelengths and in the best focus on the crystal surface through the quartz window. The images obtained by the interferometer were sent to a VTR system through a video-signal wireless transmitter.

In the temperature range 303 K $\leq T \leq$ 323 K, the differential coefficients of the refractive index of the specimen with respect to temperature and concentration were experimentally obtained as follows:

At $\lambda_1 = 680$ nm, $(\partial n_1/\partial C)_T = 3.2 \times 10^{-3}$ / mol % and $(\partial n_1/\partial T)_C = -5.9 \times 10^{-4}$ / K.

At $\lambda_2 = 780$ nm, $(\partial n_2/\partial C)_T = 2.5 \times 10^{-3}$ / mol % and $(\partial n_2/\partial T)_C = -6.7 \times 10^{-4}$ / K.

Specimen cell

Figure 5 shows the experiment cell. The molten organic compound was poured into a rectangular quartz cell, which was made of quartz glass polished to be optically flat, without a free surface of the melt. The thickness of the glass-cell wall through which incident light passed, the width of the cell, the height of the cell, and the specimen thickness were 7 mm, 34 mm, 10 mm and 1 mm, respectively. The upper and lower ends of the cell were equipped with Peltier heating and cooling units, respectively, which were controlled by thermocouples, TH and TC, via aluminum blocks attached to the units. A small reservoir of the organic compound was included in the upper metal block to accommodate volumetric changes in the melt. Thus directional solidification experiments could be performed repeatedly by changing the imposed temperature gradient in the cell and the cooling rates at the sides of the cell. The growth direction was opposite to the gravity vector to suppress thermal convection. Four calibrated chromel-alumel thermocouples, T1 - T4, of 50 μmφ with a junction of 100 μmφ, were placed at intervals of 2 mm in the cell to measure temperature distribution of the specimen. The specimen cell was set on a post-type holder, on which the cell could be rotated from 0 to 360 °.

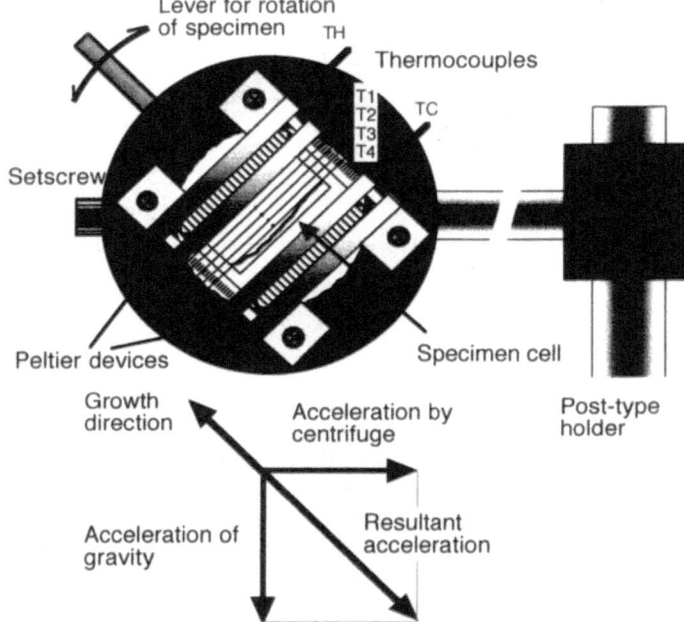

Figure 5. Schematic diagram of the experiment cell and holder.

Materials

Two types of measurements were performed during directional solidification of organic compounds: (1) temperature measurement by thermocouples, in order to observe any temperature instabilities in melt; (2) simultaneous observation of morphological change and temperature field in the melt by an interferometer. For these experiments, we used salol (phenyl salicylate, $2\text{-}(HO)C_6H_4CO_2C_6H_5$) and a mixture of thymol ($CH_3C_6H_3CH(CH_3)_2OH$) and o-terphenyl ($C_6H_4(C_6H_5)_2$). Commercial grade materials were purified by zone refining several times. The reasons why these materials were used are as follows. 1) They are often used as model materials for faceting solidification, as they have large Jackson's α factors. 2) The physicochemical properties are well established.[15-17] Figure 6 shows a portion of the phase diagram of the thymol - o-terphenyl system obtained with a DSC.

In the first stage, a single seed crystal with a favorable orientation was pre-grown from the top side of the cell through a 500 μm pore in the bottom plate of the cell. The orientation of the growth surface was adjusted to {112}. In the second stage, periodic growth and remelting of the seed crystal were performed to improve the quality of the seed. At the onset of solidification, the specimen was kept stationary in a temperature gradient for 3 hours at high acceleration, with temperatures of 58 °C and 38 °C for upper-side and lower-side of the cell, respectively. During solidification, both sides were cooled at 1 K/min.

RESULTS AND DISCUSSION

Salol

At 10 g, a fluctuation in acceleration of 2 x 10^{-2} g was measured in radial, vertical and tangential directions (cylindrical coordinates), using an accelerometer set at one position in the experiment cell. From these values, the influence of vibration on the flow-pattern may

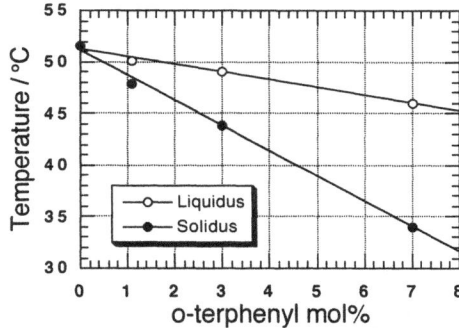

Figure 6. Thymol-rich portion of the phase diagram of the thymol - *o*-terphenyl system.

be neglected for salol, which has a high viscosity.

Temperature oscillations in molten salol were measured by the thermocouple. The amplitude of the oscillations was about 0.4 K, with a measurement error for this system of 0.1 K. The Fourier spectra of the temperature under several acceleration conditions are shown in Figure 7. Four peaks under 0.5 Hz are seen even at 2 g. Note the two peaks near 0 Hz and 0.5 Hz at 6 g, in contrast to other acceleration conditions. There is a possibility that the 6 g condition in this centrifuge system corresponded to a singular flow state.

Figure 8 shows the interference fringes, indicating how morphological change occurred and developed at 10 g. The fringes correspond to isotherms. The small fluctuations of the fringes are attributed to the flatness of the mirrors and the glass-cell.

A planar solid-liquid interface formed at the onset of cooling from 1 to 10 g. Initially, the temperature gradient was normal to the planar interface and the interval between fringes was constant, so that the gradient was considered to be uniform at this time. Macrosteps running parallel to the interface from the side to the center of the cell were observed in the bright field images using LED.

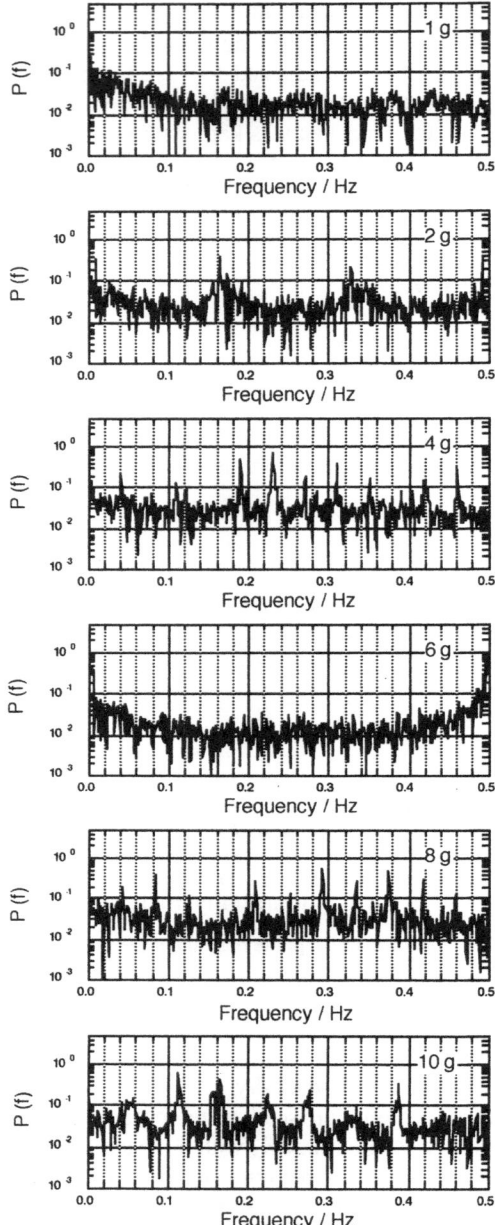

Figure 7. Power spectra of temperature oscillations in molten salol at 10 g. The temperature was measured with the T1 thermocouple.

Interface breakdown occurred at 8 minutes (left area in the bottom interferogram in Figure 8) and developed into a plateau whose orientation was the same as the seed crystal. Figure 9 shows the interface position as a function of time. The solidification rate increased as time went on.

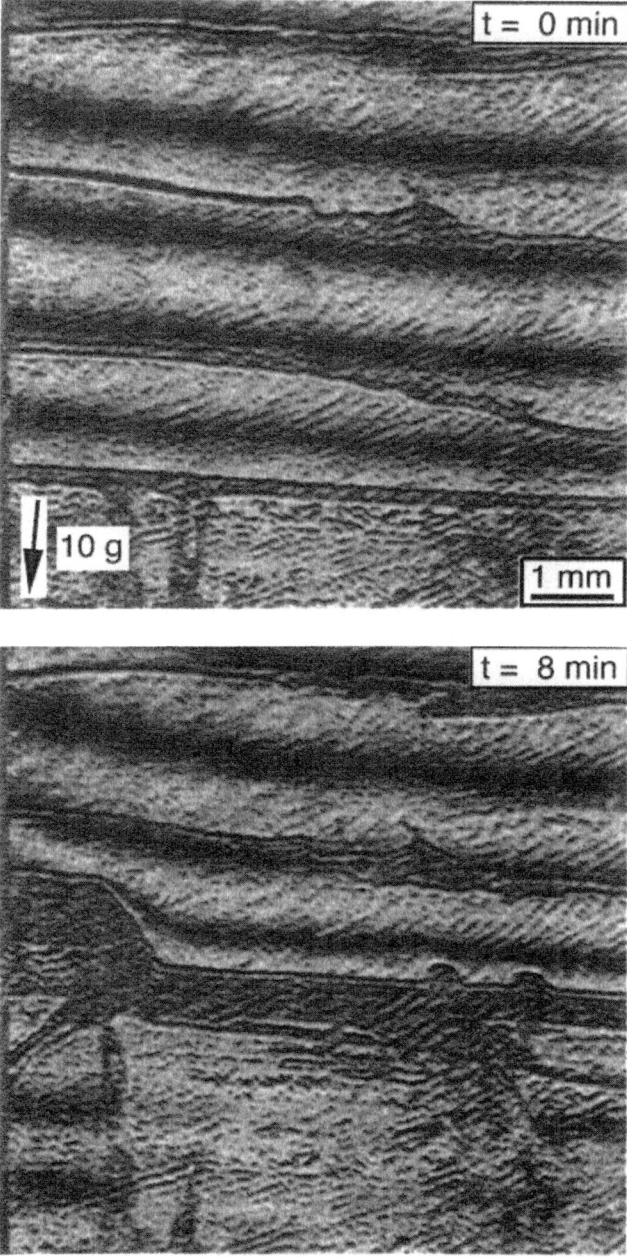

Figure 8. Fringe patterns in molten salol at the side of the cell at 10 g, initially and at 8 min.

After 8 minutes, the solidification rate became constant at about 8.3 μm/s, which was nearly same as the value defined externally by the cooling rate and the temperature gradient using the heater and the cooler, except near the side of the cell at 10 g.

Figure 10 shows the temperature profile in the melt as a function of distance from the interface. A recalescence region of about 500 μm in width is seen in this figure, except near the side of the cell at 10 g. The thermocouples indicated maximum temperatures when the interface crossed their junctions. This suggests that the latent heat released by solidification decreased the temperature gradient on the liquid side of the interface. Centrifugation increased the undercooling at the interface and then increased the solidification rate to be constant. The positive temperature gradient in front of the interface at the side of the cell at 10 g had a stabilizing effect on the interface morphology, with the interface there remaining flat even after breakdown elsewhere.

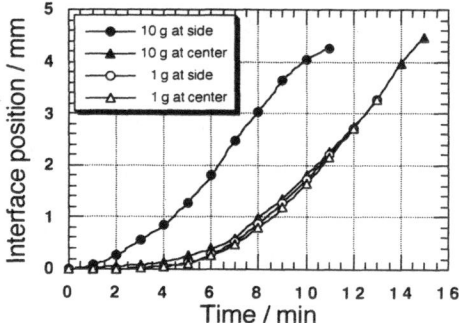

Figure 9. Freezing interface position of salol versus time, at the center and at the side of the cell.

Figure 11 shows the flow pattern in molten salol at 10 g obtained by a tracer method. Polymer particles under 10 μm in diameter were added as tracers. The direction and the magnitude of the flow velocity in this figure are averages taken over 10 minutes. Though no flow in molten salol was observed at 1 g, convection in the melt induced by high acceleration was confirmed from movement of the particles at 10 g. The tracer particles didn't enter the region about 600 μm in thickness in front of the interface at the center of the cell.

Thymol - *o*-terphenyl mixture

The maximum flow velocity in the melt measured with tracer particles was about 20% times larger than with salol, but there was no difference in the flow pattern. Figure 12 shows the temperature oscillations and the power spectrum under 0.5 Hz in molten thymol - *o*-terphenyl alloy at 10 g. The peak-to-peak amplitude of the oscillations in this mixture was about 1.0 K, which is two times higher than for salol. The peaks in the spectrum are slightly shifted from those for salol (see Figures 7 and 12b). These shifts may be attributed to the difference of fluid properties between the alloy and salol. Though the peak positions and heights in the power spectra were almost the same, the phase differences of the oscillations were approximately 6 seconds between the outputs of T1 and T2, and between T2 and T3. Since the temperature readings are out of phase and the vibrations on the table were quite small as mentioned above, the origin of the temperature oscillation is considered to be a periodic change of flow pattern at a frequency of 0.12 Hz.

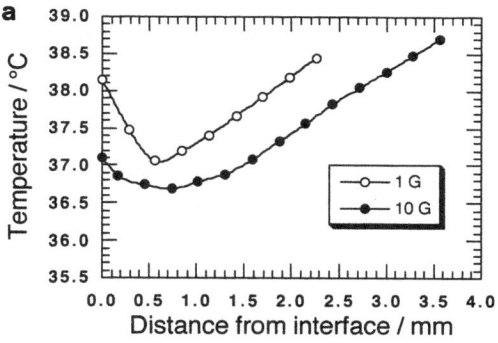

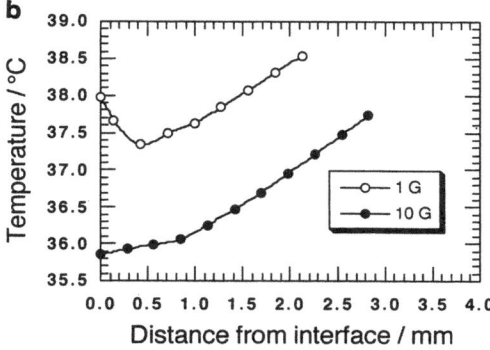

Figure 10. Temperature profile in front of the freezing interface of salol: (a) center; (b) side of the cell.

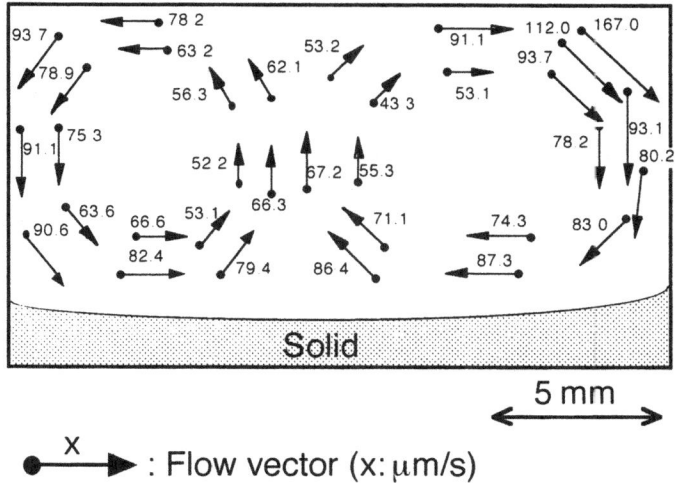

Figure 11. Flow pattern in molten salol, measured from motion of suspended particles.

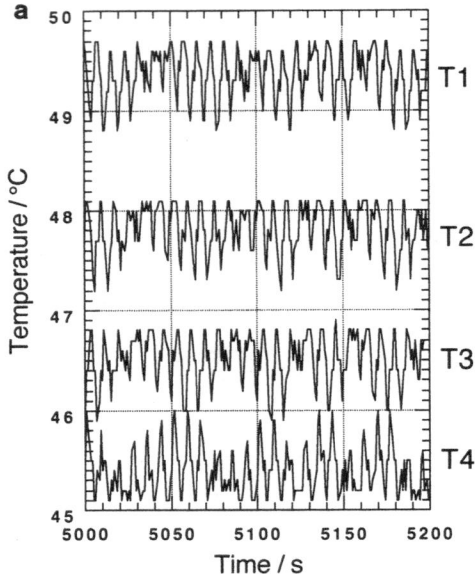

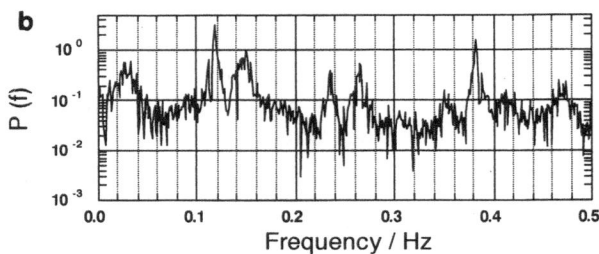

Figure 12. Temperature oscillations in molten thymol - *o*-terphenyl mixture at 10 g:
(a) measured at thermocouples T1 through T4; (b) power spectrum of temperature oscillations at T1.

Figure 13 shows the interference fringes at 10 g, indicating morphological change and distributions of temperature and concentration in the melt. The interface became concave at $t = 0$ minutes after homogenization. The fringe pattern at the center of the cell at t = 15 minutes shows close fringes near the interface compared with other conditions, i.e. the change in temperature gradient and the rejected solute in front of the interface. Figure 14 shows the interface position at the center and at the side of the cell as a function of time. The solidification rate in every condition was less than half of the value defined externally by the cooling rate and the temperature gradient. All measurements for the alloy were taken during the transient stage of solidification.

Figures 15 and 16 show the temperature and concentration profiles in front of the interface calculated from the interference fringes using equation 1. From the temperature of the interface and the phase diagram, the initial solute concentration in the melt is estimated to have been 7.8 mol%, because there had been sufficient time for homogenization of the melt.

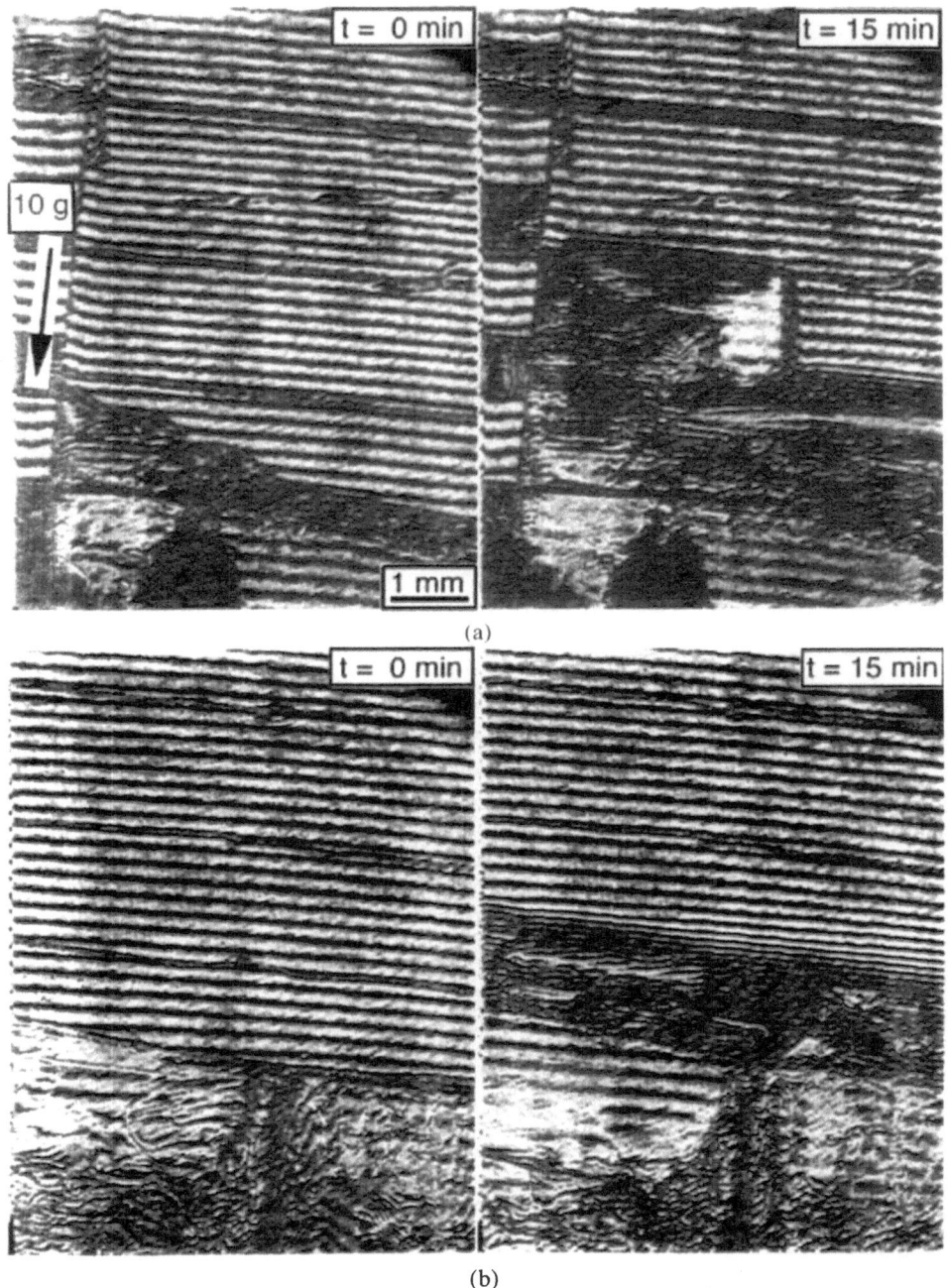

Figure 13. Interference fringe patterns in thymol - *o*-terphenyl mixture at 10 g at different times. (a) at the side of the cell; (b) at the center of the cell.

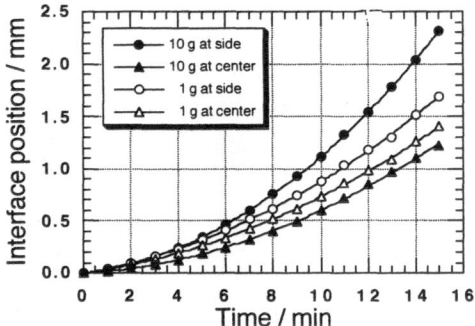

Figure 14. Interface position versus time at the center and at the side of the cell, for thymol - *o*-terphenyl.

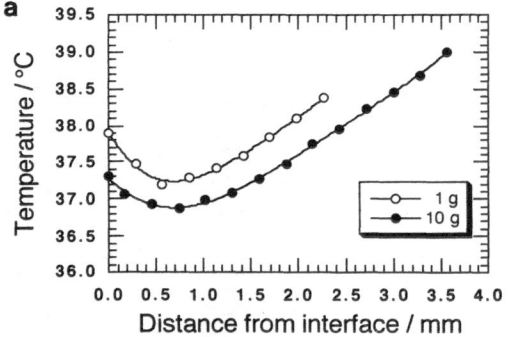

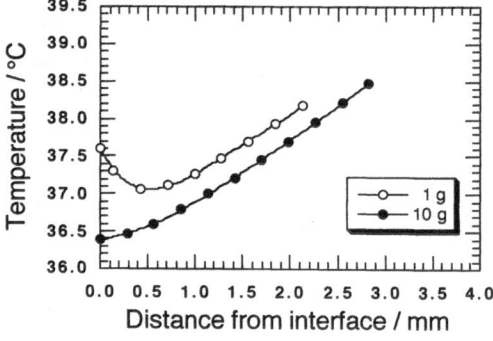

Figure 15. Temperature profile in front of the freezing interface of thymol - *o*-terphenyl at 1 g and 10 g. (a) at the center of the cell.   (b) at the side of the cell.

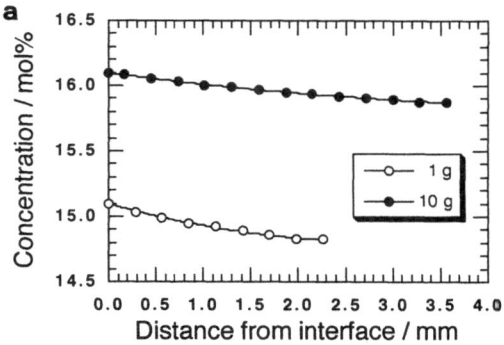

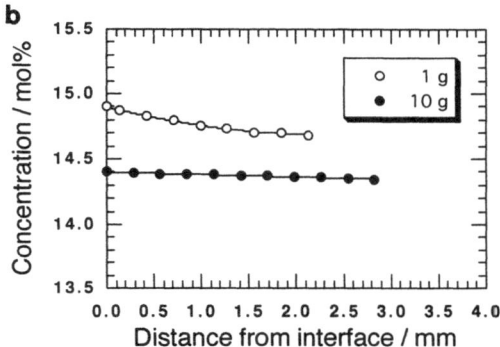

Figure 16. Concentration profile in front of the freezing interface of thymol - o-terphenyl at 1 g and 10 g. (a) at the center of the cell. (b) at the side of the cell.

Now we estimate the kinetic undercooling on the interface. The kinetic undercooling equation can be given as:

$$\Delta T_k = \Delta T - \Delta T_s - \Delta T_c = (T_m + m\,C_i) - T - \Delta T_c \qquad (2)$$

where ΔT_k, ΔT, ΔT_s, ΔT_c, T_m, m, C_i and T are the kinetic undercooling, the interface undercooling, the solute undercooling, the curvature undercooling, the melting point of thymol (51.5°C), the liquidus slope (from Figure 6, -0.77 K / mol%), the concentration of o-terphenyl in thymol at the interface in the melt, and the interface temperature, respectively. No curvature on the interface was observed, and so $\Delta T_c = 0$. The solidification rate, V, is generally a function of ΔT_k , $V = V(\Delta T_k)$. From T and C_i in Figures 15 and 16, we get a linear relationship between V and ΔT_k as follows:

$V = 2.3$ µm/s, $\Delta T_k = 2.0$ K at the center at 1 g,
$V = 2.5$ µm/s, $\Delta T_k = 2.4$ K at the side at 1 g,
$V = 2.1$ µm/s, $\Delta T_k = 1.8$ K at the center at 10 g,
$V = 4.2$ µm/s, $\Delta T_k = 4.0$ K at the side at 10 g.

This relationship explains the morphological changes not only at 1 g but also at 10 g.

From Figures 14, 15 and 16 we conclude that centrifugation induced inhomogeneous distributions of temperature and concentration during the transient stage of solidification of a high Prandtl number material.

CONCLUSION

A table-type centrifuge system was developed for materials processing experiments. This centrifuge has the capacity to load several setups on the table simultaneously. As a first step of the experiment, *in situ* observation of directional solidification was performed using transparent materials with centrifugation. Temperature oscillations were observed in melts with high Pr (~100) from 2 to 10 g resultant acceleration. Non-uniform temperature and concentration distributions in front of the solid/liquid interface were dominated by convection induced by the acceleration. Non-uniformity was considered as the driving force for interface breakdown.

REFERENCES

1. M.W. Geis, H.I. Smith, B-Y. Tsaur, J.C.C. Fan, D.J. Silversmith, and R.W. Mountain, *J. Electrochem. Soc.* 129:2813 (1982).
2. E.W. Maby, M.W. Geis, Y.L. LeCoz, D.J. Silversmith, R.W. Mountain, and D.A. Antoniadis, *Electron. Device Lett.*, edl-2:241 (1981).
3. F. Minari and B. Billia, *J. Crystal Growth* 140:264 (1968).
4. L. Pfeiffer, S. Paines, G.H. Gilmer, W. Saarloos, and K.W. West, *Phys. Rev. Lett.* 54:1944 (1985).
5. D.K. Shangguan and J.D. Hunt, *Metall. Trans.* A 23A:1111 (1992).
6. J.W. Cahn, W.B. Hillig, and G.W. Sears, *Acta Met.* 12:1421 (1964).
7. K.A. Jackson, D.R. Uhlmann, and J.D. Hunt, *J. Crystal Growth* 1:1 (1967).
8. P.J. Morris, D. Kirtisinghe, and R.F. Strickland-Constable, *J. Crystal Growth* 2:97 (1968).
9. N. Dey and J.A. Sekhar, *Acta Metall. Mater.* 41:409 (1993).
10. D.K. Shangguan and J.D. Hunt, *J. Crystal Growth* 96:856 (1989).
11. D.K. Shangguan and J.D. Hunt, *Metall. Trans.* A 22A:941 (1991).
12. T. Higashino, Y. Inatomi, and K. Kuribayashi, *J. Crystal Growth* 128:178 (1993).
13. Y. Inatomi, H. Miyashita, E. Sato, K. Kuribayashi, K. Itonaga, and T. Motegi, *J. Crystal Growth* 130:85 (1993).
14. Y. Inatomi, T. Yoshida, and K. Kuribayashi, *Microgravity Quarterly* 3:93 (1993).
15. U. Dürig, J.H. Bilgram, and W. Känzig, *Phys. Rev.* A 30:946 (1984).
16. Catalogue Handbook of Fine Chemicals, Aldrich Chemical Company Ltd. (1988).
17. Handbook of Chemistry and Physics, 68th Edition, CRC Press (1987).

IMPURITY DISTRIBUTION AND SUPERCONDUCTING PROPERTIES OF PbTe:Tl CRYSTALS GROWN IN A CENTRIFUGE

R. Parfeniev,[1] D. Shamshur,[1] L.L. Regel,[2] and S. Nemov[3]

[1]Ioffe Physical Technical Institute, RAS, St. Petersburg, Russia
[2]Clarkson University, Potsdam, NY
[3]State Technical University, St. Petersburg, Russia

ABSTRACT

We have compared crystal structure, impurity distribution and superconducting parameters of Tl-doped PbTe crystals (up to 1.5 at.% Tl) grown by the Bridgman method at different gravity levels (from 1 g_0 up to 10 g_0). The goal of these experiments was to improve our understanding of the role of centrifugation on crystallisation processes, particularly component distribution in complex semiconductor systems. The effective distribution coefficient of Tl increased with increasing acceleration. Uniform Tl-doping was achieved at 10 g_0.

INTRODUCTION

The physical properties of the narrow-gap IV-VI semiconductors, PbTe, PbSe, PbS and their solid solutions, have been studied widely (see the review in 1). Here we discuss briefly some properties of these materials relevant to the present work. The IV-VI compounds crystallise in a cubic lattice of the NaCl type, with the unit cell containing two atoms (metal and chalcogen). Crystallisation occurs with a noticeable deviation from 1:1 stoichiometry. This deviation is accompanied by a large density of intrinsic defects having a low energy of formation. These intrinsic defects are electrically active, leading to p-type material. An addition of excess Pb into the compound causes n-type conductivity, whereas excess chalcogen gives p-type conductivity. Metallic behaviour is usual for lead chalcogenides at low temperatures, but traditionally these materials are considered as semiconductors. The main extrema in the valence band and the conduction band are located at the L-points of the Brillouin zone. An extra maximum in the valence band is situated at the Σ points of the Brillouin zone; it is separated from the main extremum by an energy distance $\Delta\varepsilon_{LS} = 0.2$ eV for PbTe at 4.2 K.

The problem of impurity levels in $A^{IV}B^{VI}$ semiconductors is rather specific. In these compounds and their solid solutions, the group III impurities, Tl and In, behave in an unusual way. For PbTe doped with Tl and SnTe doped with In, the impurities form quasi-local energy levels located against the background of the allowed energy spectrum of the valence band. They lie near the maximum of the Σ-valence band. A characteristic feature of these quasi-local states is the resonant scattering of free holes by the impurity, and pinning of the Fermi level in samples if the Fermi level is located within the impurity band. An additional interest to these systems was generated by the discovery of superconductivity (SC) in PbTe doped with Tl, with a relatively high (for semiconductors) critical temperature of $T_c \geq 1$ K, by Chernik and Lykov.[2] The SC critical parameters T_c and H_{c2} (the upper critical magnetic field) depend strongly on the degree of hole occupancy of the Tl impurity band.[3,4] This hole occupancy is controlled by additional doping with an acceptor impurity, or, for a solid solution, by its composition (isovalent doping).

We note that SC in PbTe<Tl> has been observed and investigated in both bulk and thin film samples. The maximum values of the SC parameters were very similar for both types of materials.[5] We carried out an experimental investigation of the influence of centrifugation up to 10 g_0 (g_0 being terrestrial gravity) on the type of Tl segregation and on the SC transition in Tl-doped PbTe. These experiments are a continuation of the previously reported study of impurity distribution in Te-doped InSb[6] and Ag-doped PbTe[7] obtained under high gravity conditions.

EXPERIMENTAL

Semiconductor material of a definite composition was sintered from a mixture of high purity elements according to the formula $Pb_{1-x}Tl_xTe$ with $x = 0.015$. The components were melted together, shaken to ensure complete mixing, and held molten for 0.5 hr at 965°C; then the melt was rapidly quenched. The entire ingot was ground to 0.1 mm diameter powder. After mixing, the powder was loaded into a quartz ampoule, which was outgassed and sealed off at 10^{-6} Torr and placed into a pivoted "Meudon" furnace mounted on a swinging centrifuge basket. The high gravity condition was produced by a centrifuge with an arm of 5.5 m (in LCPC - Nantes, France).[8] Directional solidification was carried out by cooling the furnace at 0.5°C/ min, starting from the conical bottom end so that the resultant acceleration vector was directed along the ingot and perpendicular to the solid-liquid interface. The growth rate was estimated to be about 1 cm/hr. In this configuration, the gravity vector plays a stabilizing role on the melt. The solidification was followed by soaking at 650°C for 0.5 hr for homogenization, with the rotation being switched off. We studied the ingots of $Pb_{1-x}Tl_xTe$ grown at acceleration levels of 2.2 g_0, 5 g_0 and 10 g_0. Single crystal $Pb_{0.086}Tl_{0.014}Te$ was prepared by the Bridgman method in evacuated and sealed quartz ampoules filled with the same charge under normal laboratory conditions.

The homogeneity of the ingots was determined by electron backscattering analysis and X-ray microanalysis of sections along the sample length. In order to determine the influence of Tl and structural defects on electrical properties, we performed standard four-point probe DC measurements of resistivity versus temperature (down to 0.4 K) and magnetic field (up to 14 kOe). The SC characteristics, T_c and H_{c2}, were determined at the resistivity level $\rho = 0.5\,\rho_N$, where ρ_N is the normal-state resistivity. The SC transition width was measured between $0.9\,\rho_N$ and $0.1\,\rho_N$.

RESULTS AND DISCUSSION

Electron backscattering images and X-ray micrographs showed significant inhomogeneity of composition along the axial direction. At higher magnifications (100X and 400X), patterns of lines between blocks, dislocations and Tl-rich zones were seen on the surface of the specimen running from the bottom to the top of the ingots.

Multicomponent crystals grown by the Bridgman method tend to have a composition gradient along their length. Electron microprobe measurements demonstrated a significant Tl segregation in the 2.2 g_0 - sample, but in the 5 g_0 - ingot, the Tl content varied only slightly, as shown in Figure 1.

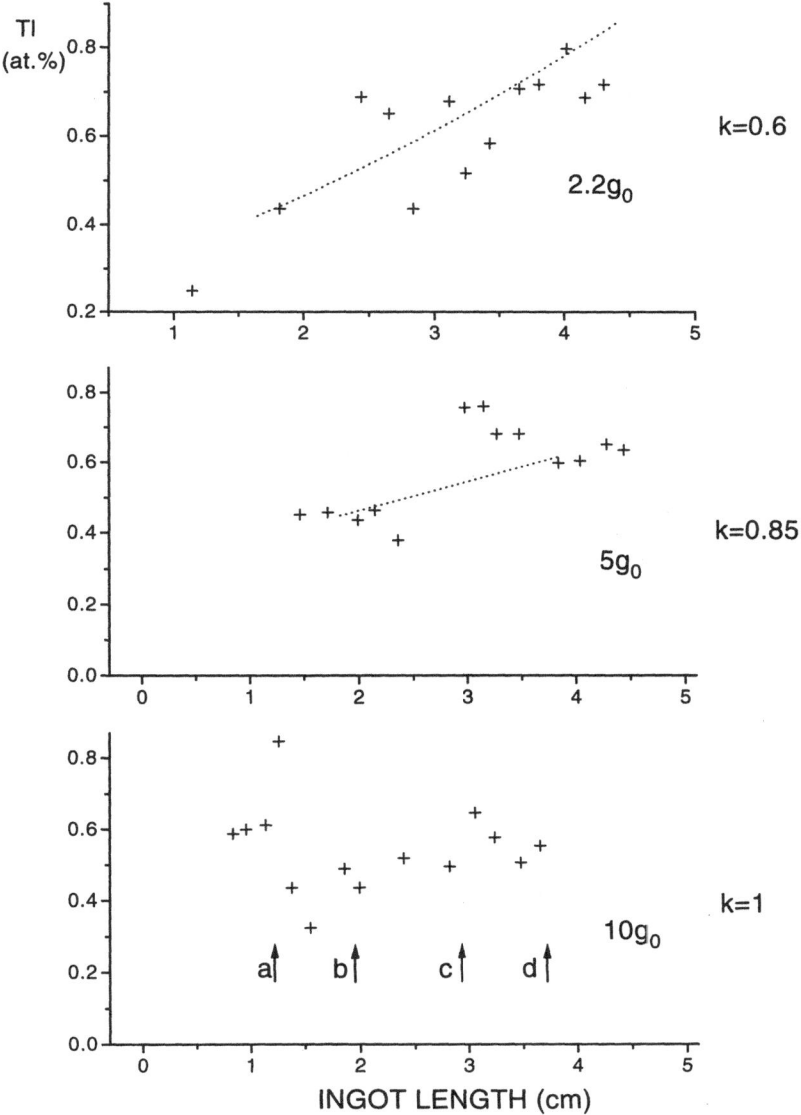

Figure 1. The influence of acceleration level on the Tl distribution in ingots of Tl-doped PbTe grown at 2.2, 5, and 10 g_0. The arrows indicate positions at which superconductivity properties were measured.

The average Tl concentration in the ingots shown in Figure 1 was about the same, but less than the Tl content in the original charge, x = 0.015. This was due to Tl-rich precipitates, which were in most parts of the crystals grown at 2.2 g_0 and 5 g_0, except the initial portion. The structure and Tl distribution in the ingot grown at 5 g_0 were very similar to those in the 2.2 g_0-ingot, i.e.:

    i) small segregation of Tl from the bottom to the top,

    ii) many Tl-rich precipitates in the middle and top parts.

The ingot grown at 10 g_0 has an irregular Tl content without noticeable segregation along its length and many Tl-rich precipitates.

Since PbTe was doped with only one acceptor impurity (Tl), the formula for the impurity concentration with partial liquid mixing and constant segregation coefficient K yields K-values increasing from K = 0.6 at 2.2 g_0, to K = 0.85 at 5 g_0, up to K ≅ 1 at 10 g_0. So, the maximum inhibition of Tl segregation in PbTe was found at an acceleration of 10 g_0, compared to 2.2 g_0 for Ag-doped PbTe ($2 \cdot 10^{18}$ cm^{-3} Ag) in the same centrifuge facilities. Based on our current understanding of convection during directional solidification on the centrifuge, this indicates that there was a more planar interface shape in the Tl-doping experiments.

Over the investigated temperature range, all Tl-doped PbTe samples had p-type conductivity. Variations in the hole concentration p calculated from the Hall coefficient at 300 K and measured in several cross-sections of the crystal, correlated with the Tl distribution shown in Figure 1. In the 2.2 g_0 sample, p changed from 3.1×10^{19} cm^{-3} in the bottom part up to p = 4.5×10^{19} cm^{-3} in the top part. The hole concentration was fixed at p = 5.3×10^{19} cm^{-3} in the 10 g_0 sample, within the accuracy of the measurements.

At liquid helium temperature, the investigation of the conductivity in the different parts of the ingot were mainly concerned with the SC transition. We found a correlation between the SC parameters and the impurity distribution in the samples. Note that the SC transition in PbTe<Tl> with T_c > 0.4 K had been observed previously only in samples in which the Fermi level lay within the energy limit of the Tl impurity band.[3] The characteristic dependence of T_c on hole concentration p was found in Reference 3 over a range of p, from the SC threshold up to the maximum T_c that could be achieved by Tl doping before the appearance of a second phase (for Tl = 2.5 at% at 650°C). The threshold character for the appearance of the SC transition is connected with entering the Tl impurity band with a high density of states (DOS) by the Fermi level. The impurity band DOS exceeds the valence band DOS at the same energy.

In our case, all samples obtained under high gravity showed the SC transition near the same T_c as in single crystal Pb$_{0.986}$Tl$_{0.014}$Te grown at 1 g_0. Figure 2 shows a comparison of the SC transitions in the resistivity of the 1 g_0 single crystal with those of parts a,b,c,d of both the 10 g_0 and 5 g_0 samples. The resistivity was measured between probes with a distance (2 ÷ 3) mm in the points indicated by the arrows in Figure 1. In the 2.2 g_0 sample from the bottom of the ingot, the SC transition was incomplete and resembled the a - curve for the 10 g_0 samples, because there were non-superconducting blocks with a small Tl concentration (N_{Tl} < 0.4 at%). Only samples from the middle part of the 2.2 g_0 ingot and above showed the complete SC transition, with T_c ≅ 1.15 K (N_{Tl} > 0.4 at.%).

The second critical field $H_{C2}(T)$ determined from the curve ρ(H) at different T (Figure 3) gives approximately equal derivatives $|dH_{C2}/dT|_{Tc}$ for the 1 g_0 single crystal and the middle part of the 5 g_0 ingot, 4.95 kOe/K and 4.8 kOe/K, respectively. This SC parameter is important because when taken together with the normal resistivity ρ_N it allows us to evaluate the DOS N(0) at the Fermi level. Comparison of T_c and $|dH_{C2}/dT|_{Tc}$ for different samples shows that the SC parameters of PbTe<Tl> reflect the distribution of the Tl concentration.

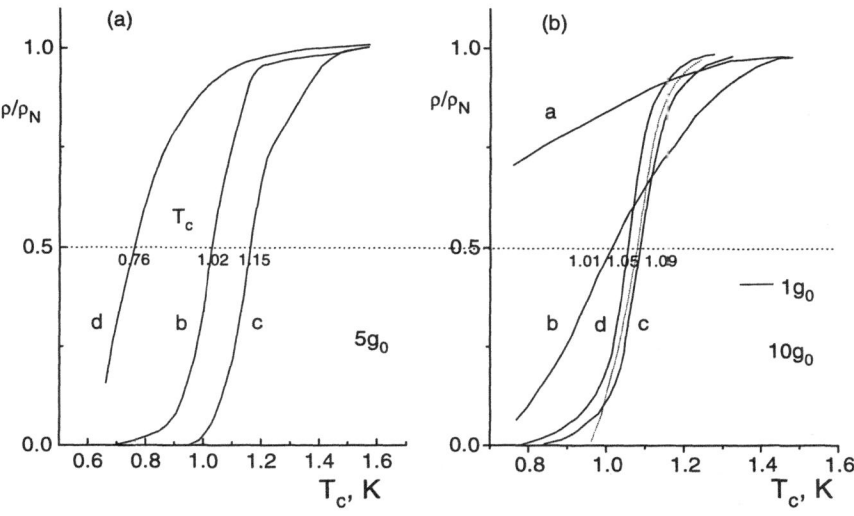

Figure 2. The dependence of resistivity on temperature for samples directionally solidified at (a) 5 g_0 and (b) 1 and 10 g_0. Measurements were taken at points a, b, c, and d along the ingots, as shown in Figure 1.

An irregular distribution of the Tl content in the bulk of the ingot could produce a step-like character or even broaden the dependence of the SC transition on temperature or magnetic field. Figure 3, for the 5 g_0-sample, shows parts with different critical SC parameters in the middle section.

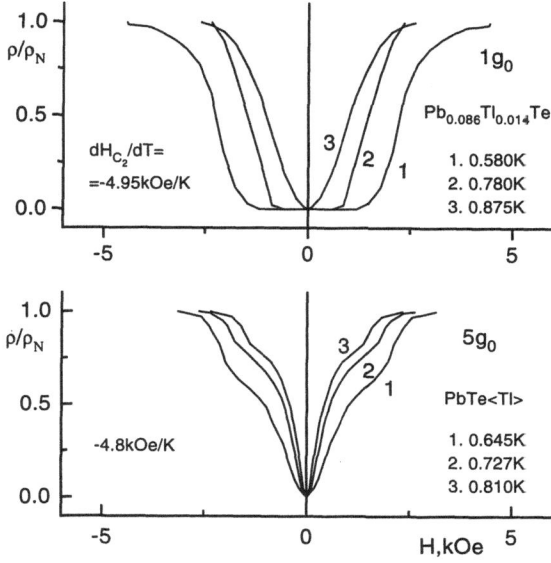

Figure 3. Comparison of the superconducting transition in a magnetic field at different temperatures $T < T_c$, for samples grown at 1 g_0 and at 5 g_0 at position b in Figure 1.

CONCLUSIONS

Suppression of Tl segregation was observed in Tl-doped PbTe crystals grown by the gradient freeze technique at high gravity (up to 10 g_0). The superconducting properties of the samples are associated with the Tl distribution, and are in accordance with the conception of the role of Tl impurity states in the occurrence of superconductivity. The suppression of impurity segregation at high gravity is of interest as a general result for doped semiconductors. The obtained data show new potential for using centrifugal acceleration to vary the critical parameters of a phase transition in electronic systems, as well as in the lattice structure.

Acknowledgments

This research was supported by the Russian Fund of Basic Research under grant N96-02-17848, the French Centre National de la Recherche Scientifique at Meudon, and the United States National Science Foundation via grant DMR-9414304. We are grateful to the Laboratoire Central des Ponts et Chaussées for permitting our solidification experiments to be performed on their 5.5 m centrifuge.

REFERENCES

1. V.I. Kaidanov and Yu.I. Ravich, *Sov. Phys. Usp.* 28:31 (1985).
2. I.A. Chernik and S.N. Lykov, *Sov. Tech. Phys. Lett.* 7:40 (1981).
3. S.A. Kaz'min, S.N. Lykov, R.V. Parfeniev, I.A. Chernik, and D.V. Shamshur, *Sov. Phys. Solid State* 24:832 (1982).
4. S.A. Nemov, R.V. Parfeniev, and D.V. Shamshur, *Sov. Phys. Solid State* 27:368 (1985).
5. V.I. Kaidanov, S.A. Kaz'min, S.A. Nemov, R.V. Parfeniev, D.V. Shamshur, and V.F. Shokh, *Sov. Phys. Solid State* 28:591(1986).
6. R. Derebail, W.R. Wilcox, and L.L. Regel, *J. Crystal Growth* 119:98 (1992).
7. H. Rodot, L.L. Regel, and A.M. Turtchaninov, *J. Crystal Growth* 104:280 (1990).
8. M. Rodot, H. Rodot, P. Williams, L.L. Regel, R.V. Parfeniev, and N.K. Shulga, book of abstracts, First International Workshop on Material Processing in High Gravity, Dubna, USSR (1991) p. 29.

A LOW COST CENTRIFUGE FOR MATERIALS PROCESSING IN HIGH GRAVITY

Y. A. Chen, L. C. Russo, M. F. Ribeiro, and I. N. Bandeira

Instituto Nacional de Pesquisas Espaciais - INPE/LAS
12227-010, São José dos Campos, SP, Brazil

ABSTRACT

A small, low cost centrifuge was constructed at the Associate Laboratory of Sensors and Materials of the Brazilian Space Research Institute (LAS/INPE) for crystal growth experiments at accelerations up to 10 g. The centrifuge and a resistance-heated tubular furnace for temperatures up to $1000^{\circ}C$ were designed and constructed at our laboratory practically from scratch, at a total cost under US$1,000. This paper presents details of the construction and performance of this system.

INTRODUCTION

The mechanisms of crystal growth and, consequently, crystal perfection are strongly influenced by acceleration. If a fluid is submitted to an acceleration produced by the angular velocity of a rotating centrifuge, several effects will appear in the growing process. For example, the increase of acceleration changes the buoyancy-driven convection, modifying the heat and mass transfer. Also, the Coriolis acceleration is introduced in the fluid, modifying the flow stability. The study of those effects can improve our understanding of crystal growth processes as much as does experiments made at low gravity.

All over the world, large centrifuges are expensive and seldom available for crystal growth experiments. In Brazil, this situation is even more problematic due to the limited number of such facilities in the country. To overcome this drawback, a small centrifuge for accelerations up to 100 m/sec^2 (about 10 times terrestrial gravity) was designed and constructed at LAS/INPE laboratories. The cost was minimized by using readily available materials. Also, a small tubular furnace (30 cm in length and 7 cm in diameter) was constructed for Bridgman and vapor growth of IV-VI compounds.

CENTRIFUGE CONSTRUCTION

The LAS centrifuge consists of an aluminum arm of 125 cm overall length, 6 cm width and 13 mm thickness. The furnace side of this arm measures 75 cm from the center of rotation, while the counter-weight side is 50 cm long. Balance of the centrifuge is obtained by adding movable weights to this short side. The arm is driven by a 1/4 HP AC motor through a gear box that reduces the motor rotation from 3600 RPM to 248 RPM. The centrifuge arm is coupled directly to the axis of the gear box. An electronic speed control drives the gears from zero to 248 RPM with ±1 RPM accuracy. The usable rotation ranges from 40 to 110 RPM, resulting in centrifugal accelerations from 1.5 to 10 g at the end of the furnace arm.

A perforated disc with 60 holes is attached to the gear box axis, allowing a photocoupling device to measure the rotation frequency.

A tubular furnace can be mounted on the centrifuge in two different ways. It can be fixed on the top of the arm by a hinge, adjustable and fixed by a screw, in the angle that aligns the resultant acceleration (vector sum of earth's gravity and centrifugal acceleration) with the axis of the furnace. Alternately, it can be held at the end of the arm by a steel pin, which gives the furnace freedom to swing out. In this case, as the centrifuge rotates the furnace aligns itself with the resultant acceleration and the distance from the end of the furnace to the centrifuge axis is augmented to 105 cm. Because the base is symmetric on which the tubular furnace is mounted, it is possible to choose which end of the furnace (hot or cold) faces toward the centrifuge center. This permits both direct and inverted configurations for Bridgman and vapor crystal growth.

The whole apparatus is mounted on an iron frame of 80 cm length x 80 cm width x 70 cm height, which is bolted to the floor. A photograph of the centrifuge and furnace is shown in Figure 1.

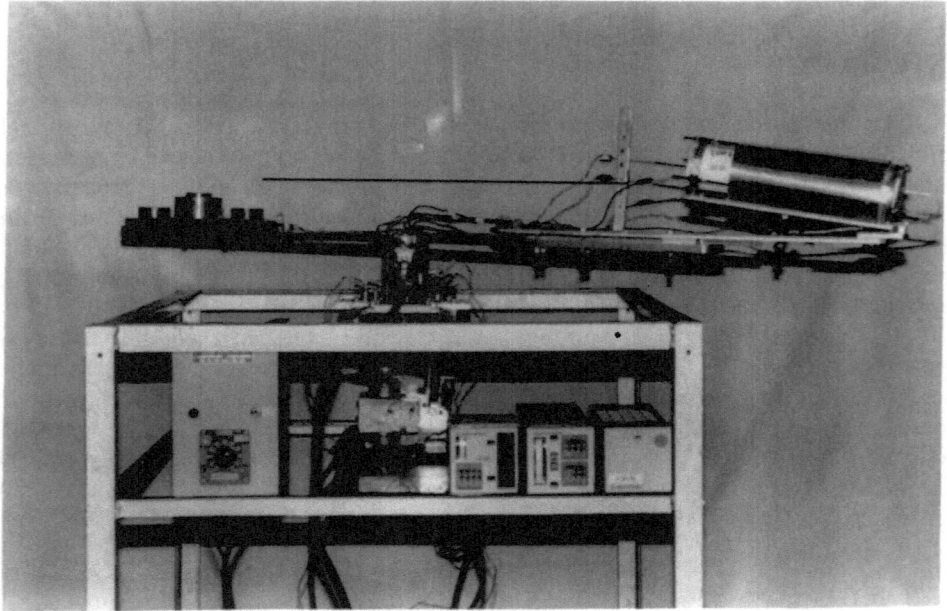

Figure 1. Photograph of LAS centrifuge with the tubular furnace mounted in the top position with its cooler end towards the center of the centrifuge.

A slip ring assembly was also constructed, consisting of eight rings; two for power and six for instrumentation. The construction consisted in inserting, as tightly as possible, a PTFE (polytetrafluoroethylene) block 65 mm diameter and 75 mm length, inside a 2 mm wall thickness brass tube. The brass tube was then machined to leave eight rings each 6 mm wide. Wires were passed through holes drilled in the top of the PTFE piece and soldered to the side of the rings. The power and instrumentation were connected by commercial automotive graphite brushes. The slip ring setup is shown in Figure 2.

Figure 2. Centrifuge slip ring assembly made of PTFE encased in machined brass tube. The brushes are commercial automotive electrical generator collectors.

FURNACE CONSTRUCTION

A quartz single-zone tubular furnace was constructed to grow crystals by the Bridgman and vapor processes in both direct and inverse configurations. As can be seen in Figure 3, the furnace consists of a quartz tube (280 mm long x 20 mm diameter) with a helical quartz rod soldered in the external wall to hold nickel-chromium resistance wire. The region covered by the electrical resistance heater is 15 cm long, leaving 10 cm of unheated quartz to establish the furnace cool end. The 20 Ω nickel-chromium heater consists of 1 mm diameter wire. To protect the resistance heater, a 35 mm diameter quartz tube was inserted; this assembly was fitted inside another 70 mm quartz tube. The gap between the two outer tubes was filled with thermal insulator powder. The whole system is held together by two aluminum end caps. Externally, in the hot region of the furnace, another thin sheet of thermal insulator was placed inside a stainless steel foil. This furnace can produce temperatures up to 1000 $^{\circ}$C and weighs only 1.6 kg.

Figure 4 shows the assembled furnace and Figure 5 the temperature profile for a typical PbSnTe Bridgman growth run. The temperature gradient in the furnace at the crystal solid-liquid interface is about 20°C/cm.

Figure 3. Disassembled furnace showing internal part details.

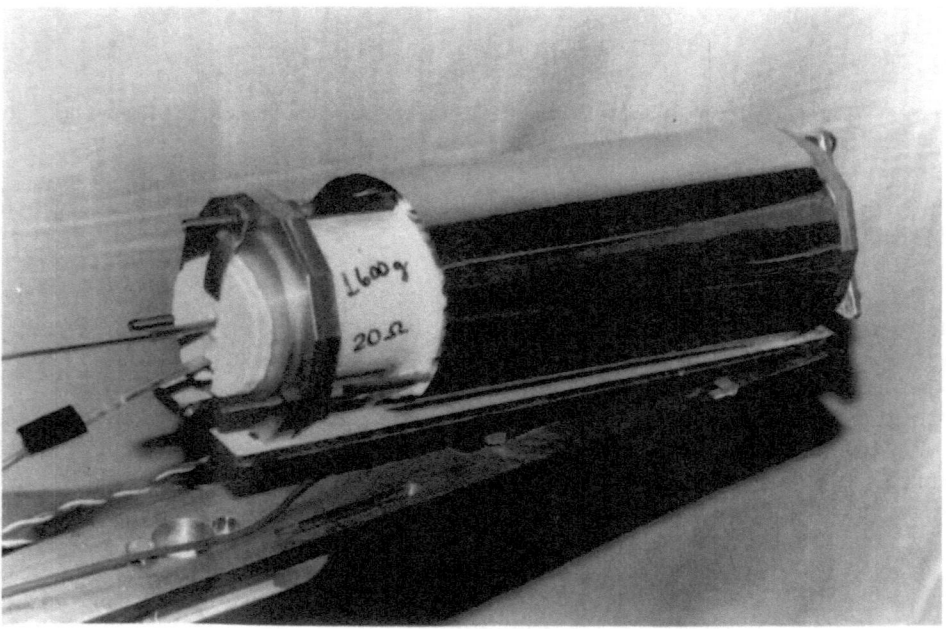

Figure 4. Assembled furnace mounted in the centrifuge.

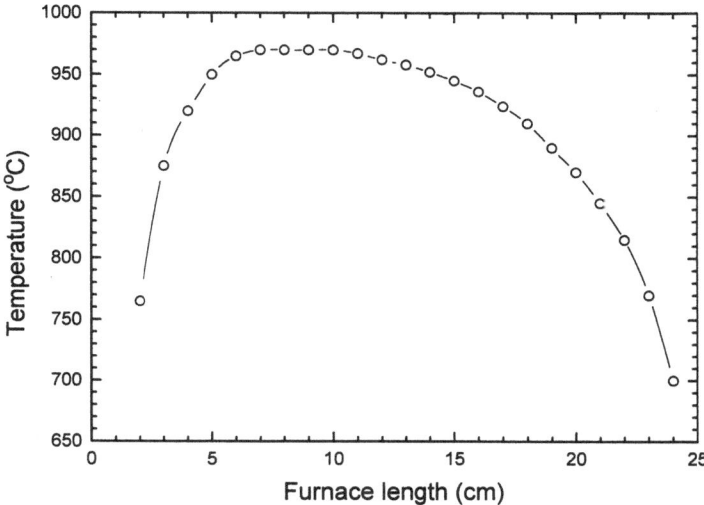

Figure 5. Furnace temperature profile when loaded with a growth ampoule inside a protective steel container.

The furnace temperature is controlled by a PID Eurotherm controller with an accuracy of ± 1°C at 1000°C, using chromel-alumel thermocouples. The six instrumentation slip rings allow the use of one thermocouple for temperature control and four for measurements. A gradient freeze growth run is made by slowly lowering the furnace temperature using an Eurotherm temperature programmer attached to the controller. The growth velocity attainable ranges from 1 to 5 mm/hour.

CONCLUSIONS

This paper describes a small, low cost centrifuge and furnace to grow crystals at accelerations up to 10 g. Despite some drawbacks, such an appreciable acceleration gradient along the crystal due to the short arm, it is a good workable choice for small research groups. The crystal growth group at LAS/INPE has been using this centrifuge and furnace for the past three years for the growth of IV-VI (PbSnTe) and II-VI (HgCdTe) semiconductor materials.[1,2]

REFERENCES

1. Y.A. Chen, I.N. Bandeira, A.H. Franzan, S. Eleutério, and R.M. Slomka, The influence of gravity on $Pb_{1-x}Sn_xTe$ crystals grown by the vertical Bridgman method, *in*: "Materials Processing in High Gravity," L.L. Regel and W.R. Wilcox, eds., Plenum Press, New York (1994).
2. Y.A. Chen, E.G. Salgado, C.R.M. Silva, and I.N. Bandeira, The influence of convection on axial and radial composition profiles in PbSnTe and HgCdTe, present volume.

THE INFLUENCE OF CONVECTION ON
AXIAL AND RADIAL COMPOSITION
PROFILES IN PbSnTe AND HgCdTe

Y. A. Chen,[1] E. G. Salgado,[2] C. R. M. Silva,[2] and I. N. Bandeira[1]

[1]Instituto Nacional de Pesquisas Espaciais - INPE/LAS
[2]Centro Técnico Aeroespacial - CTA/AMR
12227-010, São José dos Campos, SP, Brazil

ABSTRACT

Studies of Bridgman-grown lead-tin-telluride (PbSnTe) and mercury-cadmium-telluride (HgCdTe) axial and radial composition profiles are presented. For PbSnTe, both profiles were investigated for several tin-telluride initial compositions, for direct and inverted Bridgman configurations, and under high gravity conditions. Axial and radial profiles investigations were also carried out for HgCdTe. The results indicate that the radial composition homogeneity in both semiconductor alloys was enhanced by convection.

INTRODUCTION

$Pb_{1-x}Sn_xTe$ and $Hg_{1-x}Cd_xTe$ semiconductor alloys are the most common materials for the construction of detectors and arrays for the thermal infrared region. For p-n junction arrays of several elements grown on a single crystal substrate, radial concentration homogeneity is a must for focal plane detection. Although new techniques such as Molecular Beam Epitaxy (MBE) are starting to permit the growth of these materials directly on more conventional substrates such as silicon and gallium arsenide, the growth of single crystals of these IV-VI and II-VI alloys with high radial and axial homogeneity is still very important.

The behavior of the growth of PbSnTe and HgCdTe with respect to solutal stability is different. In conventional vertical Bridgman growth (VB), although both compounds are thermally stable, PbSnTe is solutally unstable while HgCdTe presents a stable solutal situation. The solutal instability of PbSnTe has lead to the search for new growth methods in order to avoid or lower natural convection and, consequently, increase compositional homogeneity. These techniques included the growth in a microgravity environment,[1] where convection is suppressed and the growth may proceed in a diffusion-controlled steady state.

Also considered has been the inverse vertical Bridgman growth method (IVB),[2] in which solidification proceeds downward under conditions of thermal instability in exchange for solutal stability. More recently, the growth under high gravity environment using centrifuge systems has been studied.[3-6]

The main focus of these studies has been, normally, the search for axial homogeneity. But, still more important for infrared devices, is the radial homogeneity across the substrate wafer where the detector array will be constructed. In this work, the radial homogeneity for both compounds was also studied, besides the measurement of axial concentration variation as a function of tin content, growth velocity and acceleration for PbSnTe and growth velocity for HgCdTe.

LEAD-TIN-TELLURIDE RESULTS

The $Pb_{1-x}Sn_xTe$ semiconductor compound was grown with the VB method in normal gravity for tin compositions ranging from $x = 0.20$ to 0.80 at the same growth velocity, with the VB and IVB methods with constant composition ($x = 0.20$) and several velocities, and in the VB configuration on a centrifuge with constant composition ($x = 0.20$) and constant growth velocity. In all this situations, both axial and radial profiles were measured and compared.

To evaluate the influence of the solutal and temperature gradients on convection, it is necessary to compare the values of their density gradients. The density gradient caused by solute accumulation at the interface at steady state is given by:

$$\left(\frac{\partial \rho}{\partial z}\right)_c = \frac{\partial \rho}{\partial C}\frac{\partial C}{\partial z} = \frac{\partial \rho}{\partial C}\left(\frac{1}{k_0} - 1\right)\frac{VC_0}{D} \tag{1}$$

while the density gradient due to heat transfer is:

$$\left(\frac{\partial \rho}{\partial z}\right)_T = \frac{\partial \rho}{\partial T}\frac{\partial T}{\partial z} = \frac{\partial \rho}{\partial T}G \tag{2}$$

where z is the growth direction, C_0 the original mole fraction of SnTe, k_0 the equilibrium segregation coefficient, V the solidification rate, D the diffusion coefficient in the melt and G the temperature gradient. The gradients are given by $\partial \rho/\partial C = -\beta_c(x)\rho(x)$ and $\partial \rho/\partial T = \beta_T(x)\rho(x)$, where $\beta_c(x)$, $\beta_T(x)$ and $\rho(x)$ are the Sn-content coefficient of bulk solutal expansion, coefficient of thermal expansion and density, respectively. Taking into account that D and k_0 also depend on Sn concentration, it is possible to calculate the dependence of density gradient on SnTe mole fraction, as plotted in Figure 1.

As can be seen from this figure, the thermal density gradient is small and practically constant within the whole range of SnTe mole fraction, whereas the solutal density gradient has a maximum at $x \approx 0.45$ (about 10 times bigger than the thermal one) suggesting that the most unfavorable cases in axial homogeneity should appear at intermediate SnTe compositions and that better results can be obtained by inverted Bridgman growth.

Figure 2 shows the axial composition for $Pb_{1-x}Sn_xTe$ crystals of different Sn compositions ($x = 0.20$, 0.40, 0.60 and 0.80) grown by VB with growth velocity (assumed equal to the crystal translation speed) of 1 mm/h in a furnace with a temperature gradient of $G = 15°C/cm$. Note that the crystals with middle SnTe compositions have less uniform axial composition and that the best results are obtained for high tin composition ($x = 0.80$).

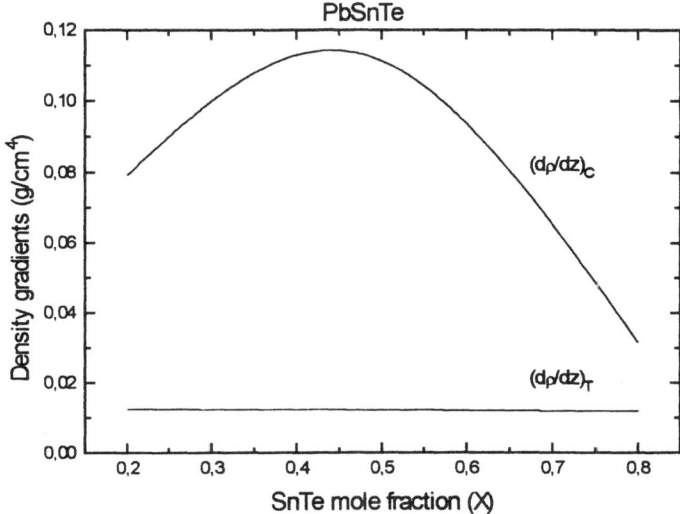

Figure 1. Calculated density gradients at the freezing interface due to composition and temperature gradients, from Equations (1) and (2), versus mole fraction SnTe.

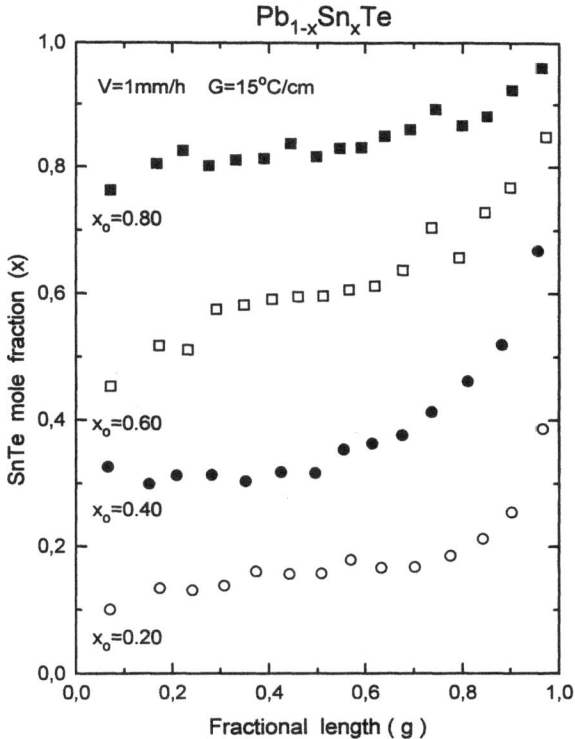

Figure 2. Axial composition profiles for $Pb_{1-X}Sn_XTe$ crystals grown by the vertical Bridgman (VB) technique for several initial SnTe compositions (x = 0.20, 0.40, 0.60 and 0.80).

For optoelectronic devices, even more important than the axial composition is the radial homogeneity. The radial variation in composition was determined for the same crystals used for the results shown in Figure 2. These crystals were 10 mm in diameter and were cut along the growth axis. A microprobe scan was employed to measured the tin content. The samples were measured on the top, middle and bottom (near the tip of the crystal) positions in two perpendicular directions (see inset in Figure 3a). The results are presented in Figures 3a, 3b, 3c and 3d. As expected, the worst radial distribution, especially at low tin concentrations, is in the crystal top. In the middle and bottom regions, the composition tends to be more homogeneous across the slice, more or less independent of the tin content.

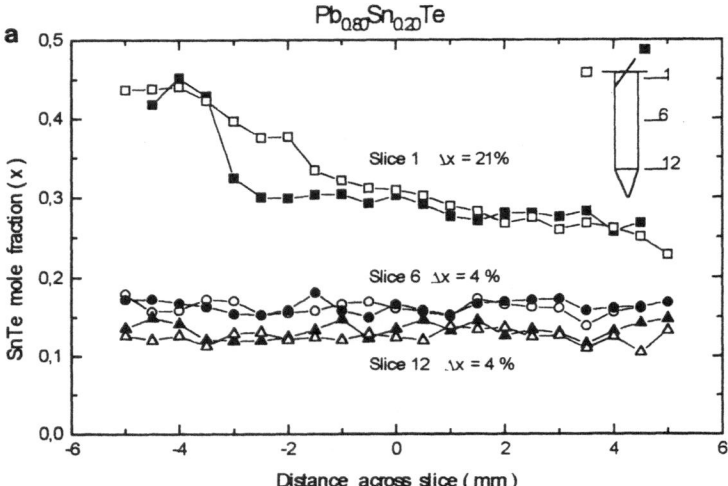

Figure 3a. Radial composition profiles for $Pb_{0.8}Sn_{0.2}Te$ at several axial positions. Here Δx is the % of solute concentration variation across the slice.

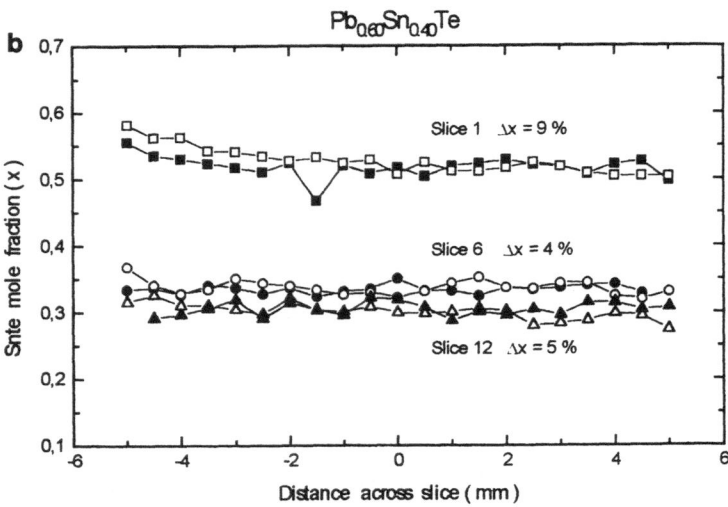

Figure 3b. Radial composition profiles for $Pb_{0.6}Sn_{0.4}Te$ at several axial positions.

Lead-tin-telluride with x = 0.20 was grown in the inverted direction (growth direction parallel to the gravity vector) in a configuration where the system is thermally unstable but solutally stable. Figure 4a shows the axial composition profile for a crystal grown at a velocity of 4 mm/h in a furnace with a temperature gradient of 15°C/cm. Figure 4b shows results for normal vertical Bridgman growth, otherwise under the same conditions. The eight-fold reduction in axial inhomogeneity for inverted growth with x = 0.20 is understandable from the exchange between thermal and solutal stability shown in Figure 1. The same tendency appears in the radial concentration profiles of both crystals, as shown in Figures 5a and 5b.

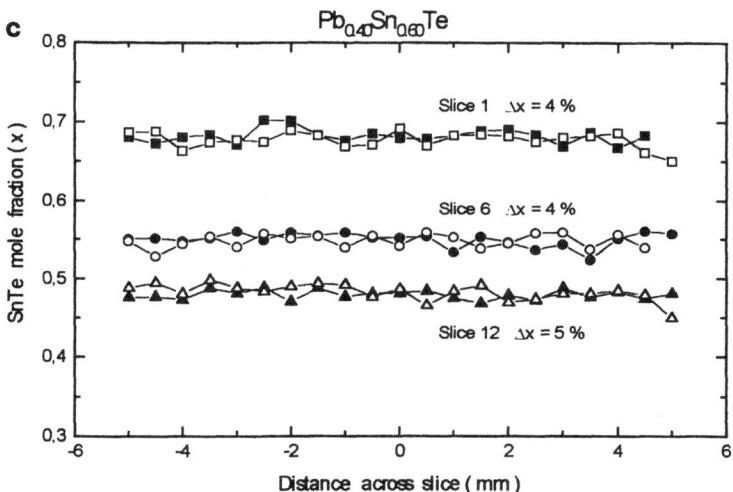

Figure 3c. Radial composition profiles for $Pb_{0.4}Sn_{0.6}Te$ at several axial positions.

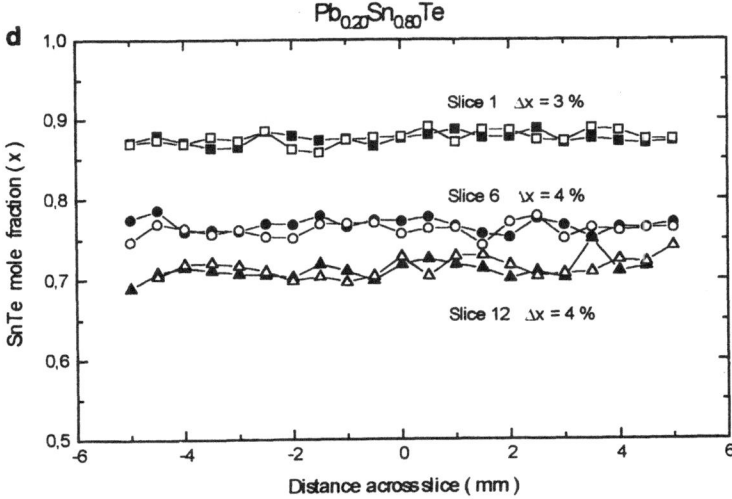

Figure 3d. Radial composition profiles for $Pb_{0.8}Sn_{0.2}Te$ at several axial positions.

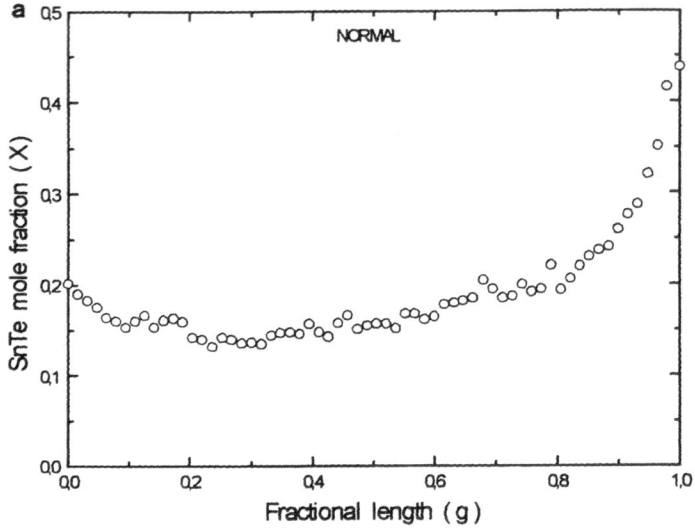

Figure 4a. Axial composition profile for $Pb_{0.8}Sn_{0.2}Te$ grown upward at 4 mm/hr.

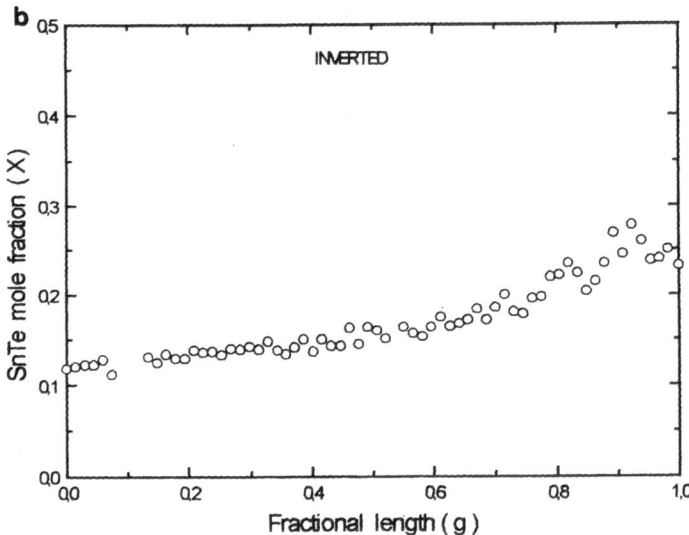

Figure 4b. Axial composition profile for $Pb_{0.8}Sn_{0.2}Te$ grown downward at 4 mm/hr.

Similar growth experiments were carried out at several accelerations with the same material (x = 0.20) on a centrifuge made at LAS/INPE laboratories.[6] Figure 6 presents the axial (a) and radial (b) SnTe composition profiles for vertical Bridgman growth at an acceleration of seven times Earth's gravity (7 g). The axial profile is similar to that obtained at 1 g. The percent radial variations in composition, measured at top, middle and tip of the crystal, also are near those for non-accelerated growth (average of 5% dispersion). On the other hand, the average composition at these three positions lies closer to the initial alloy concentration (x = 0.20) for the sample grown at high gravity than for the other growth conditions, especially compared to vertical Bridgman growth at 1 g. This can be attributed to enhanced convection and the Coriolis force. Although the axial homogeneity can in this situation be somewhat spoiled, the change in convection due to centrifugation appears to benefit the radial concentration distribution and lead to more homogeneous crystals.

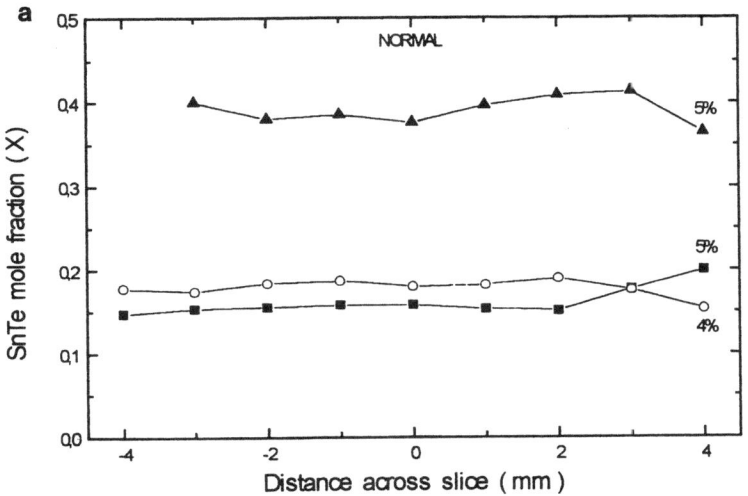

Figure 5a. Radial composition profiles for $Pb_{0.8}Sn_{0.2}Te$ grown upward at 4 mm/hr. The percentages represent the variation in Sn composition across the slice.

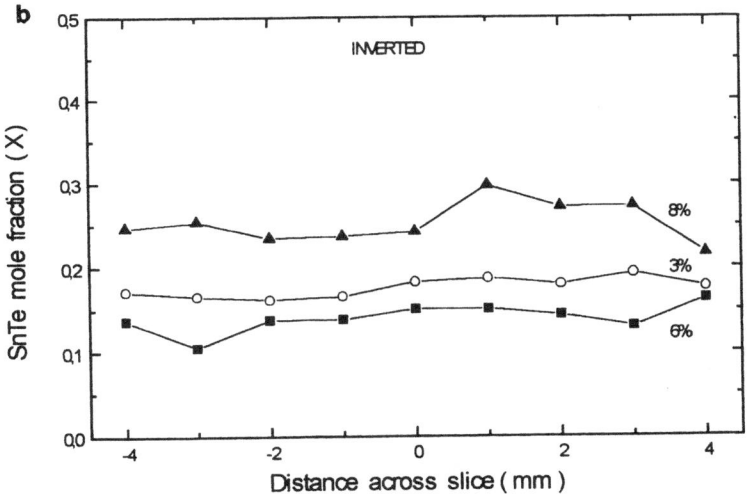

Figure 5b. Radial composition profiles for $Pb_{0.8}Sn_{0.2}Te$ grown downward at 4 mm/hr.

MERCURY-CADMIUM-TELLURIDE RESULTS

With the same purpose, to improve the radial composition profile for materials suitable for infrared detector arrays, crystals of $Hg_{0.80}Cd_{0.20}Te$ were made at several growth velocities. As HgCdTe is both thermally and solutally stable for solidification upward anti-parallel to the gravity vector, the method employed was vertical Bridgman growth. The temperature gradient of the furnace was 28°C/cm, and the axial concentration profile followed the expect D/V < L prediction, approaching the diffusion-controlled steady state. Figure 7 shows the CdTe mole fraction variation for crystals grown at speeds varying from 0.4 to 4.0 mm/h.

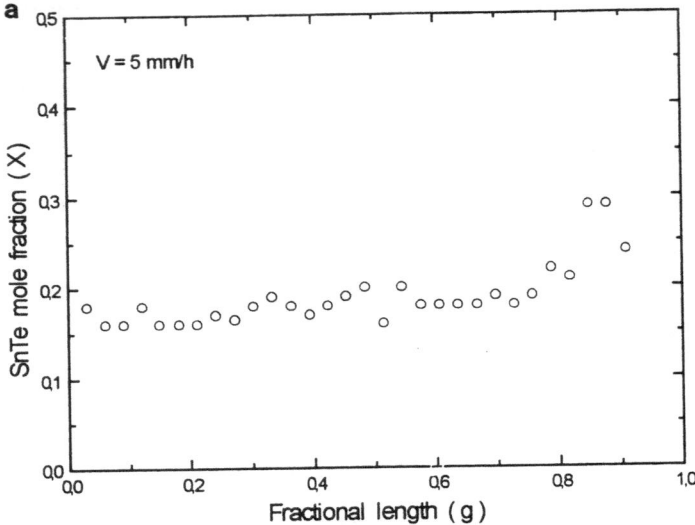

Figure 6a. Axial concentration profile in $Pb_{0.8}Sn_{0.2}Te$ grown at 5 mm/hr in a centrifuge at 7 g.

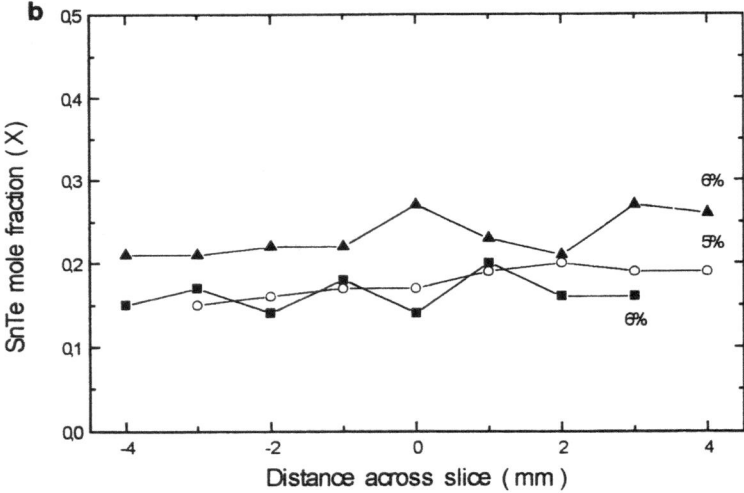

Figure 6b. Radial concentration profiles in $Pb_{0.8}Sn_{0.2}Te$ grown at 5 mm/hr in a centrifuge at 7 g.

Radial concentration profiles were determined in the same way as for PbSnTe, i.e., the 6 mm diameters crystals were cut into two halves in the growth direction and scanned for cadmium with a microprobe. The results are shown in Figures 8a to 8d. As can be seen from these figures, the crystal grown at the lowest speed has the most homogeneous radial solute distribution. The percent variation in composition is plotted versus freezing rate in Figure 9.

For all growth rates, the last to solidify region (crystal end) presents a more constant concentration of CdTe radially, with a variation of about 5%. On the other hand, the first to freeze region (crystal tip) had a speed-dependent compositional variation, reaching about 30% for growth velocities higher than approximately 0.6 mm/h and dropping sharply to around 5% variation for a velocity of 0.4 mm/h. The smaller radial variation in composition at low growth speeds (in contrast to the axial behavior) is attributed to the long time available for homogenization by diffusion and/or convection. Growth in centrifuges, where convection should be enhanced, would probably yield higher radial homogeneity even at high growth velocities.

CONCLUSIONS

The principal aim of this work was to determine the influence of growth conditions on the radial composition profiles of semiconductor compounds useful for thermal infrared devices. Small concentration variations across the device provoke changes in wavelength radiation detection, diminishing the detector array's figure of merit.

In the case of vertical Bridgman (VB) grown lead-tin-telluride, the composition varied more axially for intermediate starting concentrations of tin ($0.4 \leq x \leq 0.6$) than for low ($x \approx 0.2$) and high ($x \approx 0.8$) initial concentrations. The radial variation in composition was almost independent of the initial content of tin, but was strongly affected by axial position. For inverted vertical Bridgman (IVB) growth, the exchange between thermal and solutal stability helped the homogenization, not only in the axial direction but also radially. Growth under high gravity conditions (7 g) in the VB configuration produced percent variations in

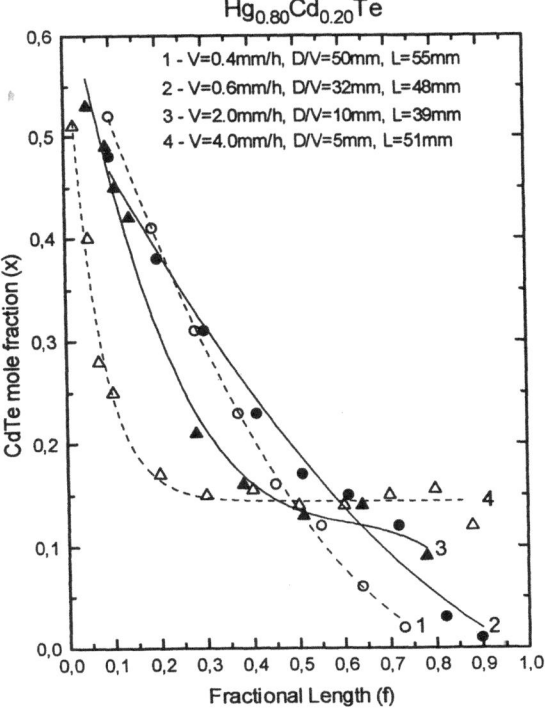

Figure 7. Cadmium concentration versus fractional length for $Hg_{0.8}Cd_{0.2}Te$ grown at a velocity V ranging from 0.4 to 4.0 mm/h. Temperature gradient $G = 28°C/cm$, diffusion coefficient $D = 5.5 \times 10^{-5}\ cm^2/s$.

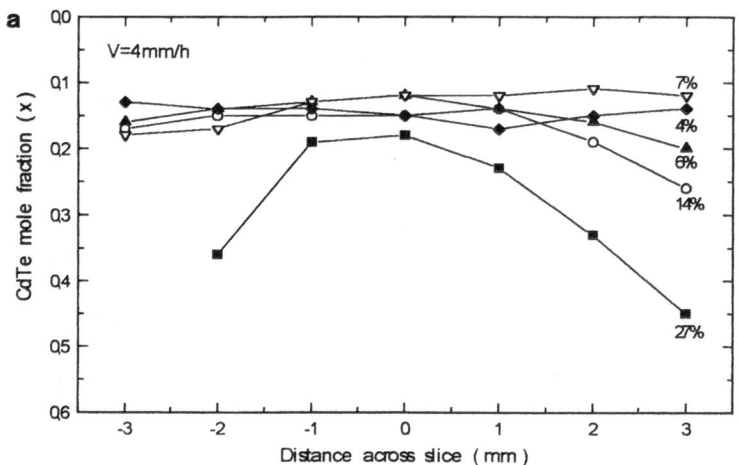

Figure 8a. Radial concentration profiles for $Hg_{0.8}Cd_{0.2}Te$ crystals solidified upward at 4 mm/h. The top curve corresponds to the end of that crystal.

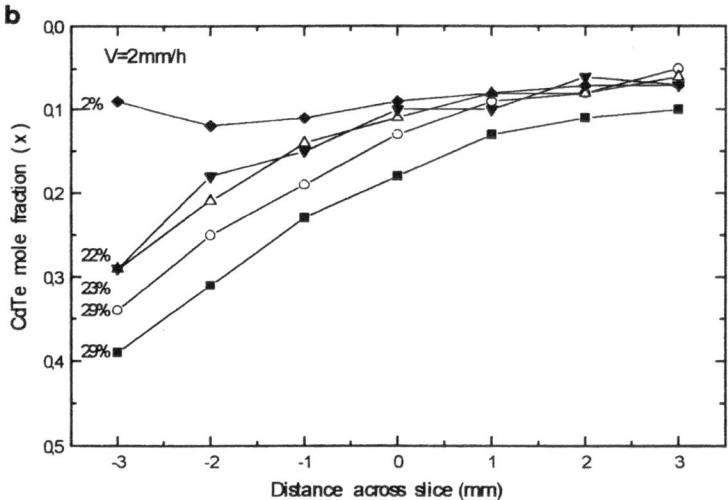

Figure 8b. Radial concentration profiles for $Hg_{0.8}Cd_{0.2}Te$ crystals solidified upward at 2 mm/h.

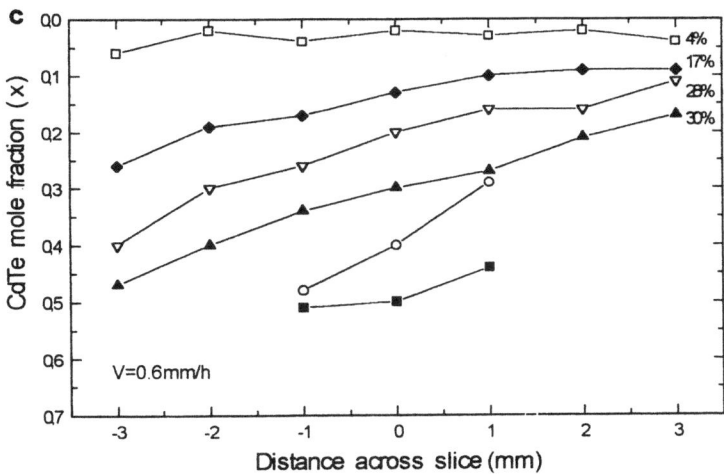

Figure 8c. Radial concentration profiles for $Hg_{0.8}Cd_{0.2}Te$ crystals solidified upward at 0.6 mm/h.

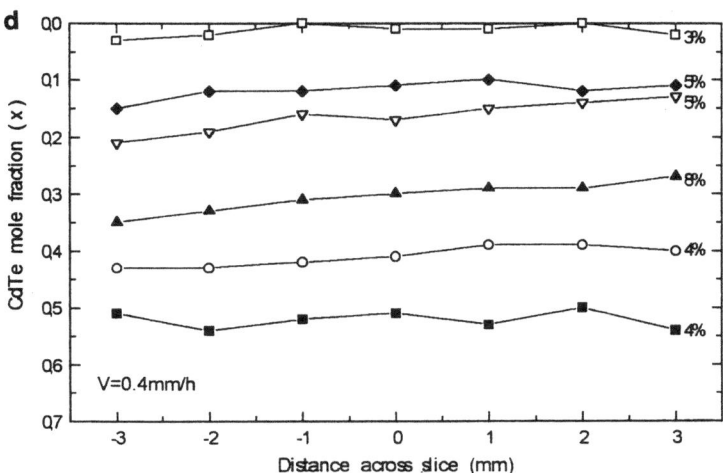

Figure 8d. Radial concentration profiles for $Hg_{0.8}Cd_{0.2}Te$ crystals solidified upward at 0.4 mm/h.

the axial and radial Sn concentration profiles similar to growth by VB and IVB at 1 g. However, the radial variation in composition in crystals grown at high-g was less dependent on axial position. This result may be explained by the enhancement of convective fluxes that mix the solute in the lateral direction, facilitating radial homogenization.

Mercury cadmium telluride was grown at several velocities. The axial composition profile was the expected one, tending to the diffusion-controlled steady state growth condition when the ratio D/V diminished with respect to the crystal length. Near the end of the crystal, all experiments gave a more or less constant low radial variation in composition, whereas near the tip the radial variation went as high as 30%. At low growth rates, however, there was a smaller variation across the slice even near the tip of the crystal, in contrast to the axial concentration behavior. This seems to indicate that convection is an important factor for radial compositional homogeneity, and that better results can be attained with the use of centrifuges.

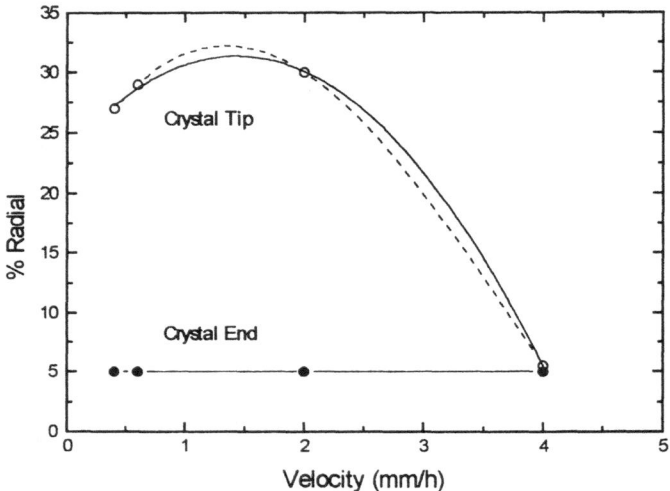

Figure 9. Percent variation in CdTe radial distribution of $Hg_{0.8}Cd_{0.2}Te$ crystals versus growth velocity.

REFERENCES

1. L.L. Regel, N.T. Nghi, and O.I. Rachmatov, Directional crystallization of PbTe under microgravity conditions, *in:* "Proceedings of the 4th European Symposium on Materials Science under Microgravity," European Space Agency, Paris, p 429 (1983).

2. K. Grasza and U. Zuzga-Grasza, Temperature field computations in $Pb_{1-x}Sn_xTe$ crystal grown by inverted Bridgman method, *J. Crystal Growth* 16:139 (1992).

3. H. Rodot, L.L. Regel, and A.M. Turtchaninov, Crystal growth of IV-VI semiconductors in a centrifuge, *J. Crystal Growth* 104:280 (1990).

4. G. Müller, G. Neumann, and W. Weber, The growth of homogeneous semiconductor crystals in a centrifuge by the stabilizing influence of Coriolis force, *J. Crystal Growth* 119:8 (1992).

5. L.L. Regel, A.M. Turchaninov, O.S. Shumaev, I.N. Bandeira, Y.A. Chen, and P.H.O. Rappl, Growth of lead-tin-telluride crystals in high gravity, *J. Crystal Growth* 119:94 (1992).

6. Y.A. Chen, I.N. Bandeira, A.H. Franzan, S. Eleutério, and M.R. Slomka, The influence of gravity on $Pb_{1-x}Sn_xTe$ crystals grown by the vertical Bridgman method, *in:* "Materials Processing in High Gravity," L.L. Regel and W.R. Wilcox, eds., Plenum Press, New York (1994).

DIRECTIONAL SOLIDIFICATION OF $Cd_{1-x}Zn_xTe$ AT HIGH GRAVITY

L.O. Ladeira[a], J. Shen[b], L.L. Regel, and W.R. Wilcox

International Center for Gravity Materials Science and Applications
Clarkson University
Potsdam, New York 13699-5814

ABSTRACT

$Cd_{0.96}Zn_{0.04}Te$ was directionally solidified by the gradient freeze technique in a furnace mounted on Clarkson's centrifuge, at accelerations of 1, 2, 3, 4, and 5 times earth's gravity g. The concentration of tellurium inclusions decreased markedly with increasing acceleration during solidification. The polycrystallinity of the initial portion of the ingots increased as the acceleration during solidification was increased from 3 g to 5 g. The maximum grain size and minimum infrared transmission were at 2 g. The dislocation density was unaffected by centrifugation. The axial concentration profile of zinc appears to have been unaffected by centrifugation, although the scatter in the data is large. The radial variation in zinc concentration was a maximum at 2 g and increased down the ingot.

Our interpretation of our experimental results is that the primary effect of centrifugation during solidification of Zn-doped CdTe was to float away foreign particles that caused heterogeneous nucleation of solid. As the acceleration was increased, this removal of nucleating particles led to increased supercooling prior to nucleation, an increased initial nucleation of grains, an increased initial freezing rate, and reduced nucleation of new grains later in the solidification. The radial variation in zinc concentration indicates a minimum in convection at 2 g, while the large scatter in the axial data can be explained by unsteady convection. Dislocation generation was unaffected by acceleration because of the dominance of thermal stress and differential thermal contraction with the ampoule. The reduction in the concentration of large inclusions with increasing g might have been caused by the floating away of micro bubbles that aid in trapping melt at the freezing interface. The reduction in the concentration of small precipitates with increasing g indicates that the accepted mechanism of precipitation involving the retrograde solubility of Te may not be complete or even correct.

---

[a] Permanent address: Departamento de Fisica, Universidade Federal de Minas Gerais, Minas Gerais, Brazil.

[b] Present address: Rodel, Inc., Newark, Delaware 19713.

INTRODUCTION

Zinc-doped cadmium telluride ($Cd_{1-x}Zn_xTe$) is grown primarily as a substrate for $Hg_{1-y}Cd_yTe$ films for fabrication of infrared focal plane arrays. Both vertical and horizontal Bridgman techniques are utilized for commercial production. As with nominally pure CdTe, the resulting ingots suffer from a variety of defects. These defects include a non-uniform Zn concentration, deviations from stoichiometry, electrically active impurities, dislocations, grain boundaries, twins, inclusions and precipitates.[e.g.,1,2]

The initial motivation for this research was the hypothesis that because CdTe is soft near its melting point,[3-6] it deforms during growth under its own weight. The argument was that this self-deformation made it impossible to produce low-dislocation density CdTe on earth, especially long crystals by a vertical Bridgman technique. If this hypothesis were true, then growth in a centrifuge should increase the dislocation density.

Another motivation for centrifugation during solidification of $Cd_{1-x}Zn_xTe$ stemmed from problems with axial and radial variation of Zn concentration.[7-11] It had been shown experimentally and theoretically that centrifugation during gradient freeze growth can reduce buoyancy-driven convection, and thereby increase doping homogeneity.[12-18] Thus we thought it possible that centrifugation could produce a more uniform Zn concentration in $Cd_{1-x}Zn_xTe$.

Finally, we noted that complete characterization had not yet been performed on crystals grown at different accelerations under otherwise identical conditions. Thus our intent was to test the hypothesis for self-deformation, while determining the influence of centrifugation on doping homogeneity and defects in $Cd_{1-x}Zn_xTe$.

EXPERIMENTS

Gradient freeze experiments were performed in the centrifuge facility developed at Clarkson University.[18,19] This centrifuge has a 1.5 m radius arm, maximum rotation rate of 99 rpm, and a maximum acceleration of 14 g (where g is the acceleration due to earth's gravity). A gradient freeze furnace was placed in a basket that was attached by a hinge to the end of the centrifuge arm, so that the resultant acceleration was always aligned with the furnace tube and the growth ampoule. Figure 1 shows the temperature profile in the furnace during the soak period prior to solidification. Note that the temperature gradient was about 3°C/cm in the region where the growth ampoule was located, with the temperature increasing with height (thermally stable configuration).

All 14 experiments were performed using the same ampoule design and growth conditions, with the exception of acceleration. Each carbon-coated quartz ampoule was 11 mm in inside diameter, had a tapered conical tip, and contained about 40 g of thoroughly-cleaned chunks of $Cd_{0.96}Zn_{0.04}Te$. No excess Cd was added. The ampoule was evacuated to about 10^{-6} torr and sealed with a vapor space approximately 1 cm high above the CdTe.

A thermocouple was placed about 1 mm below the ampoule tip for control of the furnace power. After placing the ampoule in the furnace, the temperature was raised until the ampoule tip reached 1100°C. This temperature was held for 24 hr before beginning centrifugation. The centrifuge was rotated at the desired rate for 4 hours before initiating controlled cooling of the ampoule tip at 1.5°C/hr. Numerical simulation of the heat transfer in the ampoule gave an interfacial temperature gradient of about 1.9°C/cm, so that the freezing rate was approximately 8 mm/hr. After solidification was complete, the ampoule was cooled at 20°C/hr to 900°C, 50°C/hr to 500°C, and then the furnace power was turned off. A minimum of two runs was performed at each acceleration, from 1 g to 5 g.

All 14 ingots readily slid out of their ampoules after cooling. All had a shiny and smooth surface with no bubbles or pits. A wire saw was used to cut each ingot longitudinally into two halves. One half was used for zinc composition profiling, while the other half was cut into (111) wafers for measurement of precipitate density, etch pit density and infrared transmission. After mechanical and chemical polishing, these wafers were about 1.2 mm thick. The characterization methods are described in reference 6.

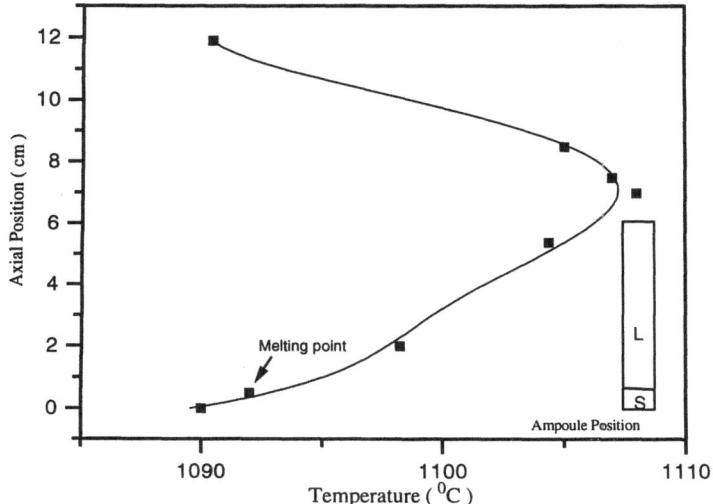

Figure 1. Temperature profile in the furnace during the soak period. Measured with a thermocouple in an empty quartz ampoule without centrifugation. Centrifugation changed the profile negligibly.

MICROSTRUCTURE

Figure 2 shows several of the resulting ingots. Notice that those solidified with centrifugation had rough tops, suggestive of extrusion of melt during rapid freezing. All ingots contained gas bubbles with diameters from 50 to 100 μm. The concentration of bubbles was greatest near the bottoms and tops of the ingots, indicating that the freezing rate was greater near the two ends.

As the acceleration was increased from 3 to 5 g, the resulting ingots were increasingly polycrystalline near their bottom ends. Increasing polycrystallinity indicates increased supercooling prior to nucleation of the solid. Two of the three ingots solidified at 2 g had the largest grain sizes, with one of these having a single grain over much of its length.

Rudolph[21-23] experimentally demonstrated a connection between supercooling, grain nucleation, and the superheating of molten CdTe prior to solidification. Heating the melt to 10°C or more above the melting point dramatically increased the supercooling required to nucleate CdTe, increased the initial polycrystallinity of the ingots, and reduced subsequent nucleation of new grains. The supercooling for nucleation jumped abruptly at a critical superheating of about 10 K. Rudolph attributed this phenomenon to the destruction of pre-nucleation clusters in the melt by superheating. We have an alternate explanation that is supported by the present results.

If pre-nucleation clusters are destroyed by heating, one would expect them to reform in a short time when the temperature is lowered again, i.e. the phenomenon should be reversible. However, Rudolph's experiments showed this behavior to be highly irreversible.[22] After holding the melt above the critical superheating for 10 minutes, "the supercooling effects was not a function of the chosen cooling rate." It should be noted that viscosity versus temperature data for molten CdTe showed no sudden change, only a steady decrease indicating gradual breaking of atomic bonds within the melt as the temperature was raised.[24]

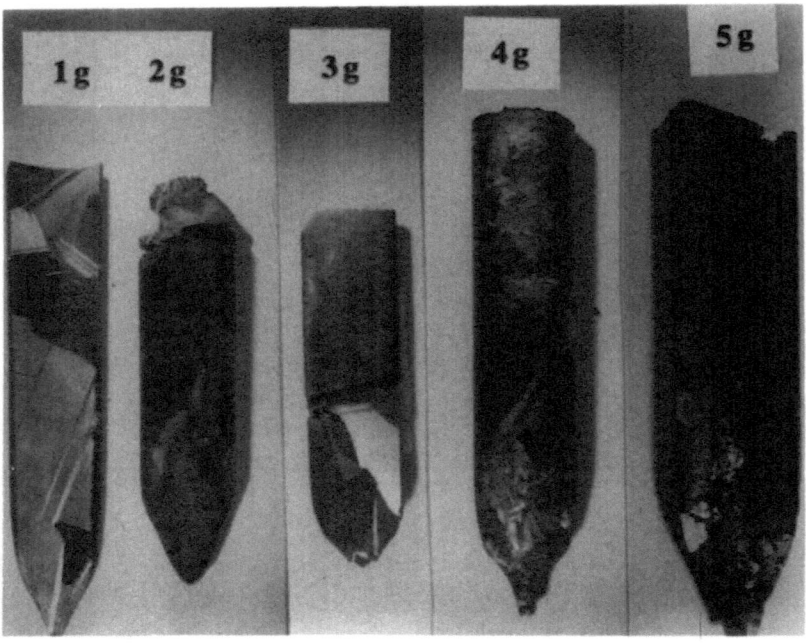

Figure 2. Photographs of several of the 14 ingots produced in these experiments. The numbers are the accelerations at which they were solidified, where $g = 9.8$ m/s^2.

An alternate explanation for Rudolph's superheating/supercooling/nucleation results is the dissolution of second-phase material that causes heterogeneous nucleation. The paper by Kanter[25] is instructive. Supercooling for nucleation of PbTe was measured versus prior superheating. Up to a critical superheating of about 12 K, the supercooling for nucleation was negligible. Above this critical superheating, the supercooling depended on the treatment of the melt and the presence of second-phase material. For example, PbTe prepared in air from the elements gave a maximum supercooling of only about 5 K. Filtration of the melt increased this value to 15 K. In the presence of molten B_2O_3, which is known to dissolve oxides, the maximum supercooling reached 60 K. Interestingly, the maximum supercooling was 50 K in the presence of the fused silica (quartz) ampoule wall, but only 5 K with fine SiO_2 particles added to the melt. Kanter comments that the same supercooling versus superheating behavior had been observed for many metals and indicates heterogeneous nucleation.

Thus we believe that Rudolph's experimental results were due to the presence of foreign particles in the molten CdTe that acted as good heterogeneous nucleation sites, that these particles dissolved in the melt at superheats of 10 K and above, and that they did not themselves nucleate and regrow upon cooling. These particles were probably composed of oxides. Annealing of CdTe in hydrogen prior to melting and growth greatly reduced nucleation of grains and twins, while controlled oxidation yielded a low quality ingot.[26] Silicon-rich inclusions are another possible nucleating agent.[27,28] In support of the notion of nucleating particles is the relationship between wetting of molten CdTe, adhesion of solid CdTe, and grain nucleation.[29] In progressing from uncoated quartz growth ampoules, to carbon coated, to pyrolytic boron nitride coated, wetting decreased, adhesion decreased, and the apparent supercooling required for nucleation increased. On coated quartz, the initial material was polycrystalline, while on uncoated quartz the solid began as single crystals. On the other hand, there was little nucleation of new grains during growth in PBN-coated quartz, while continued nucleation occurred with uncoated ampoules.

Our experimental results on the centrifuge are consistent with the presence of hetero-nucleating particles that were less dense than the melt. We propose the following hypothesis to explain our experimental results. At 1 g, nucleating particles were present both at the ampoule tip and dispersed throughout the melt. These particles caused nucleation of solid at a small supercooling, such that the number of grains nucleated at the ampoule tip was small. Nucleation of new grains also occurred on particles later during solidification. At 2 g, sufficient particles remained at the tip to permit ready nucleation there, but most rose so that little nucleation occurred during subsequent solidification. At 3 g, nearly all nucleating particles had been removed from the tip and substantial supercooling was required to cause the initial nucleation. As g was increased further, fewer and fewer nucleating particles remained, greater supercooling was required to initiate nucleation, greater polycrystallinity occurred at the beginning, the subsequent freezing rate was larger, and the ingots became more polycrystalline. We suggest that the balance between these opposing effects was responsible for achieving the largest grain size at 2 g.

ZINC DISTRIBUTION

For composition analysis, the surface of each longitudinal slice was mechanically polished and then chemically polished in Br-methanol. Both K and L x-ray emissions of zinc were measured in the scanning electron microscope using energy dispersive spectrometry (SEM-EDX). Measurements were made at 2 mm intervals along the centerline, and were repeated once or twice at each point.

The logarithmic form of the normal distribution equation was used to interpret the axial composition data:

$$ln\,C \;=\; A + (k - 1)ln(1 - f) \tag{1}$$

where C is the zinc concentration in atom %, A is a constant, k is the effective distribution coefficient, and f is the mass fraction solidified, approximated here as the fractional distance d down the ingot. Figure 3 shows typical results plotted as lnC versus ln(1-d). A least squares fit to each such plot yielded a line, the slope of which was (k-1). Because of the large scatter in the data, the 95% confidence limits for k ranged from ± 0.07 to ± 0.15, independent of g. The resulting values are plotted versus acceleration in Figure 4, from which it can be concluded only that k was greater than 1 and was not appreciably influenced by centrifugation.

On the same longitudinal slices, we also measured the radial variation in zinc concentration in ingots grown at 1 g, 2 g, and 3 g. Figures 5 and 6 show typical results. Some of the concentration profiles appear symmetric, while others are less so. A symmetric profile indicates an axi-symmetric convection cell during solidification. Figure 7 compares the standard deviation in radial concentration data taken at different values of acceleration and lengths down the ingots. Under the present growth conditions, reduced convection should cause an increase in the radial variation in concentration. Thus our results indicate that convection near the freezing interface reached a minimum at about 2 g, and increased with decreasing height of the melt. This conclusion is consistent with prior experiments and with theory for a concave interface shape,[14-18] which is expected for gradient freeze growth.

We need to address the large scatter in our axial concentration data. If this were merely experimental error, we would have a similar scatter in our radial concentration data. We do not. Furthermore, others have obtained good results for Zn in CdTe using an electron microprobe with wave-dispersive spectrometry. We suggest that the scatter in our axial concentration data arises from meandering convection, as has been observed in organic melts in a "thermally stable" environment with temperature increasing with height.[30-33]

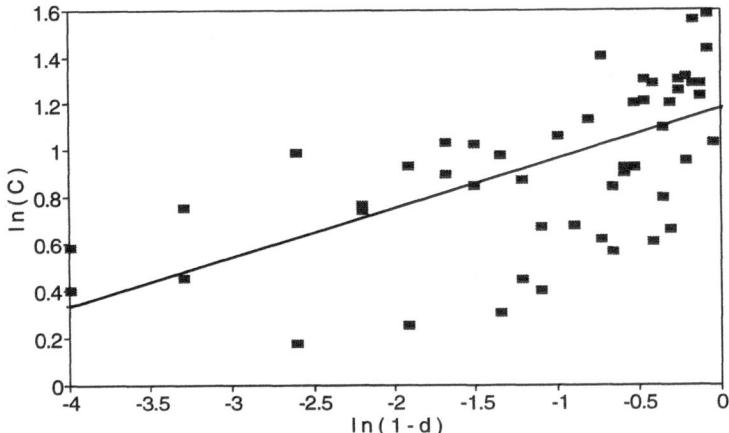

Figure 3. Zinc concentration C in atom percent versus fractional distance d down an ingot solidified at 1 g (without centrifugation). The zinc concentration was determined using energy dispersive x-ray spectroscopy in a scanning electron microscope, using the zinc L emission. The line is a least squares fit to the data. From the slope, k = 1.2.

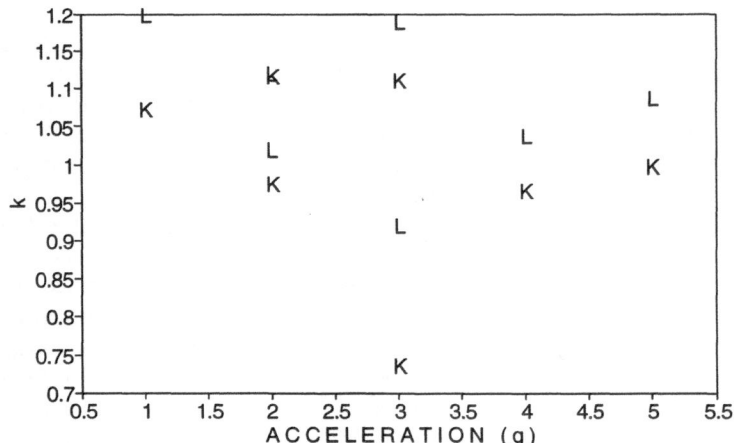

Figure 4. Effective distribution coefficient k obtained from the slopes of ln C versus ln(1-d) data. Here L represents a value obtained using the L emission line of zinc, while K is a value obtained using the K line. The 95% confidence limits are approximately ± 0.1.

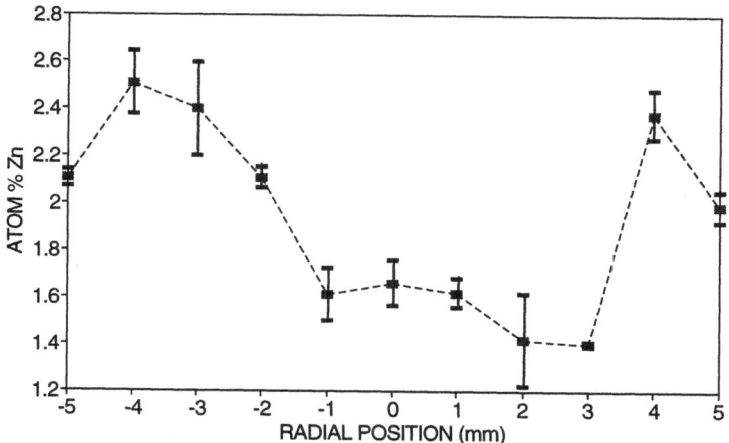

Figure 5. Radial variation in zinc concentration 37 mm down an ingot solidified at 2 g. The error bars are ± one standard error, from the three measurements taken at each position.

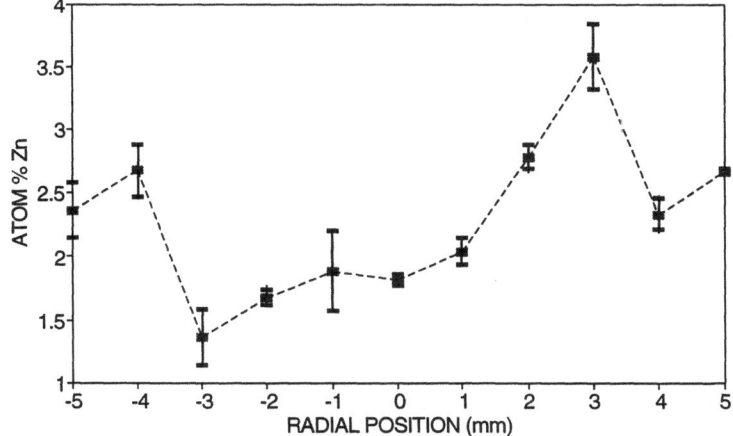

Figure 6. Radial variation in zinc concentration at 15 mm down an ingot solidified at 3 g. The error bars are ± one standard error, from the three measurements taken at each position.

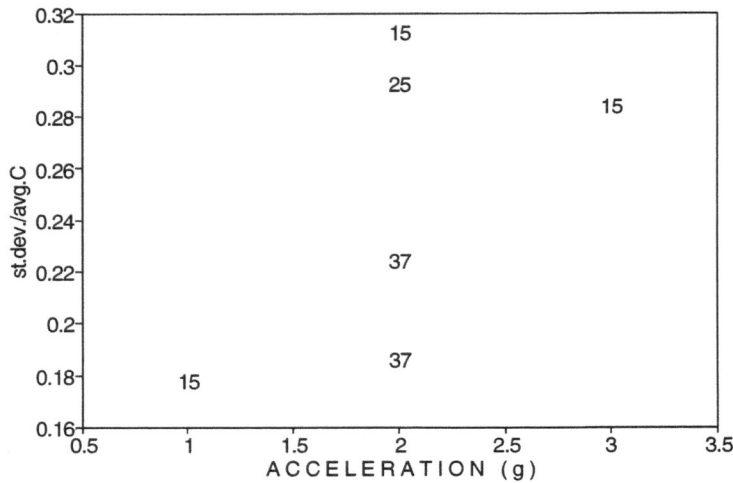

Figure 7. Standard deviation in zinc concentration divided by the average zinc concentration. Each point represents a set of measurements taken across one diameter at the distance from the front (bottom) of the ingot indicated by the number (in mm).

INCLUSIONS

Infrared microscopy revealed inclusions throughout all ingots. Using an SEM-EDX technique,[6,34,35] we found these inclusions were primarily Te. Those larger than 5 μm tended to be associated with bubbles, which were concentrated near the head and tail ends of the ingots. Previously, we had reported a close association between bubbles and large inclusions in CdTe:[35] "Most of the polyhedral-shaped Te precipitates with a size range from 3 to 20 μm had voids inside." We suggested that bubbles may have formed first at the freezing interface and aided in trapping melt containing excess tellurium or cadmium. Some researchers have proposed that voids found in CdTe result from Cd evaporation. Although this is a possibility for a Cd-rich melt, due to rejection of excess Cd by the freezing interface, it is not possible for a Te-rich melt as in the present experiments. The most likely origin of bubbles and voids is residual gas, which would dissolve in the melt and be rejected by the freezing interface. Although our ampoules were sealed in a vacuum, this vacuum was not perfect. Additional possible sources of gas are out gassing from the quartz ampoule, hydrogen produced by reaction of residual water vapor with CdTe, and carbon monoxide from reaction of the carbon coating with the quartz ampoule.

Figure 8 shows the density of large inclusions in the central portion of the ingots versus acceleration. This density appears to decrease with increasing acceleration, at least up to 3 g. This probably reflects the increased tendency of bubbles to float away from the freezing interface as acceleration is increased.

Infrared microscopy also revealed smaller precipitates, between 1 and 5 μm. Most of these were randomly distributed inside grains, but some were on grain and twin boundaries. Figure 9 shows the acceleration-dependence of the density of these precipitates in the central portions of the ingots. The precipitate density declined linearly with increasing acceleration, by a factor of over 3 in going from 1 g to 5 g.

140

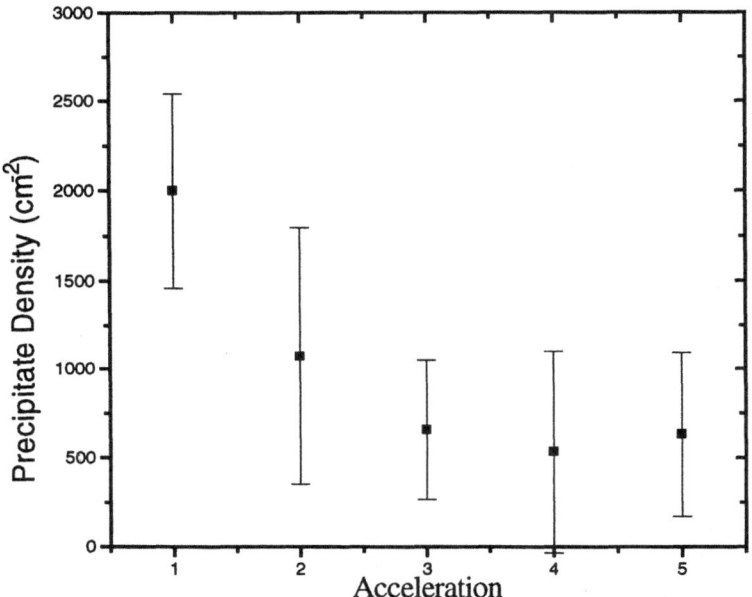

Figure 8. Density of tellurium-rich inclusions over 5 μm in size in the central portion of ingots, versus the acceleration in g at which the ingots were solidified. Error bars are one standard error. The precipitates in a given area were counted by infrared microscopy of a 1.2 mm thick (111) slice.

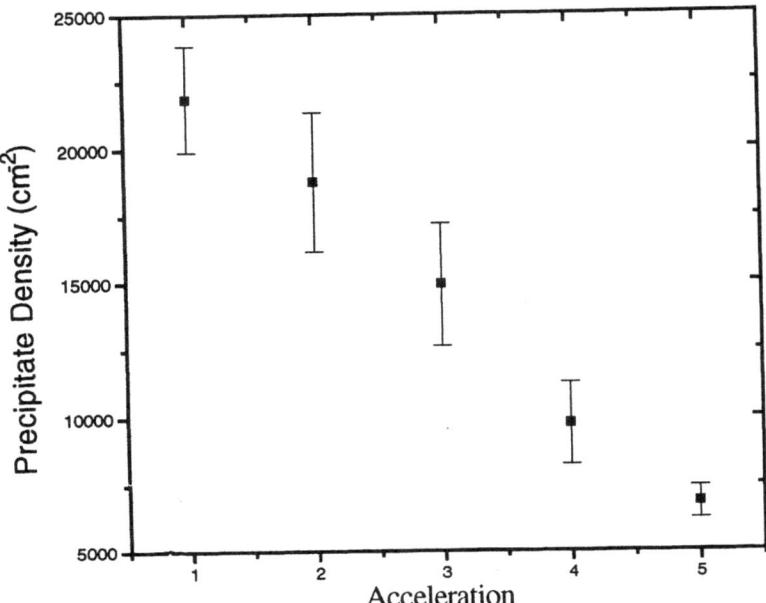

Figure 9. Density of tellurium-rich precipitates between 1 and 5 μm in size.

Large inclusions in CdTe have been attributed to trapping of melt at the interface, while small precipitates are thought to stem from retrograde solubility.[21,36-40] The CdTe lattice can tolerate a much higher deviation from 1:1 stoichiometry at high temperature than at room temperature. Thus, if CdTe is grown from a Te-rich melt, the crystal will initially contain excess Te, which forms precipitates as the crystal cools to room temperature (via condensation of cadmium vacancies[21,36-38] or tellurium interstitials[6,39]). Centrifugation should not influence this precipitation process. The strain field at 5 g is not sufficient to influence the thermodynamics or kinetics of precipitation (see the results below for plastic deformation). Another possibility is that centrifugation changes the temperature gradient in the ingot during cooling from the freezing point. However, our measurements in an empty ampoule showed that centrifugation caused only a only a slight increase in the axial temperature gradient, insufficient to alter precipitation kinetics due to retrograde solubility.

What centrifugation at this level does strongly influence are convection and second-phase sedimentation[c] in the melt. One possibility is that the melt is trapped initially at grain boundary grooves. Since the melting point of tellurium is only about 450°C, the resulting inclusions would remain molten and would migrate out of the grain boundaries due to the temperature gradient that exists during cooling.[e.g.,37,40] Convection would influence this trapping process; Yuferev[41] has shown, for example, that centrifugation can increase morphological stability.

Because Te is less dense than CdTe, centrifugation may increase convective removal of the Te enriched melt adjacent to the freezing interface.

We might also hypothesize the existence of Te-rich second-phase droplets, such as TeO_2, in the melt adjacent to the freezing interface. Engulfment of such droplets would be influenced by centrifugation, both because of changes in convection and because the drops would experience a force pushing them away from the freezing interface.

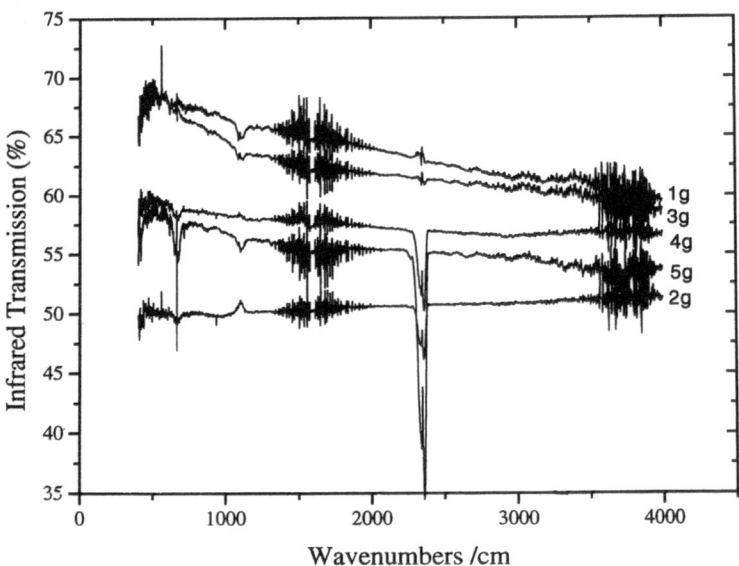

Figure 10. Infrared transmission through 1.2 mm thick slices taken from the middle portions of ingots.

---

[c] At 5 g, sedimentation within a single-phase mixture is negligible.

INFRARED TRANSMISSION

In all ingots, the infrared transmission was greatest in the central portion of the ingot, where the densities of bubbles, precipitates and grain boundaries were least. Figure 10 shows the infrared transmission in the central portions of ingots solidified at different accelerations. The transmission was greatest in the ingot solidified at 1 g, and lowest in that solidified at 2 g.

ETCH PIT DENSITY

Dislocation etch pits were produced on (111)A surfaces of the wafers by Nakagawa etchant.[6] As predicted in a theoretical model for temperature and thermal stress,[10,42-44] the etch pit density was highest in the shoulder region[d] of the ingots (see Figure 11). As shown in

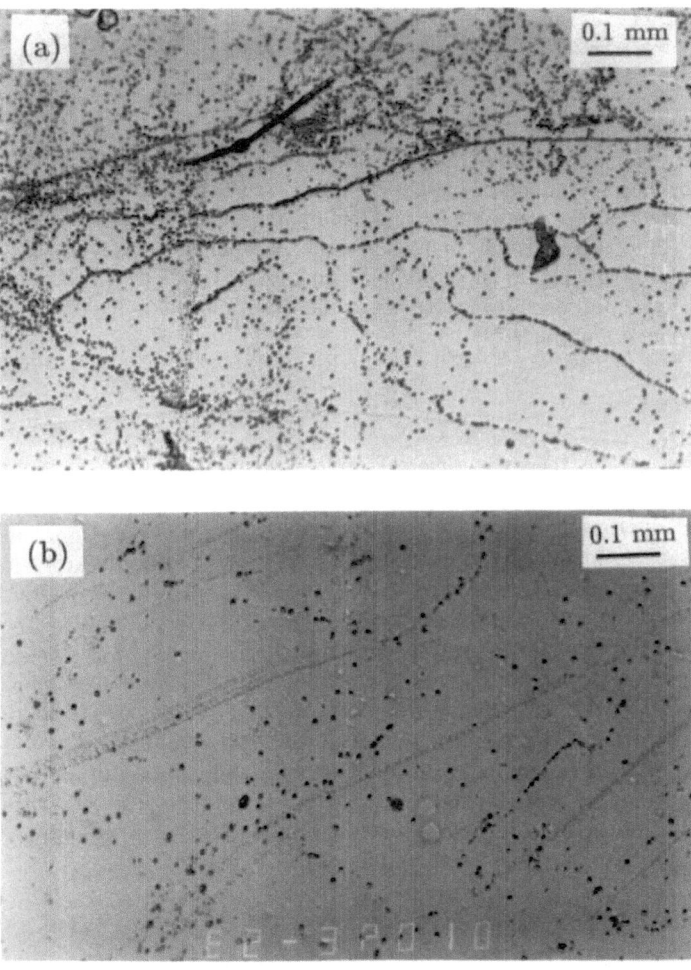

Figure 11. Etch pits in an ingot solidified at 2 g, produced by Nakagawa etchant.[6]
(a) top: shoulder.   (b) bottom: middle.

---

[d] The shoulder is the region where the initial conical section meets the cylindrical part of the ingot.

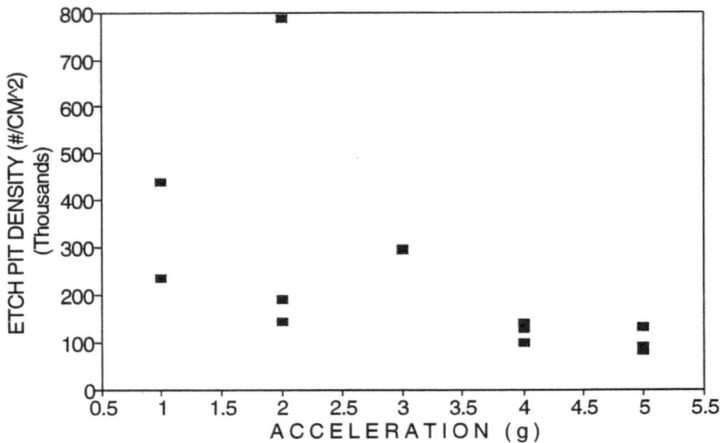

Figure 12. Etch pit density versus acceleration at which the ingots were solidified. Slices were taken from the middle portions of the ingots.

Figure 12, the etch pit density did not depend on the acceleration during solidification. This is consistent with the predictions of theoretical models showing that the hydrostatic stress is negligible compared to the thermal stress, and that the thermal stress is negligible compared to the stress due to differential thermal contraction when the solid sticks to the ampoule wall.[42,43,45] The importance of sticking to generation of dislocations in CdTe was demonstrated by comparing experimental results obtained using uncoated, carbon-coated, and pyrolytic boron nitride-coated quartz ampoules.[29]

CONCLUSIONS

Our interpretation of these experimental results is that the primary effect of centrifugation during solidification of Zn-doped CdTe was to float away foreign particles that caused heterogeneous nucleation of solid. As the acceleration was increased, this removal of nucleating particles led to increased supercooling prior to nucleation, an increased initial nucleation of grains, an increased initial freezing rate, and reduced nucleation of new grains later in the solidification. Consequently, the grain size was a maximum in ingots solidified at 2 g. The radial variation in zinc concentration indicates a minimum in convection at 2 g, while the large scatter in the axial data can be explained by unsteady convection. Dislocation generation was unaffected by acceleration because of the dominance of thermal stress and differential thermal contraction with the ampoule. The reduction in the concentration of large inclusions with increasing g might be attributed to the floating away of micro bubbles that aid in trapping melt at the freezing interface. The reduction in the concentration of small precipitates indicates that the generally accepted mechanism of precipitation, involving the retrograde solubility of Te, is incorrect or at least incomplete.

Acknowledgments

This work was supported by the National Science Foundation under grant DMR-9414304, the Consortium for Commercial Crystal Growth, and the Center for Advanced Materials Processing. L.O. Ladeira is grateful to the Conselho Nacional de Desenvolvimento Cientifico de Tecnologico, Brazil, for a faculty fellowship. The II-VI Corporation of Saxonburg, Pennsylvania donated the high purity ZnCdTe used for the experiments. Our thanks to R.

Derebail for assistance with the growth experiments, to T.P. Lee for the simulation of heat transfer in the growth ampoule, and to W. Plunkett for help in the SEM-EDX analysis.

REFERENCES

1.   R. Triboulet, W.R. Wilcox, and O. Oda, eds., "CdTe and Related Cd Rich Alloys," North-Holland, Amsterdam (1993).
2.   P. Rudolph, Fundamental studies on Bridgman growth of CdTe, *Prog. Cryst. Growth Charact. Mater.* 29:275 (1995).
3.   R. Balasubramanian and W.R. Wilcox, Mechanical properties of CdTe, *Mat. Sci. Engin.* B16:1 (1993.
4.   J.C. Moosbrugger and A. Levy, Constitutive Modeling for CdTe single crystals, *Metallurgical & Materials Transactions* 26A:2687 (1995).
5.   J.C. Moosbrugger, Continuum slip viscoplasticity with the Haasen constitutive model: application to CdTe single crystal inelasticity, *Internat. J. Plasticity* 11:799 (1995).
6.   J. Shen, "Microstructural Imperfections and Characterization of CdTe and CdZnTe crystals," Ph.D. Dissertation, Clarkson University (1993).
7.   M. Azoulay, A. Raizman, G. Gafni and M. Roth, Crystalline perfection of melt-grown CdTe, *J. Crystal Growth* 101:156 (1990).
8.   M. Muhlberg, P. Rudolph, C. Genzel, B. Wermke and U. Becker, Crystalline and chemical quality of CdTe and $Cd_{1-x}Zn_xTe$ growth by the Bridgman method in low temperature gradients, *J. Crystal Growth* 101:275 (1990).
9.   P. Capper, J.E. Harris, E.S. O'Keefe, and C.L. Jones, Macro- and microsegregation of Zn in Bridgman-grown CdZnTe, *Adv. Mat. Opt. Electron.* 5:101 (1995).
10. D.J. Larson, Jr., L.G. Casagrande, D.D. Marzio, A. Levy, F.M. Carlson, T. Lee. D. Black, J. Wu, and M. Dudley, Producibility improvements suggested by a validated process model of seeded CdZnTe vertical Bridgman growth, *in:* "Producibility of II-VI Materials and Devices," *Proc. SPIE - Int. Soc. Opt. Eng.* 2228:11 (1994).
11. Q. Hou, J. Wang, J. Deng, M. Li, and F. Tao, Vertical gradient freeze growth of $Cd_{1-x}Zn_xTe$ with Cd/Zn reservoir, *Rare Met. (Beijing)* 13:287 (1994).
12. L.L. Regel, M. Rodot, and W.R. Wilcox, eds., "Materials Processing in High Gravity," special issue of *J. Crystal Growth* 119:1-167 (1992).
13. L.L. Regel and W.R. Wilcox, eds., "Materials Processing in High Gravity," Plenum Press, New York (1994).
14. W.A. Arnold and L.L. Regel, Thermal stability and the suppression of convection in a rotating fluid on earth, *ibid.*
15. V.A. Urpin, Convective flows during crystal growth in a centrifuge, *ibid.*
16. W. Arnold, "Numerical Modeling of Directional Solidification on a Centrifuge," Ph.D. Dissertation, Clarkson University (1993).
17. J. Friedrich and G. Müller, Segregation in crystal growth under high gravity on a centrifuge: a comparison between experimental and theoretical results, *in present volume.*
18. J. Friedrich and G. Müller, Convection in crystal growth under high gravity on a centrifuge, *in present volume.*
19. R. Derebail, W.A. Arnold, G.J. Rosen, W.R. Wilcox, and L.L. Regel, HIRB - the centrifuge facility at Clarkson, *in ref 13.*
20. R. Derebail, "Directional Solidification of Indium Antimonide under High Gravity in Large Centrifuges," PhD Dissertation, Clarkson University (1994).
21. P. Rudolph and M. Mühlberg, Basic problems of vertical Bridgman growth of CdTe, *Mater. Sci. Eng.* B16:8 (1993) 8.
22. M. Mühlberg, P. Rudolph, M. Laasch, and E. Treser, The correlation between superheating and supercooling in CdTe melts during unseeded Bridgman growth, *J. Crystal Growth* 128:571 (1993).
23. P. Rudolph, K. Umetsu, H.J. Koh, and T. Fukuda, The correlation between growth stability and superheating of the melt in semiconductor compounds, *J. Jap. Assoc. Crystal Growth* 21:38 (1994).
24. V.M. Glazov, S.N. Chizhevskaya, and N.N. Glagoleva, "Liquid Semiconductors," Plenum Press, NY (1969)
25. Yu.O. Kanter, Heterogeneous nucleation in PbTe alloys, *Crystal Res. Technol.* 16:1333 (1981).
26. L.O. Ladeira, private communication, Universidade Federal de Minas Gerais, Belo Horizonte, Brazil (1996).
27. V.V. Krapukhin, I.V. Perepelkin, T.V. Kul'chitskaya, N.A. Kul'chitskii, and A.A. Glebkin, Real structure of single crystals of cadmium telluride, *Sov. Phys. Crystallogr.* 36:581 (1992).

28. L.P. Shcherbak, I.M. Fodchuk, and O.M. Tikhonova, Impurity structure defects in CdTe single crystals, *Sov. Phys. Crystallogr.* 36:862 (1991).

29. R. Shetty, W.R. Wilcox, and L.L. Regel, Influence of ampoule coatings on cadmium telluride solidification, *J. Crystal Growth* 153:103 (1995).

30. H. Potts and W.R. Wilcox, Chaotic asymmetric convection in the Bridgman-Stockbarger technique, *J. Crystal Growth* 74:443 (1986).

31. G.T. Neugebauer and W.R. Wilcox, Convection in the vertical Bridgman-Stockbarger technique, *J. Crystal Growth* 89:143 (1988).

32. G.T. Neugebauer and W.R. Wilcox, Experimental observation of the influence of furnace temperature profile on convection and segregation in the vertical Bridgman crystal growth technique, *Acta Astronautica* 25:357 (1991).

33. G. Neugebauer, "The Influence of Convection on Radial Segregation in the Vertical Bridgman-Stockbarger Crystal Growth Technique," Ph.D. Dissertation, Clarkson University (1990).

34. J. Shen, D.K. Aidun, L.L. Regel, and W.R. Wilcox, Etch pits originating from precipitates in CdTe and $Cd_{1-x}Zn_xTe$ grown by the vertical Bridgman-Stockbarger methods, *J. Crystal Growth* 132:351 (1993).

35. J. Shen, D.K. Aidun, L. Regel, and W.R. Wilcox, Characterization of precipitates in CdTe and $Cd_{1-x}Zn_xTe$ grown by vertical Bridgman-Stockbarger technique, *J. Crystal Growth* 132:250 (1993).

36. P. Rudolph, M. Neubert, and M. Mühlberg, Defects in CdTe Bridgman monocrystals caused by nonstoichiometric growth conditions, *J. Crystal Growth* 128:582 (1993).

37. P. Rudolph, A. Engel, I. Schentke, and A. Grochocki, Distribution and genesis of inclusions in CdTe and (Cd,Zn)Te single crystals growth by the Bridgman method and by the traveling heater method, *J. Crystal Growth* 147:297 (1995).

38. H. Zimmerman, R. Boyn, P. Rudolph, C. Albers, K.W. Benz, D. Sinerius, C. Eiche, B.K. Meyer, and D.M. Hoffman, State and distribution of point defects in doped and undoped Bridgman-grown CdTe single crystals, *J. Crystal Growth* 128:593 (1993).

39. R.D.S. Yadava, R.K. Bagai, and W.N. Borle, Theory of Te precipitation and related effects in CdTe crystals, *J. Electron. Mat.* 21:1001 (1992).

40. T.S. Lee, Y.T. Jeoung, H.K. Kim, J.M. Kim, I.H. Park, J.M. Chang, S.U. Kim, and M.J. Park, Study on tellurium precipitates in (Cd,Zn)Te crystals grown by seeded vertical Bridgman method, *Proc. - Electrochem. Soc.* 94-30:242 (1995)("Long Wavelength Infrared Detectors and Arrays: Physics and Applications").

41. V.S. Yuferev, Morphological stability of directional solidification in a centrifugal field, *in ref. 13*.

42. T. Lee, "Finite Element Analysis of the Thermal and Stress Fields during Directional Solidification of Cadmium Telluride," Ph.D. Dissertation, Clarkson University (1996).

43. T. Lee, J.C. Moosbrugger, F.M. Carlson, and D.J. Larson, Jr., The role of thermal stress in vertical Bridgman growth of CdZnTe crystals, *in ref 13*.

44. D.J. Larson, Jr., L.G. Casagrande, D. Di Marzio, A. Levy, F.M. Carlson, T. Lee, D. Black, J. Wu, and M. Dudley, Producibility improvements suggested by a validated process model of seeded CdZnTe vertical Bridgman growth, *SPIE* 2228:11 (1994).

45. W. Rosch and F. Carlson, Computed stress fields in GaAs during vertical Bridgman growth, *J. Crystal Growth* 109:75 (1991).

LIQUATION PHENOMENA IN MOLTEN CMT AND VOLUMETRIC PROPERTIES OF HgTe AND CMT

Vasilii M. Glazov,[1] Lidya M. Pavlova,[1] and Sergei V. Stankus[2]

[1]Department of Physical Chemistry
Moscow Institute of Electronic Engineering (Technical University)
103498 Moscow, Russia.
[2]Department of Radiation and Matter
Siberian Branch of RAS, Institute of Thermophysics
630090 Novosibirsk, Russia

INTRODUCTION

Our report is devoted to the investigation results for the volumetric properties of mercury telluride and CMT (cadmium mercury tellurium) alloys in a liquid phase by the gamma radiation attenuation method. These substances are widely used as materials in electronic engineering and are popular objects for experiments in space material science. Wider use of electronic devices based on CMT is being retarded by the difficulties of homogeneous material production. Different techniques are being used to solve the question of single bulk uniform crystals, but they haven't yet succeeded completely.[1,2] Chemically homogeneous single crystals of solid solution couldn't have been obtained even under zero gravity, although the attempt to solve that question by this technique is being continued. It should be noted that we do not know all of the mechanisms by which gravitation influences the degree of homogeneity of solid solutions obtained by unidirectional solidification.

The behavior of HgTe and CMT at high gravity is not directly considered in the present report. Our investigations were connected with the determination of structural CMT melt inhomogeneity mentioned by Holland and Chandra.[3,4] We believe that information on the structural peculiarities of these substances in the liquid phase will permit us to predict their behavior at high gravity.

EXPERIMENTAL METHODS

We investigated the temperature dependence of the density of HgTe and $Hg_xCd_{1-x}Te$ alloys at x=0.2 by the penetrating γ-radiation method.[5,6] The mercury telluride synthesis was performed by direct alloying of the pure components. The impurity content of the initial reagents was no more that 10^{-5} %. HgTe and CdTe compounds were synthesized beforehand to prepare CMT alloys. The crystal production of solid solutions was performed by the heater traveling method.[2] The quality of the synthesized samples was determined by X-ray diffraction, X-ray microscopy, chemical analysis, and picnometric density measurements.

The procedure for measuring density by means of penetrating radiation and the experimental setup were given in detailed elsewhere.[5] The sample under consideration was placed in an evacuated, sealed quartz container 25 mm in i.d. and 60 mm high, which was placed inside a bulk copper block. Temperature was measured by a platinum/platinum-rhodium thermocouple embedded into a block wall. Corrections for the difference in temperature between the container and the block were determined by special calibration experiments using a sample simulator with an auxiliary thermocouple in the center. Studying the density of mercury telluride was mainly carried out in a dynamic mode at heating and cooling rates of no more than 2 K/min. In addition, at some temperatures the sample density was determined under static conditions after being held 20-30 min at constant temperature. No significant discrepancies were observed for the data obtained under different conditions of measurements.

A specific feature of studying mercury telluride is that thermal dissociation of a sample at high temperature can result in a loss of uniformity and in the formation of gas inclusions at the walls of the container. Such defects are particularly serious when they occur in the zone of the γ-quantum beam. Uniformity of each sample was monitored by measuring the attenuation coefficient of radiation at different heights. This procedure made it possible to locate the defects and avoid their occurrence in the γ-ray measurement zone.

RESULTS AND DISCUSSION

Mercury Telluride

The results for the temperature dependence of mercury telluride density, in both liquid and solid states, are shown in Figure 1. Mercury telluride is seen to melt like water, that is, with increased packing. Earlier density measurements also are given. A comparison indicates that our data for the liquid phase coincide, within experimental error, with the data of Regel and Mockrovsky,[7] and noticeably exceed the data from Chandra and Holland.[3,4] Following are two possible reasons for underestimation of densities in both liquid and solid phases. Perhaps Chandra and Holland did not avoided the gas bubble effect in the melt due to mercury evaporation during the measurements. Moreover, they did not synthesize mercury telluride *in situ* in the measuring cell. This means that the quality of the sample was not verified either before or after the density measurements. In the present investigation, the quality of the mercury telluride was verified by X-ray diffraction both before and after the density measurements. The measured lattice constant coincided with the reference data[7] to the fourth decimal place.

Note that the solid-phase density exhibits a rather weak dependence on temperature. For example, changing the temperature by 650 K results in a density change of only 120 kg/m^3. Most likely, this can be caused by the dissociation of the solid compound, resulting in the emergence of a factor that balances the thermal expansion and, hence, the decrease in density. Dissociation in the solid is accompanied by the formation of the Frenkel defects and, in the given case, by migration of the mercury atoms into the interstices and filling of the tetrahedral voids.

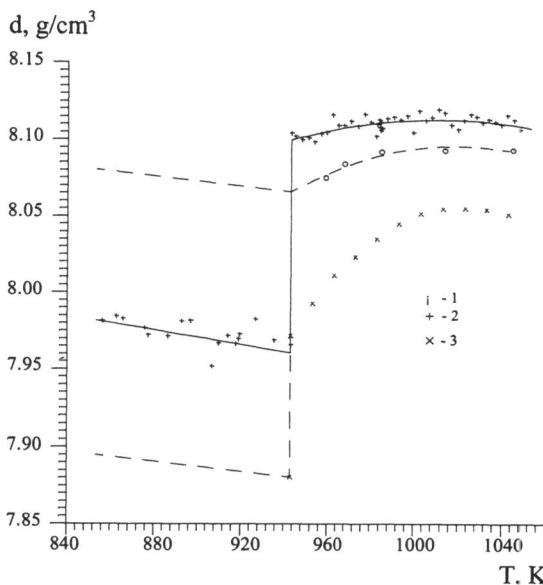

Figure 1. Density versus temperature for mercury telluride in the vicinity of the crystal-melt phase transition. ○ - Data of Regel and Mockrovsky[7] ; + - Present results; x - Data of Chandra and Holland[3,4]

This should bring about a certain increase in compactness of the lattice, smoothing over the effect of the increase in interatomic distances and loosening of the lattice due to thermal expansion.

Our results unambiguously indicate an increase in density in the course of the solid-liquid phase transition. Hence, the volume change ΔV is negative and, according to the Clapeyron-Clausius equation,

$$\frac{dT}{dP} = \frac{\Delta V}{\Delta S} \tag{1}$$

increasing the pressure P should result in a decrease of the melting point T of mercury telluride. (Here ΔS is the entropy of melting, which is always positive.) It has been shown[9,10] that the baric coefficient of the HgTe melting temperature is negative. According to the Clapeyron-Clausius equation, this can be true if $\Delta V = (V_l - V_s) < 0$. The volume difference ΔV was estimated[9,10] to be -1.94×10^{-6} and -1.08×10^{-6} m^3. Our result here gives the value -0.71×10^{-6} m^3. The volume difference was determined[3,4] to be -0.47×10^{-6} m^3, which is considerably lower than our value. The discrepancy between these values of the experimental volume effect of melting is caused by the difference in the measured densities of molten HgTe noted above. Note that the values of the volume effect obtained from the experimental values of the melting baric coefficient[9,10] by Equation 1 are of the same sign. The quantitative discrepancies can be rationalized by the error in determining the entropy of melting, which enters into the Clapeyron-Clausius equation, because of the influence of the thermal dissociation[6] of HgTe and errors in constructing a p-T plot. Summarizing the discussion of the volume effect values, we believe that its previous estimation[11] is likely to be invalid. Our results for the density of mercury telluride at the melting temperature in the solid and liquid states are summarized below:

Density, kg/m³			Volume, (m³/mol)•10⁻⁶		
ρ_s	ρ_l	$\Delta\rho$	V_s	V_l	ΔV
7965	8103	137	41.19	40.49	-0.7

The observed abnormal change in density upon heating the melt can be rationalized in terms of the concept of a postmelting effect.[12] When heating a melt immediately after melting, an increase in density can be caused by an intensive decay of clusters that inherited a relatively loose structure from the crystalline phase. In the temperature range from T_m to T_{max} in the plot of temperature dependence of density, mercury telluride can be considered to be a two-structure melt.

A quantitative description of the temperature dependence of HgTe density could also be successfully made on the basis of the equilibrium associated solution theory. According to Chandra,[4] the anomaly in the temperature dependence of density cannot be satisfactorily explained by means of a model based upon the dissociation of mercury monotelluride. We suggest a chemical structure model including, in addition to the HgTe associate, the possibility of forming the more complex Hg_2Te_3 associate. This possibility is suggested by the presence of a minimum on the curve of the concentration dependence of excess entropy of mixing of Hg-Te liquid alloys (Figure 2).

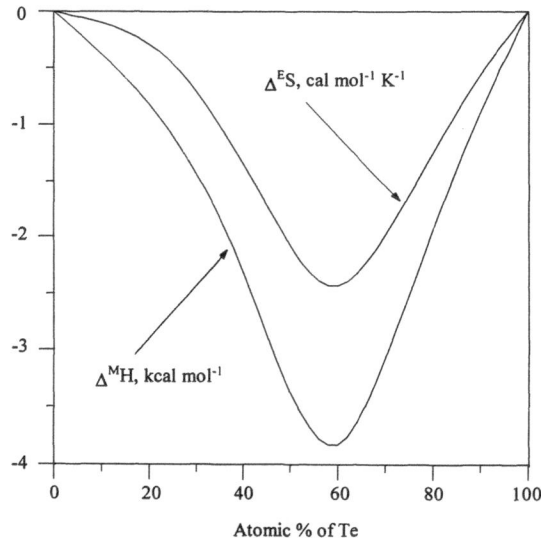

Figure 2. Enthalpy of mixing and excess entropy of mixing for molten Hg-Te alloys versus atomic percent tellurium calculated on the basis of a consistent approach by Marbeuf et al.[15]

It is seen that the minimum position on the curve of excess entropy of mixing corresponds to the composition of the Hg_2Te_3 associate. Consequently, in molten HgTe the following chemical processes relate to thermal dissociation of the suggested associates:

$$Hg_2Te_3 = 2HgTe + Te \tag{2}$$

$$HgTe = Hg + Te \tag{3}$$

The enthalpies and entropies of reactions (2) and (3) were found by fitting calculated and experimental partial pressures of Hg and Te_2 along the three-phase line HgTe(c). In this procedure, we utilized the experimental data of Brebrick and Strauss[16] obtained by an optical density measurement technique. Figure 3 shows that the resulting calculated temperature dependence of the density agrees well with experimental data obtained by the penetrating radiation technique.[12] The complex structure of molten HgTe was confirmed by the direct neutron diffraction results of Gaspard et al.[17] According to their data, the structural factor for molten HgTe melt has a characteristic shoulder, which indicates the availability of loosely-structured clusters with a primarily covalent type of interatomic bond.

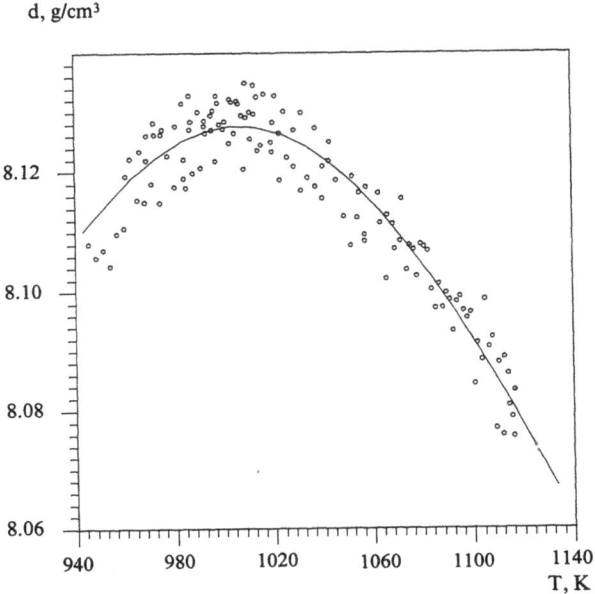

Figure 3. The temperature dependence of the density of molten HgTe. The solid line represents the results of our calculation, while the circles are our experimental data.

Cadmium Mercury Telluride Results

A new effect was detected when investigating the temperature dependence of the density of $Cd_xHg_{1-x}Te$ melts containing less than 0.19 mol fraction of CdTe, in addition to the HgTe anomaly described earlier. While investigating the temperature dependence of the densities of molten CdHgTe having 0.1 and 0.192 CdTe mol fractions, we discovered a density gradient with height, caused by a CdTe concentration gradient.

In the gamma-technique, direct measurements of a series of μ and ρ value were made, where μ is the mass coefficient of radiation attenuation and ρ is the density of the sample. Both parameters correspond to a point on the gamma-quanta beam's path. The value of μ was estimated as follows: $\mu = \mu_{CdTe}x + \mu_{HgTe}(1-x)$, where x is the mole fraction of CdTe, and μ_{CdTe} and μ_{HgTe} are the mass coefficients of radiation attenuation of the components determined separately. When a concentration gradient x(h) takes place in the melt, the values of μ and ρ must change along the sample's height, $\mu\rho = f(h)$, where h is the height of the gamma rays beam above the bottom of the sample's container. The concentration profiles for $Hg_{0.9}Cd_{0.1}Te$ and $Hg_{0.808}Cd_{0.192}Te$ melts at different time intervals are shown in Figure 4. Within measurement error limits, the dependence is linear, and the slope of the straight line slowly decreases with time. The intersection of the straight lines (isochrones) is around the point h_0, i.e. at the height corresponding to about half of the sample's height. In the vicinity of this point $\mu\rho = $ const, i.e. the composition in the middle of the sample does not change with time, and, evidently, equals the mean composition x_0 of the sample. Using

x_0 and the relations above help find the melt density corresponding to the initial composition. Using the values of μ for HgTe and CdTe and this value of density, one can establish a dependence, $\rho = \rho(h)$.

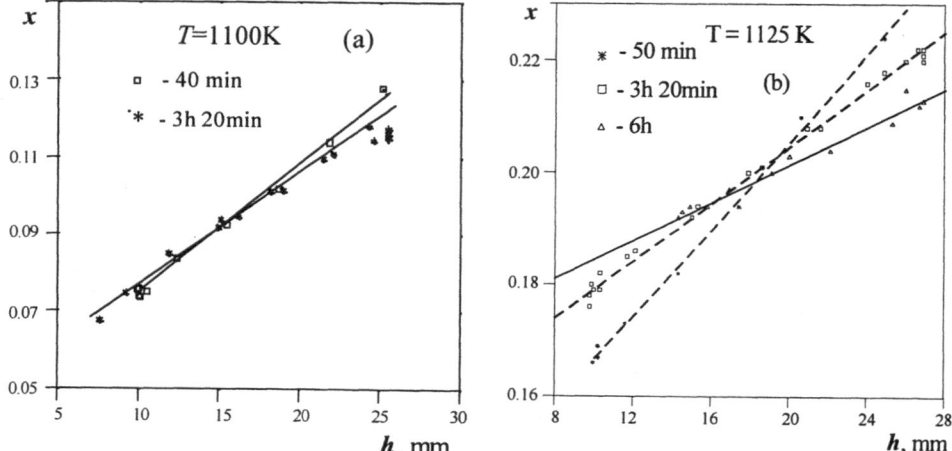

Figure 4. Concentration profiles at different time intervals since melting.
(a) $Hg_{0.808}Cd_{0.192}Te$ (b) $Hg_{0.9}Cd_{0.1}Te$

The leveling-up of the composition with time observed in these experiments is related to diffusion, because convection is not expected due to the isothermal conditions under which the measuring cell was being held. We can approximate the density gradient's change with time by the relation:

$$\frac{d\rho}{dh} = \left(\frac{d\rho}{dh}\right)_{t=0} \exp\left(-\frac{t}{t_{char}}\right) \tag{4}$$

The data for $Cd_xHg_{1-x}Te$ with x=0.192 give a value of t_{char} = 4 hours 40 min. Note that in the $Cd_xHg_{1-x}Te$ melt with x = 0.1, the concentration gradient and the density gradient practically did not change with time, at least within the time of 3-5 hours.

To determine the density it is not required that the sample composition be uniform with height. It is sufficient to make measurements at the height of the gamma beam equal to h_0, i.e. where the composition does not change with time. Figure 5 shows our results for the temperature dependence of $Cd_xHg_{1-x}Te$ melt density, for x = 0.1 and 0.192, and compares the with the prior data.[3,4] Our measurements and the prior results coincide within the limit of experimental error.

The occurrence of concentration and density gradients in molten $Cd_xHg_{1-x}Te$ may be due to the formation of complex polyanions, similar to Hg_2Te_3. Some prior work[17] indicates the possibility of such polyanion formation derived from HgTe-alloys by the effect of different organic substantion.

The formation of both a density gradient and a concentration gradient in CMT melts with x<0.2 would be expected to be absent under low gravity conditions. *Vice versa,* the formation of a prominent concentration gradient must occur very effectively and exhibit stability during the time period under increased gravity conditions (centrifugation).

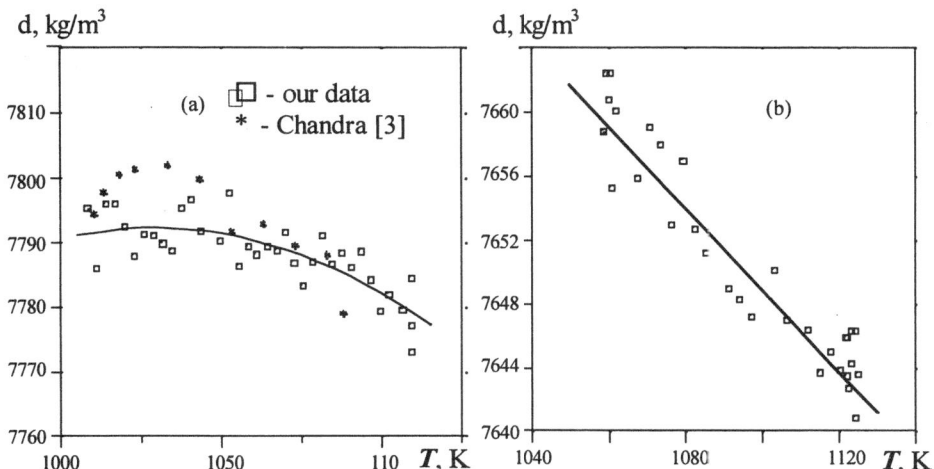

Figure 5. The temperature dependence of density.　(a) $Hg_{0.9}Cd_{0.1}Te$　(b) $Hg_{0.808}Cd_{0.192}Te$

REFERENCES

1.   V.N. Romanenko, Upravlenie sostavom poluprovodnikovyh sloev, *Metallurgiya* 192 (1978).
2.   I.B. Mizetskaya, G.S. Oleynik, L.D. Budennaya, V.N. Tomashik and I.D. Oleynik, Fiziko-khimicheskie osnovy sinteza monokristallov poluprovodnikovyh tverdyh rastvorov soedinenii A^2B^6, *Kiev: Naukova Dumka* 160(1986).
3.   D. Chandra and L.R. Holland, *J. Vac. Sci. Technol.* A11:1620(1983).
4.   D. Chandra, *Phys. Rev. B.* 31:7206(1985).
5.   S.V. Stankus and R.A. Hairulin, *Teplofizika Vysokih Temp.* 30:487(1992).
6.   V.M. Glazov, M. Vobst, and V.I. Timoshenko, Metody issledovaniya svoistv zhidkih metallov I poluprovodnikov, *Metallurgiya* 184(1989).
7.   N.P. Mokrovsky and A.R. Regel, *Zh. Tehnicheskoi fiziki* 22:1281(1952).
8.   A V. Novoselovoi and V.B. Lazareva, "Fiziko-Khimicheskie Svoistva Poluprovodnikovyh Veschestv. Spravotchnik Pod. Redaktsiei," Nauka, Moscow 340(1979).
9.   J.C. Tedenac, M.C. Record, R.M. Ayral-Marin and G. Brun, *Japan J. Appl. Phys.* 32:26(1993).
10. A.V. Omel'chenko and V.I. Soshnikov, *Izvestiya AN SSSR. Neorg. Mat.* 18:685(1982).
11. A. Nasar and M. Shamsuddin, *J. Less-Common Metals* 161:87(1990).
12. V.M. Glazov, L.M. Pavlova, S.V. Stankus and R.A. Khairulin, *Dokl. Akad. Nauk.* 347:202(1996).
13. A.R. Regel and V.M. Glazov, "Zakonomernosti Formirovaniya Struktury Elektronnyh Rasplavov," Nauka, Moscow 320(1982).
14. A.R. Regel and V.M. Glazov, *Fizika i Technika Poluprovodnikov* 17:1729(1983).
15. A. Marbeuf, M. Feran, E. Janik and A. Heurtel, *J. Crystal Growth* 72:126(1985).
16. R.F. Brebrick and A.J. Strauss, *J. Phys. Chem. Sol.* 26:989(1965).
17. J.P. Gaspard, J.Y. Raty, R. Ceolin and R. Bellissent, Local order in II-VI liquid compounds, LAM9 Conference, Chicago (August 1995).

ESTIMATION OF THE SELF AND MUTUAL
DIFFUSION COEFFICIENTS IN MOLTEN CdTe
BY A MOLECULAR DYNAMICS TECHNIQUE

Vasilii M. Glazov, Lidya M. Pavlova, and Kirill V. Rezontov

Department of Physical Chemistry
Moscow Institute of Electronic Engineering (Technical University)
Moscow 103498, Russia

INTRODUCTION

Diffusion processes play an important role during semiconductor material production by crystallization techniques. The experimental determination of diffusion coefficients becomes difficult due to convection, and therefore diffusion processes are being studied in microgravity. However, measurement in space is rather restricted due to the large cost. Thus, it is attractive to estimate diffusion coefficients by the computational techniques of statistical physics.

The present paper is devoted to the calculation of component diffusion coefficients in molten cadmium telluride by a molecular dynamics technique. There are no reliable data on component diffusion coefficients in liquid cadmium telluride. Taking into account the covalent bonding in molten cadmium telluride,[1] an empirical potential of the Stillinger-Weber type[2] was used to make the calculation. Wang and his colleagues were the first who calculated the surface tension of liquid cadmium telluride using this potential by a Monte-Carlo method, and showed the possibilities of its use.[3]

It is impossible to compare the results of diffusion coefficient calculation in liquid cadmium telluride with experimental data. Therefore, the reliability of our results was indirectly controlled by comparison of the calculated structural characteristics of cadmium telluride melts with data from neutron diffraction investigations.[4]

CALCULATION METHODS, RESULTS AND DISCUSSION

The partial velocity autocorrelation functions $Z_\alpha(t)$ were calculated by the formula:

$$Z_\alpha(t) = < \bar{v}_{i_\alpha}(t)\bar{v}_{i_\alpha}(t_0) > = \frac{1}{N_\alpha}\sum_i \bar{v}_{i_\alpha}(t)\bar{v}_{i_\alpha}(t_0) \qquad (1)$$

where N_α is the number particles of the α type in the system, $\bar{v}_{i_\alpha}(t)$ and $\bar{v}_{i_\alpha}(t_0)$ are the velocity vectors of the i_a - particles at the given and initial moments of time. Diffusion coefficients using expression (1) were calculated by the following formula:

$$D_\alpha = \frac{1}{3}\int_0^{k\Delta t} Z_\alpha(t)\,dt \qquad (2)$$

where Δt is a time step in the molecular dynamics calculation, and k is the number of steps necessary for $Z_\alpha(t)$ to have an oscillatory decay to zero.

Cd and Te diffusion coefficient calculations were done for stoichiometric, molten cadmium telluride for an *NVT* assemblage. The particle number of both types equaled 512 and was located in a cubic box subjected to the conventional periodic boundary conditions. The length of the cubic box edge *L* corresponded to the real density of liquid cadmium telluride at the given temperature. The averaging of the vector product in formula (1) was done on set out of 100 lengths of phase trajectories. The calculated temporal picture of the behavior of partial velocity autocorrelation functions of components in cadmium telluride at 1373K is shown in Figure 1.

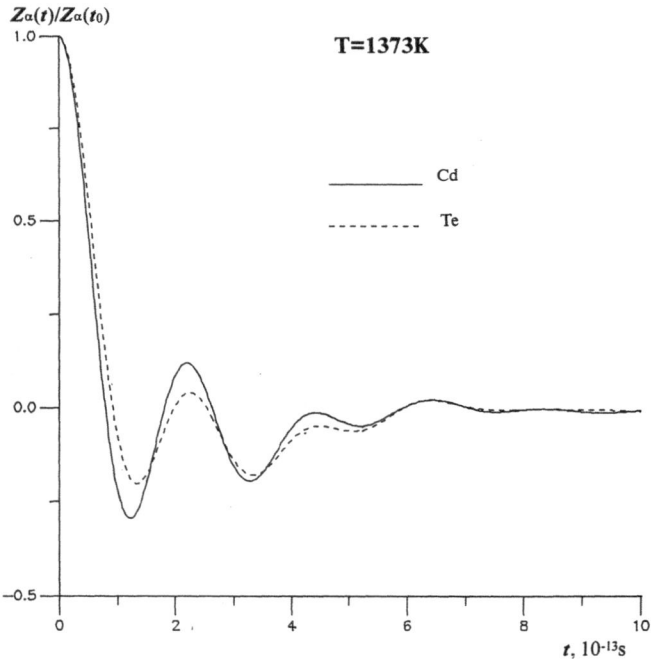

Figure 1. Partial velocity autocorrelation functions for liquid cadmium telluride, calculated by molecular dynamics.

It is necessary to pay attention to the similarity of the individual dynamics of Cd and Te atoms in the CdTe melt. The negative oscillations observed both in the given case and in all molecular dynamic calculations point to the exhibition of back scattering. This effect might result in lower values of the diffusion coefficient in the molecular dynamic calculation.

The values of self-diffusion coefficients of cadmium and tellurium obtained by Equation 2 were used to calculate the so-called interdiffusion coefficients from:

$$D_{12} = \sum_{\alpha} x_{\alpha} D_{\alpha} , \qquad (3)$$

where x_{α} and D_{α} are atomic fraction and diffusion coefficient of α-type atoms. The data obtained for diffusion coefficients in molten cadmium telluride at 1373 K are presented in the table below.

Table 1. Calculated Cd and Te diffusion and mutual diffusion coefficients in stoichiometric molten cadmium telluride at 1373 K.

$D_{Cd} \cdot 10^9$, ı^2/ñ	$D_{Te} \cdot 10^9$, ı^2/ñ	$D_{12} \cdot 10^9$, ı^2/ñ	Technique
0.87	1.14	1.01	MD-calculation
0.13	0.82	0.47	Thermodynamic evaluation

Values also are presented in this table that were evaluated approximately on the basis of data on the structural factors in the long-wavelength limit by the formula:

$$D_i = D_i^0 \left(1 + \frac{\Delta V^0}{V_j^0} x_i \right) \frac{S_{XX}(0)}{S_{XX}^{id}(0)} \qquad (4)$$

where D_i^0 is the self diffusion coefficient in the corresponding simple ith substance, V_j^0 is the molar volume of the jth component, $S_{XX}(0)$ and $S_{XX}^{id}(0)$ are the concentration - concentration correlation function in the long-wavelength limit in real and ideal solutions, respectively, and $\Delta V^0 = V_i^0 - V_j^0$.

The results from the thermodynamic evaluation agree with those from molecular dynamic calculations within one order of magnitude. This can be considered good agreement for such properties as diffusion coefficient.

For additional testing of the reliability our results, we simulated structural characteristics using the same potential of atomic interaction and compared the results with data obtained by neutron diffraction.[4] Three-pair correlation functions were calculated characterizing the pair interactions Cd-Cd , Te-Te and Cd-Te in a liquid cadmium telluride. The calculation was done by:

$$g_{ij}(r) = \frac{2V}{N} \left[\frac{\langle n(r) \rangle}{4 \pi r^2 D r} \right] \qquad (5)$$

where $\langle n(r) \rangle$ is the average number of j-type particles in a layer of Δr thickness at distance r from an i-type particle. The calculation results are presented in Figure 2, which shows the three radial distribution functions $g_{Cd-Cd}(r)$, $g_{Cd-Te}(r)$ and $g_{Te-Te}(r)$ for liquid in the vicinity of the melting temperature (T = 1373 K). Here $g_{Cd-Cd}(r)$, for example, represents the probability density for finding a Cd atom at r, given that there is a Cd at the origin. All three functions are normalized to unity at large separations. As can be seen, there is an excess of Cd-Te nearest-neighbor bonds in the liquid state.

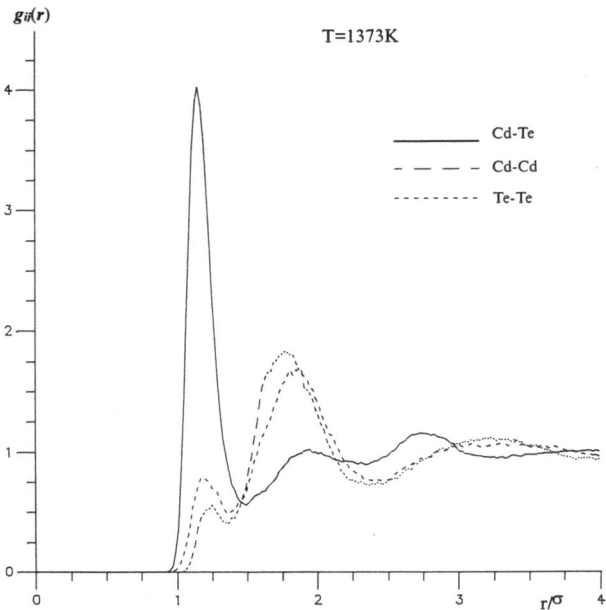

Figure 2. Partial radial correlation functions $g_{ij}(r)$ for stoichiometric liquid CdTe as calculated by molecular dynamics.

This behavior represents a residue of the perfect short-range order that exists in the ordered solid; our model predicts that there is still considerable short-range order in the liquid state.

The common pair correlation function for liquid cadmium telluride was calculated on the basis of the partial pair correlation functions shown in Figure 2. This was determined by a linear combination of partial pair correlation functions according to the following expression:

$$g(r) = \sum_{i,j} W_{ij} g_{ij}(r) \qquad (6)$$

$$W_{ij} = x_i x_j b_i b_j / \langle b \rangle^2 \qquad (7)$$

where $\langle b \rangle = \sum_i x_i \cdot b_i$ and (i, j) is (Cd, Te). Numerical values of neutron scattering amplitudes[5] are $b_{Cd} = (3.8 + i1.2) \cdot 10^{-15}$m and $b_{Te} = 5.43 \cdot 10^{-15}$m. There appear to be no experimental data available for partial pair correlation functions for molten cadmium telluride. The results of our calculations practically coincide with data obtained by a Monte-Carlo method.[3]

The common pair correlation function for molten cadmium telluride melt obtained by the pointed method using relationships (5) to (7) is given in Figure 3, where it is compared with experimental data[4].

A quantitative characteristic comparison of calculated and experimental pair correlation functions for liquid cadmium telluride is given in Table 2.

The positions of all maxima on both curves almost coincide. The position of the first maximum correlates with the interatomic distance in a crystalline lattice of cadmium telluride (2.809A).

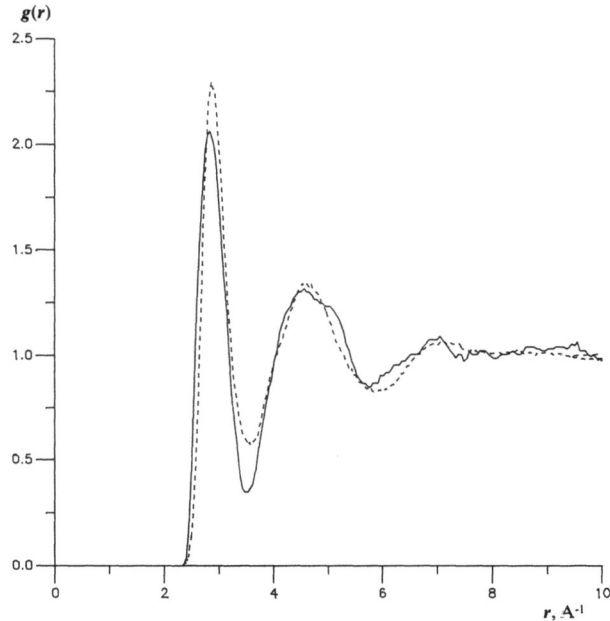

Figure 3. Total pair correlation function for molten CdTe at 1373 K. The results of our molecular dynamic calculations are given by the solid curve. The dashed curve is from experimental data.[6]

Table 2. $g(r)$-function maximum position compared with experimental data obtained by neutron diffraction.

maximum position, A			Data source
r_1	r_2	r_3	
2.86	4.63	7.15	our results
2.83	4.6	7.1	[4]

Next we present calculations for the coordination number in liquid cadmium telluride. Two techniques were used to evaluate it. The first technique consisted in a direct calculation of the number of nearest neighbours at a distance R_m from the central particle, followed by averaging the coordination number over all the system.

The second technique to calculate the coordination number used the equation:

$$N_{av} = \frac{4\pi N}{L^3} \int_0^{r_m} g(r) r^2 dr \qquad (8)$$

where r_m is the cutoff distance. As a result, we obtained coordination number values in liquid cadmium telluride of 3.31 and 4.35 using the first and second techniques. Experimental values[4] of the coordination number lie on the curve of the pair correlation function at the first minimum and equal 3.3 and 4, which agree well with the results of our calculations.

Besides the characteristics mentioned above, we also calculated partial and total structural factors using:

$$S_{ij}(k) = 1 + \frac{4\pi}{L^3} N \int_0^{6\sigma} \left[g_{ij}(r) - 1 \right] \frac{Sin\ kr}{kr} r^2 dr \qquad (9)$$

$$S(k) = \sum_{i,j} W_{ij} S_{ij}(k) \qquad (10)$$

Integration was done by Filon's method. The results of these calculations are given in Figures 4 and 5. Table 3 shows a comparison of the maxima on the calculated curve of the total structural factor with experimental neutron diffraction data.[4,6] The main characteristics of the calculated curve differ from the experimental ones. However, the tendency to the agreement improvement of the pointed characteristics obtained in our work with the experimental data should be noted, in particular that is referred to the relationship of parameters k_1 and k_2.

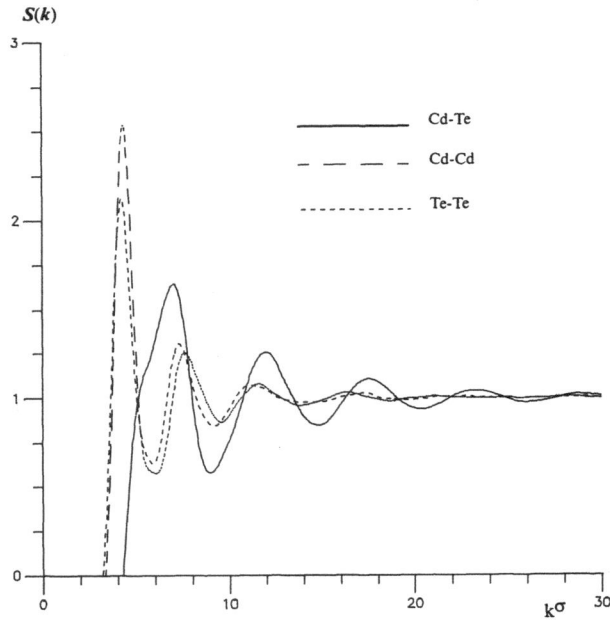

Figure 4. Partial structure factors for stoichiometric molten CdTe at 1373 K calculated by molecular dynamics.

Table 3. Calculated maxima of total structural factor compared with the results of neutron diffraction experiments.[4,6]

k_2, Å^{-1}	k_1, Å^{-1}	k_3, Å^{-1}	k_4, Å^{-1}	k_2/k_1	Data source
2.88	1.8	4.73	6.96	1.6	Calculation with suggested potential parameters.
2.84	1.97	4.72	6.89	1.44	Calculation with potential parameters suggested by Wang and Stroud.[3]
3.1	1.7	5.2	8.5	1.82	Experimental results of Gaspard et al.[4]
3.1	1.9	5.1	7.5	1.63	New results[6]

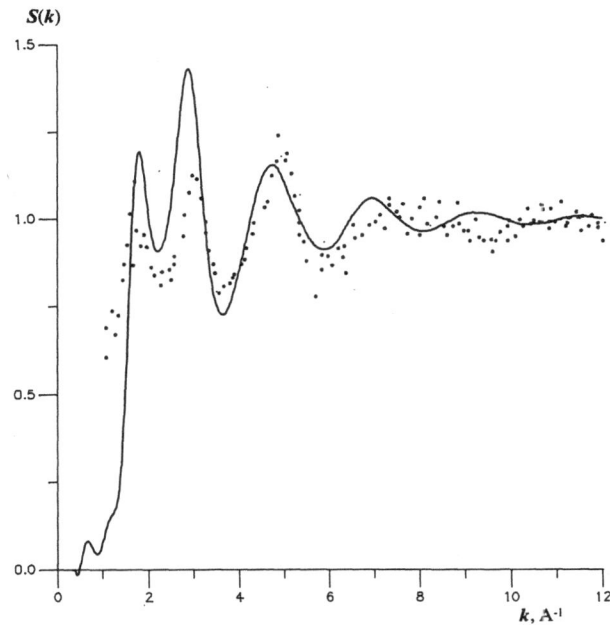

Figure 5. Total structure factor for molten CdTe. Full line is from molecular dynamic simulation Data points are from neutron scattering experiments.[6]

REFERENCES

1. V.M. Glazov, S.N. Chizhevskaya, and N.N. Glagoleva, "Liquid Semiconductors," Plenum, New York (1969).
2. F.H. Stillindger and T.A. Weber, *Phys. Rev. B* 31:5262(1985).
3. Z.Q. Wang, D. Stroud, and A.J. Markworth, *Phys. Rev. B* 40:3129(1989).
4. J.P. Gaspard, C. Bergman, C. Bichara, R. Bellissent, P. Chieux, and J. Goffard, J. Non-Cryst. Solids 97/98:1283(1987).
5. L. Koster, "Neutron Scattering Length and Fundamental Neutron Interaction," Springer Verlag, Berlin (1977).
6. J.P. Gaspard, J.Y. Raty, R. Ceolin, and R. Bellissent, Local order in II-VI liquid compounds, *in:* LAM9 Conference, Chicago (August 1995).

CHIMNEY FORMATION AND FLOW INSTABILITIES IN A MUSHY LAYER WITH DEFORMED BOUNDARIES

C.F. Baker and D.N. Riahi

Mechanical and Industrial Engineering, and
Department of Theoretical and Applied Mechanics
University of Illinois
Urbana, Illinois 61801

ABSTRACT

Three-dimensional linear flow instabilities and initial formation of chimneys in a mushy region during solidification of a binary alloy are investigated. Chimneys are vertical channels that can lead to freckles in the solidified material. Freckles are imperfections that interrupt the uniformity of the solidified material and cause areas of mechanical weakness. A stability analysis and numerical computations for deformed liquid-mush and mush-solid interfaces provide information on the neutral stability curve, velocity data and density data for the solid fraction perturbation in the mushy layer that leads to local regions of chimney formation. The results for the flow pattern in the form of hexagons indicate two distinct hexagonal modes of convection. The importance and roles of these two modes are discussed in relation to chimney formation in the mushy layer, as well as to future nonlinear instability and high gravity research with application to freckles elimination.

INTRODUCTION AND GOVERNING SYSTEM

Recently, Sayre and Riahi[1] investigated two-dimensional linear instabilities of the liquid and mushy layers with flat mush-solid interface during solidification of binary alloys under a high gravity environment. It was assumed that the solidification system was placed in a centrifuge basket[2] whose rotation axis was inclined with respect to the high gravity vector. The high gravity vector was also assumed to be anti-parallel to the direction of the solidification velocity. They developed a numerical code, which was used to determine various two-dimensional flow features due to the preferred primary instability modes of their stability system. However, there were several problems associated with that study.[1] First, it was assumed that the mush-solid interface is flat, although the liquid-mush interface was considered to be deformed. But, in general, both of these interfaces are non-planar.

Secondly, linear stability analysis was performed wherein only two-dimensional disturbances of infinitesimal amplitude were taken into account. Due to the two-dimensionality of the resulting flow, the Coriolis force was not included in the high gravity system and, hence, the results were somewhat restrictive. In addition, it is known[3] that nonlinear flow prefers three-dimensional hexagonal planform for sufficiently small amplitude of the flow due to the existence of asymmetries in the flow of the solidification problem.

One major goal of the present research work has been to determine the nonlinear evolution of the solidification system in order to understand more fully the formation of freckles in the mushy layer. The reason lies in the fact that freckles are intrinsically a finite amplitude phenomenon, since there is no mechanism for localization of the flow in the linear regime.

For the reasons stated above, we decided to undertake the present investigation of three-dimensional flow instabilities in a mushy layer bounded above and below by two deformed interfaces. Linear results under normal gravity are reported here, and it is hoped that both nonlinear evolution and high gravity extensions will be done soon. It should be noted that the present linear study is an essential and pre-requisite part of a weakly nonlinear study of the problem. In particular, the regions showing a tendency for chimney formation from linear analysis could also serve as initial locations of such chimneys in a nonlinear evolution study. As stated earlier, freckles originate from vertical chimneys or channels filled mostly with convective flow,[4] and are imperfections that reduce the quality of the solidified materials.

We consider a thin layer of a binary alloy melt of some constant composition C_0 and temperature T_∞ that is solidified at a constant rate V_0, with eutectic temperature T_e at the solid-mush interface at an average position $z = 0$ held fixed in a frame moving at the solidification speed in the vertical z-direction (anti-parallel to the gravity vector). Within the layer of melt, there is a very thin mushy layer adjacent to the solidifying surface of thickness $h(x,y,t)$, where t is time, and x and y are the horizontal coordinates.

The governing system of equations is based on the equations for momentum, continuity, heat and solute transport for both the liquid region $(z > h)$ and the mushy region $(l < z < h)$ in the moving frame oxyz, where $l(x,y,t)$ is the solid-mush interface function. These equations are non-dimensionalized using V_0, K/V_0, K/V_0^2, $\beta\Delta C\varphi_0 g K/V_0$, ΔC and ΔT as scales for velocity, length, time, pressure, solute and temperature, respectively. Here K is the thermal diffusivity, φ_0 is a reference (constant) density, β is the expansion coefficient for solute, $\Delta C \equiv C_0 - C_e$, C_e is the eutectic concentration of the alloy, $\Delta T \equiv T_L - T_e$, and T_L is the local liquidus temperature. Following Worster[5] and Sayre and Riahi,[1] we treat the mushy layer as a porous medium where Darcy's law applies. We assume that the temperature contribution to the buoyancy term in the momentum equation is negligible and $K \gg d$, where d is the solute diffusivity.

Now, we consider a linear stability system for three-dimensional disturbances superimposed on the base flow. The base flow solutions are functions of z in a motionless state and are given in Worster.[5] Their expressions will not be repeated here. Considering the vertical component of the double curl of the momentum equations for both liquid and mushy zones, and making using of the divergence-free velocity vector expressed by the continuity equation, we end up with equations for vertical component w of disturbance velocity vector, disturbance solute concentration S, disturbance temperature θ, and disturbance local solid fraction ϕ of the mushy layer. We then seek solution of the form:

$$(w, S, \theta, \phi) = \left[w_1(z), S_1(z), \theta_1(z), \phi_1(z)\right]f(x, y), \tag{1a}$$

164

where the planform function $f(x, y)$ is that assumed for hexagonal cells[3] given by:

$$f(x, y) = 2\left[\cos(\alpha x) + \cos\left(-\alpha x/2 + \sqrt{3}\alpha\, y/2\right) + \cos\left(\alpha x/2 + \sqrt{3}\alpha\, y/2\right)\right]\Big/\sqrt{6}. \tag{1b}$$

where α is the horizontal wave number. Using the above equations and the appropriate boundary conditions[1,5], we find the governing system. In the liquid region, the equations and the boundary conditions are:

$$\left[\left(D^2 - \alpha^2\right)^2 + \left(D^2 - \alpha^2\right)D/P_r\right]w_1 + R_1\alpha^2 S_1 = 0,\ D \equiv \frac{d}{dz}, \tag{2a}$$

$$\left(D^2 + D - \alpha^2\right)\theta_1 = w_1 D\theta_0, \tag{2b}$$

$$\left(\varepsilon D^2 + D - \varepsilon\alpha^2\right)S_1 = w_1 DS_0, \tag{2c}$$

$$\theta_1 - S_1 = [w_1] = Dw_1 = D(S_1 - \theta_1) + (\varepsilon - 1)h_1 D\theta_0 = 0 \text{ at } z = h_0, \tag{2d}$$

$$\theta_1 = S_1 = w_1 = Dw_1 = 0 \text{ at } z = \infty, \tag{2e}$$

where θ_0 and S_0 are the basic temperature and basic solute concentration, respectively, $\Pr = v/K$ is the Prandtl number, v is the kinematic viscosity, $R_1 = \beta\Delta C g K^2/(V_0^3 v)$ is the liquid solutal Rayleigh number, $\varepsilon = d/K$ is the inverse of the Lewis number, h_0 is the unperturbed mush-liquid interface (constant), $h_1 = (h - h_0)/f$, and the square brackets denote the jump in the enclosed quantity across the interface. In the mushy region, the equations and the boundary conditions are:

$$\left(D^2 - \alpha^2\right)w_1 = R_m\alpha^2\theta_1, \tag{3a}$$

$$\left\{D^2 + \left[1 + S_t(C_r - \theta_i)/(C_r - \theta_0)^2\right]D + 2S_t(C_r - \theta_i)D\theta_0/(C_r - \theta_0)^3 - \alpha^2\right\}$$

$$\theta_1 = S_t\xi_1 D\theta_0/(C_r - \theta_0)^2 + w_1 D\theta_0 + S_t D\xi_1/(C_r - \theta_0), \tag{3b}$$

$$D\xi_1 = w_1 D\theta_0, \tag{3c}$$

$$[\theta_1] = [D\theta_1] + S_t h_1 D\theta_0/(C_r - \theta_i) = \xi_1 - \theta_1 - h_1 D\theta_0 = R_1 Dw_1\big|_{\text{mush}} +$$

$$R_m\left(D^3 + D^2/P_r\right)w_1\big|_{\text{liquid}} = 0 \text{ at } z = h_0, \tag{3d}$$

$$w_1 = Dw_1 = [\theta_1] = [D\theta_1] + S_t l_1 D\theta_0/(C_r + 1) = 0 \text{ at } z = 0, \tag{3e}$$

where $\theta_i = \varepsilon\theta_\infty/(\varepsilon - 1)$, θ_∞ is the non-dimensional T_∞, $S_1 = \theta_1$,[1,5] $\xi_1 = (1 - \phi_1)S_1 + \phi_1 C_r$, $S_t = L/(C\Delta T)$ is the Stefan number, C is the specific heat per unit volume, L is the latent heat of solidification per unit volume, $C_r = (S_s - C_0)/\Delta C$ is a concentration ratio, C_s is the composition of the solid phase forming the dendrites, $R_m = \beta\Delta C g \Pi_0/(V_0 v)$ is the mush solutal Rayleigh number ($R_1/R_m = H$ is a large parameter[1,5]), Π_0 is a reference value of the permeability and $l_1 = l/f$. The boundary conditions (3e) correspond to conditions at a deformed solid-mush interface and, thus, are different from those for the flat interface case applied by Worster[5] and by Sayre and Riahi.[1] Equations (2a) and (3a), together with the last boundary condition given in (3d), are different from those used in previous two-dimensional studies,[1,5] and they are, together with the rest of the equations and the boundary conditions given by (2)-(3), applicable to general three-dimensional flow cases.

RESULTS

Due to the complexity of the system (2)-(3), we used a modified version of the numerical code developed by Sayre and Riahi[1] that applies shooting and iterative techniques to find the solution. It should be noted that the third and fourth boundary conditions given in (3e) required solution for the temperature variations in the solid. This solution was found by solving the steady state heat diffusion equation in the solid. The bounded solution in the $(-z)$ direction for the temperature θ_1 in the solid was then used in these boundary conditions, leading to satisfactory boundary conditions at $z = 0$ involving only dependent variables in the mushy layer. For details regarding the numerical approach the reader is referred to Worster[5] and to Sayre and Riahi.[1]

Following Worster[5], we set $S_t = C_r = \theta_\infty = 1$ and $P_r = 10$. Such a value of the Prandtl number is representative for aqueous solutions We plan to consider cases for small Prandtl number, say $P_r = 0.02$, which are representative for metallic alloys, in the near future. The numerical procedure leads to an eigenvalue relation which provides a marginal stability curve $R_1(\alpha)$ for each choice of the parameters ε and H. The parameter ε is the inverse of the Lewis number and is typically very small. The parameter $H \equiv R_1/R_m$ is a representative of the square of the ratio of the thermal length scale, on which the depth of the mushy layer depends, to the average spacing between dendrites within the mushy layer.[5] This parameter is typically very large. The values chosen for $\varepsilon = 0.025$ and $H = 10^5$ are the same as those chosen in earlier studies.[1,5]

The results for the neutral stability curve, R_1 versus α, are shown in Figure 1.

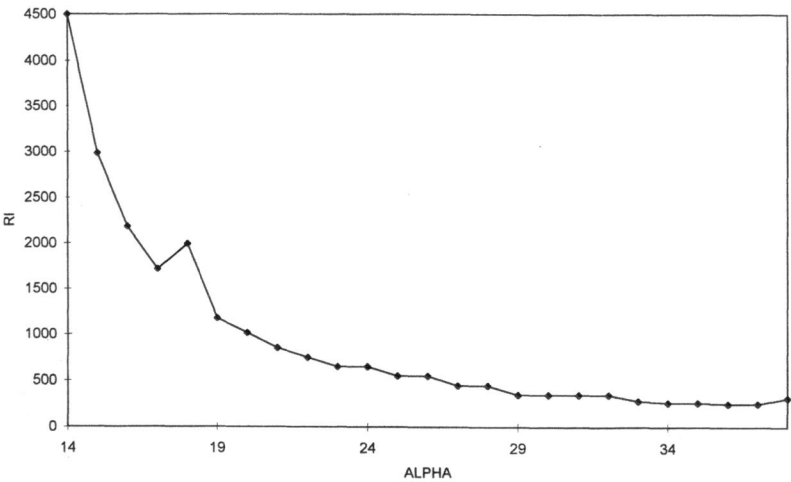

Figure 1. Marginal stability curve for $\varepsilon = 0.025$ and $H = 10^5$.

The system is unstable in the region above the curve and is stable below the curve. Just as in the case of non-deformed solid-mush interface,[1,5] the marginal curve has two minima, corresponding to two distinct hexagonal modes of convection. The first mode, corresponding to the first minimum with smaller α but larger R_l, is referred to here as the first mode. Its wavelength is comparable to the depth of the mushy layer and causes flow throughout the mushy layer. The second mode, corresponding to the second minimum with larger α but smaller R_l, is more unstable than the first mode. Its horizontal wavelength is smaller than that of the first mode and it causes stronger up and down flow in the mushy layer, as indicated by the velocity data obtained from the computation. The critical values of R_l at the two minima of the neutral stability curve are smaller than the corresponding ones for a non-deformed solid-mush interface.[1,5] This is due to the fact that for a non-deformed solid-mush interface, a more restricted isothermal condition for the temperature was imposed, which led to higher values of the critical R_l. The second mode is more unstable, carries stronger flow and is more associated with solid fraction perturbations than the first mode. The second mode is more associated with a decrease in the solid fraction in the interior of the mushy layer in regions of up flow, which is indicative of a tendency to form chimneys. Convective up flow results in a local decrease of the solid fraction and an elevation of the mush-liquid interface. Our numerical results support these conclusions.

We determined numerically the vertical and horizontal components of velocity vector in both liquid and mushy regions and for both of the convective modes. The magnitude of the flow velocity for the second mode is more significant in the mushy layer than for the first mode, while the opposite is true in the liquid layer. The magnitude of the velocity in the liquid region is generally smaller than the corresponding one in the mushy region, while the opposite was generally found to be true for a non-deformed solid-mush interface with isothermal condition.[1]

Density data for the perturbation to the solid fraction in the mushy layer for both convective modes were determined. It was concluded from these results that the first mode causes less perturbation to the solid fraction than the second mode. The second mode is, thus, mostly associated with solid fraction perturbations. For this mode, there is a substantial decrease in the solid fraction in the interior of the mushy layer in regions of up flow, which indicates a tendency to form chimneys and subsequently freckles in the solidified material. The expectation that $\phi_1(z)$ changes with rotation rate in a high gravity environment, can indicate that the spatial locations in the mushy region with a tendency to form chimneys, that is regions corresponding to negative perturbations to the solid fraction that represent local melting of the dendrites, may change as rotation rate increases or decreases. This result suggests that an important operational procedure for possible elimination of chimneys would be to rotate the solidifying system. It should also be noted that hexagonal modes of convection provide fewer regions with a tendency to form chimneys than two-dimensional modes of convection rolls. The spatial locations that tend to form chimneys for hexagonal modes of convection are generally different from those for two-dimensional modes of convection.

DISCUSSION

In the present study, we investigated the three-dimensional linear primary instability of motionless z-dependent basic state of a solid-melt system with deformed interfaces and a horizontal flow structure in the form of hexagons. We determined two distinct convective hexagonal modes and discussed their description and roles in relation to chimney formation in the mushy layer and subsequent freckles in the solidified material.

Hexagonal convection is expected to be preferred in the weakly nonlinear development of the primary instability modes for sufficiently small amplitude of convection leading to subcritical instability.[3] Nonlinear evolution of the convective modes is expected to provide deeper information on the chimney convection and on the formation of such channels.

A high gravity environment can be beneficial. The second author plans to study such an environment in the near future using a variable rotation rate to see if chimney formation is halted or at least reduced.

The present study was a necessary first step to investigate three-dimensional instabilities in the liquid and mushy layers for several different values of the Prandtl number in a high gravity environment, in order to determine the effects of both Coriolis and centrifugal (gradient acceleration[1]) terms in the momentum equations on various flow features, particularly on the formation of chimneys in the mushy layer.

Acknowledgment

This research was supported by a grant from the UIUC Research Board (grant number 1-2-68608) and by the UIUC College of Engineering.

REFERENCES

1.   T.L. Sayre and D.N. Riahi, Effect of rotation on flow instabilities during solidification of a binary alloy, *Int. J. Engin. Sci.* (in press).
2.   W.A. Arnold, W.R. Wilcox, F. Carlson, A. Chait, and L.L. Regel, Transport modes during crystal growth in a centrifuge, *J. Crystal Growth* 119:24 (1992).
3.   D.N. Riahi, Solutal convection in the melt during solidification of a binary alloy, *Phys. Fluids* 31:27 (1988).
4.   D.G. Neilson and F.P. Incropera, Unidirectional solidification of a binary alloy and the effects of induced fluid motion, *Int. J. Heat Mass Transfer* 34:1717 (1991).
5.   M.G. Worster, Instabilities of the liquid and mushy regions during solidification of alloys, *J. Fluid Mechanics* 237: 649 (1992).

EFFECT OF HIGH GRAVITY ON FRECKLE FORMATION AND CONVECTION IN A MUSHY LAYER

D.N. Riahi

Department of Theoretical and Applied Mechanics
University of Illinois
Urbana, Illinois 61801

ABSTRACT

Nonlinear, time-dependent natural convection in a mushy layer during centrifugal solidification of binary alloys is investigated for low, moderate and high rates of rotation. Asymptotic and scaling analyses are applied to axisymmetric convection within the mushy layer and in vertical chimneys with small radius. These chimneys produce freckles in the final solidified material. Freckles are imperfections that reduce the quality of solidified materials. Results are obtained for several distinct regimes, corresponding to high or low Prandtl number melts and strongly or weakly time-dependent flow. In particular, conditions are determined under which convection and freckle formation can be eliminated, or at least reduced, by the rotational component of a high gravity environment.

INTRODUCTION AND FORMULATION

Recently, Riahi and Sayre[1] investigated nonlinear steady natural convection at infinite Prandtl number in a mushy layer of a solidification system under a high gravity environment with the rotation axis inclined to the acceleration vector. They applied asymptotic and scaling analyses to convective flow within the mushy layer and in vertical chimneys. They found that, for some particular moderate rotation range, the vertical velocity in the chimneys is significantly reduced. These chimneys are characterized by having zero solid fraction, and they are assumed to be in the vertical direction. These channels subsequently become locations of severe compositional inhomogeneity, which in the final solidified form are called freckles. Freckles are imperfections that interrupt the uniformity of the solidified material cause mechanical weakness.

There is some experimental evidence[2,3] that inclined rotation at an appropriate rate can reduce the number of channels substantially and under some conditions eliminate them completely. Other experimental results[4] indicate that rapid rotation is harmful. The main purpose of the present study was to investigate theoretically the effects of inclined rotation on chimneys within the mushy layer using analyses of the type of Riahi and Sayre,[1] only with realistic values of the Prandtl number and taking into account the time-dependent behavior of convection that can be important for the range of parameters values considered. In regard to the Prandtl number and time-dependent effects not considered in Riahi and Sayre,[1] Riahi[5] developed recently a time-dependent model with inclusion of Prandtl number effects under normal gravity conditions and zero rotation. In the present study, we extend the models due to Riahi and Sayre[1] and Riahi[5] to cases where the effects of time dependence and Prandtl number are taken into account in a high gravity environment under low, moderate or rapid rotation.

We consider a thin layer of a binary alloy melt of some constant composition C_0 and temperature T_∞ being solidified at a constant rate V_0, with eutectic temperature T_e at position $z = 0$ held fixed in a frame moving with the solidification speed in the z-direction. The z-axis is assumed to be anti-parallel to the acceleration vector, as described below. Within the layer of melt, there is a very thin mushy layer adjacent to the solidifying surface and of thickness h. We assume that the solidifying system is placed in a centrifuge rotating at some constant angular velocity Ω about the centrifuge axis which makes an angle γ with respect to the z-axis. The centrifuge axis is assumed to be anti-parallel to earth's gravity vector.

Next, we consider the equations for momentum, continuity, heat and solute for both liquid and mushy layers in the moving frame oxyz whose origin 0 is centered on the solid-mush interface. The governing system of equations for a solidifying system rotating with the centrifuge basket[6] and translating with the solidification front at speed V_0 is non-dimensionalized using V_0, K/V_0, K/V_0^2, $\beta\Delta C\varphi_0 g K/V_0$, ΔC and ΔT as scales for velocity, length, time, pressure, solute and temperature, respectively. Here K is the thermal diffusivity, φ_0 is a reference (constant) density, β is the expansion coefficient for solute, $\Delta C = C_0 - C_e$, C_e is the eutectic concentration of the alloy, $\Delta T = T_L - T_e$,, and T_L is the local liquidus temperature.

We assume axisymmetric flow and do not include the effect of Coriolis force.[1] The Coriolis effect would be included naturally in the fully non-axisymmetric flow case planned for investigation soon. The centrifugal acceleration term in the momentum equation, for either liquid or mushy layer, is split into an average term, which is superimposed on the gravity term, and a so-called gradient acceleration term.[6] At a high rotation rate, the non-dimensional parameters representing the modified gravity term can be significantly larger than the corresponding one due to earth's gravity alone. We shall treat the mushy layer as a porous medium[1,7-8] in which Darcy's law is valid. We use a cylindrical coordinate system to analyze the convective flow in vertical and cylindrical chimneys[1] within the mushy layer. We consider a strong buoyancy force due to solute concentration, in the limit of sufficiently large Lewis number K/D and zero thermal buoyancy.[8] Here D is the solute diffusivity.

The non-dimensional form of the equations and the boundary conditions in the rotating frame for both liquid and mushy layers are given below, based on the assumptions and approximations described above. In the liquid layers the equations for momentum, continuity, temperature and solute concentration are:

$$\frac{1}{P_r}\left(\frac{\partial}{\partial t} - \frac{\partial}{\partial z} + \underset{\sim}{u}\cdot\nabla\right)\underset{\sim}{u} = -R_1\left(\nabla P + S\hat{K}\right) + \nabla^2\underset{\sim}{u} + A_1 S\left(r\hat{r} + z\sin^2\gamma\,\hat{K}\right), \tag{1a}$$

$$\nabla\cdot\underset{\sim}{u} = 0, \tag{1b}$$

170

$$\left(\frac{\partial}{\partial t} - \frac{\partial}{\partial z} + \underset{\sim}{u}\cdot\nabla\right)\theta = \nabla^2\theta,$$

(1c)

$$\left(\frac{\partial}{\partial t} - \frac{\partial}{\partial z} + \underset{\sim}{u}\cdot\nabla\right)S = 0,$$

(1d)

where $\underset{\sim}{u}$ is the velocity vector, P is pressure, S is the solute concentration, θ is the temperature, $\hat{K}$ is a unit vector along the (z) axis, $\hat{r}$ is a unit vector in the radial (r) direction, $P_r = \mu/K$ is the Prandtl number, μ is the kinematic viscosity, $R_l = \beta\Delta C N_g K^2/\left(V_0^3\mu\right)$ is the liquid solutal Rayleigh number, $N_g = \left(g^2 + \Omega^4 R_0^2\right)^{1/2}$ is the total acceleration, Ω is the rotation rate, R_0 is the perpendicular distance from the center of gravity of the centrifuge basket to the rotation axis, g is the acceleration due to gravity, t is time and $A_l = \beta\Delta C\Omega^2 K^3/\left(V_0^4\mu\right)$ is the liquid gradient acceleration parameter. The boundary conditions for the liquid layer are:

$$\theta - S = \frac{\partial}{\partial z}(\theta - S) = \left[\hat{n}\cdot\underset{\sim}{u}\right] = \underset{\sim}{u} - \hat{n}\cdot\underset{\sim}{u} = 0 \text{ at } z = h,$$

(2a)

$$\theta \to \theta_\infty, \; S \to 0, \underset{\sim}{u} \to 0 \text{ as } z \to \infty,$$

(2b)

where θ_∞ is the non-dimensional form of T_∞, the square brackets denote the jump in the enclosed quantity across the interface, and $\hat{n}$ is a unit vector normal to the interface, which is non-planar, in general. The non-dimensional form of the equations for momentum, continuity, temperature and solute concentration in the mushy layer are:

$$\underset{\sim}{u}/\Pi = -R_m\left(\nabla P + S\hat{K}\right) + A_m S\left(r\hat{r} + z\sin^2\gamma\,\hat{K}\right),$$

(3a)

$$\nabla\cdot\underset{\sim}{u} = 0,$$

(3b)

$$\left(\frac{\partial}{\partial t} - \frac{\partial}{\partial z} + \underset{\sim}{u}\cdot\nabla\right)\theta = \nabla^2\theta + S_t\left(\frac{\partial}{\partial t} - \frac{\partial}{\partial z}\right)\phi,$$

(3c)

$$\left(\frac{\partial}{\partial z} - \frac{\partial}{\partial t}\right)[(1-\phi)(C_r - S)] + \underset{\sim}{u}\cdot\nabla S = 0,$$

(3d)

where ϕ is the local solid fraction, $S_t = L/(C\Delta T)$ is the Stefan number, C is the specific heat per unit volume, L is the latent heat of solidification per unit volume, $C_r = (C_S - C_0)/\Delta C$ is a concentration ratio, C_S is the composition of the solid phase forming the dendrites, Π is the permeability of the mushy layer (generally a function of ϕ), $R_m = \beta\Delta C N_g \Pi_0/(V_0\mu)$ is the mush solutal Rayleigh number, Π_0 is a reference value of the permeability, and $A_m = \beta\Delta C\Pi_0 K\Omega^2/(V_0^2\mu)$ in the mush gradient acceleration parameter. It turns out that $R_l/R_m = A_l/A_m = H$, where the constant H is assumed to be large.[1,5,8] The boundary conditions for the mushy layer are:

$$\theta + 1 = \underset{\sim}{u}\cdot\hat{K} = 0 \text{ at } z = 0,$$

(4a)

$$[\theta] = [\hat{n} \cdot \nabla \theta] = [P] = 0 \text{ at } z = h. \tag{4b}$$

Earlier analyses[1,5,8] indicated that $\theta = S$ in the mushy layer; this condition is also valid here.

We shall apply scaling analysis for (1)-(4) to determine some qualitative results for the strongly nonlinear axisymmetric convection in the mushy layer, and mainly the convection in the chimneys once these channels are fully developed within the mushy layer, in the asymptotic limit of sufficiently large R_m.[1,5,8]

HIGH P_r AND WEAK TIME DEPENDENCE CASE

Let us designate $a(r,t)$ as the radius of a chimney under consideration, whose axis coincides with the z-axis. It is assumed that a is small $(a \ll 1)$. We define a stream function $\psi(r,z,t)$ for flow in the chimney:

$$\underset{\sim}{u} = (u,w) = \left(-\frac{1}{r}\frac{\partial \psi}{\partial z}, \frac{1}{r}\frac{\partial \psi}{\partial r} \right), \tag{5}$$

where $\underset{\sim}{u} \equiv u\hat{r} + w\hat{K}$. We assume that the orders of magnitude of r and z are a and 1, respectively. Assuming $|\underset{\sim}{u}| \sim 1$ in the mushy layer, then (3a) implies that to the leading terms the pressure field is hydrostatic and $\theta = S$ is independent of r. Equations (3c)-(3d) then imply that $w = w_0(z,t), \phi = \phi_0(z,t)$ and $\theta = \theta_0(z,t)$ at most, where w_0, ϕ_0 and θ_0 are the leading order terms for w, ϕ and θ, respectively. It is also assumed that $C_r \gg \theta$ in the mushy layer.[1,5,8]

Next, consider the flow in the chimney, as described by (1a)-(1d) since $\phi = 0$ is assumed in the chimney.[1,5,8] The analysis presented in this section is for the case where the nonlinear and time derivative inertia terms in (1a) can be, at most, as significant as the viscous term in (1a). Using (5) and assuming $S \sim 1$ and $w \gg 1$,[8] equation (1a) implies that:

$$w \sim R_1 a^2, u \sim R_1 a^3, \psi \sim R_1 a^4. \tag{6}$$

Now, we note that θ_0 is the leading order temperature solution in the mushy zone outside the chimney and its surrounding boundary layer.[8] We designate $\theta_1(r,z,t)$ as the deviation of θ from θ_0. From (1) and the condition:

$$1/a^2 \ll R_1 \ll 1/a^4, \tag{7}$$

we find $\theta_1 \ll 1$. Using these results, (1a)-(1d) in the chimney are reduced and, in particular, (1c) becomes:

$$\frac{1}{r}\frac{\partial}{\partial r}\left(r\frac{\partial \theta_1}{\partial r} \right) = \frac{1}{r}\frac{\partial \psi}{\partial r}\frac{\partial \theta_0}{\partial z} + \frac{\partial \theta_0}{\partial t}. \tag{8}$$

Integrating (8) in r from $r = 0$ to $r = a$ and following Worster,[8] we find:

$$\theta_1 \sim \psi_a \frac{\partial \theta_0}{\partial z} \ln r + \frac{\partial \theta_0}{\partial t} r^2/4, \tag{9a}$$

where
$$2\pi\psi_a = \int_0^a 2\pi\, r\, w\, d\, r \tag{9b}$$

is the vertical volume flux in the chimney. Using (5), we have:

$$u \sim (1/r)\frac{\partial \psi_a}{\partial z} \quad \text{as } r \to a. \tag{10}$$

From (3a), we have:

$$u/\Pi \sim -R_m \frac{\partial P}{\partial r} + aA_m S. \tag{11}$$

Using (10) in (11), we find:

$$\Delta P \sim \left(\frac{\partial \psi_a}{\partial z}\right) \ln a/R_m + A_m a^2 \theta_0/R_m, \tag{12}$$

where it is assumed that $\Pi = 0(1)$ and ΔP represents the pressure near the chimney. Using (9a) and (3a), we find that:

$$w \sim R_m \psi_a 1_{na}(1 - A_m z \sin^2 \gamma/R_m), \tag{13}$$

holds near the wall of the chimney.

Applying the results for the order of magnitudes of the flow variables,[1,5] and using a Polhausen type method as suggested by Lighthill[9] and following Worster,[8] we find that the total volume flux in the chimney, due to upward flow,[8] is given by:

$$2\pi\psi_a = 2\pi\lambda a^4 R_l [1+\theta_0(z,t)](1 - A_m z \sin^2 \gamma/R_m), \tag{14}$$

where λ depends generally on z, a, h and P_r. To satisfy mass conservation, the downward flow through the mushy zone $w_0(z,t)$ must be equal to the total upflow through all of the chimneys per unit horizontal area. Thus:

$$w_0 = -2\pi\psi_a N, \tag{15}$$

where N is the number density of chimneys.

It can be seen from (12)-(14) that rotational effects can be non-zero only for $\gamma \neq 0$, and the case of weak rotation can correspond to the condition:

$$A_m \ll R_m. \tag{16}$$

This condition can be satisfied for Ω of order, say, 0.1 rad/sec or less. In such cases, N_g is hardly different from 1_g for realistic centrifuges,[6] and the results are essentially those for normal gravity conditions in the absence of inclined rotation. For moderate rotation:

$$A_m \sim R_m, \tag{17}$$

and the results (11)-(14) indicate the potential influence of the inclined rotational constraint upon convection. The results (13) and (14) imply that the vertical flow velocity in the chimney decreases with increasing rotation rate. For some particular limit, where $A_m z \sin^2 \gamma \to R_m$ for some critical value $z = z_c$, $w \to 0$ based on (13) and the next leading order terms determine the order of magnitude of w, which turns out to be much less than that for weak rotation. However, for $z > z_c$, (14) indicates that ψ_a has the opposite sign to that for $z < z_c$. Furthermore, in such a limiting situation, a smaller z_c corresponds to a larger $A_m \sin^2 \gamma / R_m$ and, *vice versa*, a larger z_c corresponds to a smaller $A_m \sin^2 \gamma / R_m$. For given R_m and γ, these results indicate the stabilization of convection in the chimney by the rotational constraint. These results, together with those due to (11)-(12), indicate also some possible double-cell formation and circulation within the chimney due to the effect of a moderate rotation rate. The condition (17) can be satisfied for Ω, say, of order of 6 rad/sec, which implies $N_g > 1_g^1$ so that the results for (17) can be applicable in a high gravity environment. For rapid rotation:

$$A_m \gg R_m, \tag{18}$$

and the results (11)-(14) indicate that the order of magnitudes of pressure and convection velocity in the chimney increase with rotation rate. Hence, rotational effects are destabilizing for the case (18), and can lead quickly to channel formation in the mushy zone.

LOW P_r AND WEAK TIME DEPENDENCE CASE

In this section we assume that the order of magnitude of the nonlinear inertial term in (1a) is larger than that of the viscous term in (1a), and the order of magnitude of the time derivative inertial term in (1a) is at most as large as the nonlinear inertial term in (1a). The order of magnitudes of the velocity components and the stream function in the chimney are now:

$$w \sim (P_r R_1)^{\frac{1}{2}}, u \sim a(P_r R_1)^{\frac{1}{2}}, \psi \sim a^2 (P_r R_1)^{\frac{1}{2}}. \tag{19}$$

From (1) and the condition:

$$1 \ll (P_r R_1)^{\frac{1}{2}} \ll 1/a^2, \tag{20}$$

we find $\theta_1 \ll 1$. Using these results, (1a)-(1d) in the chimney are reduced, but (8)-(13) are found to be still valid.

The total volume flux in the chimney is determined by using a procedure similar to that described in the previous section. The results (14)-(15) then follow, provided $R_1 a^4$ on the right-hand-side of (14) is replaced by $a^2 \sqrt{R_1 P_r}$ and where now λ is found to be independent of P_r. The results regarding (16)-(18) and rotational effects are then found to be qualitatively the same as those described in the previous section, and so will not be repeated here.

HIGH P_r AND STRONG TIME DEPENDENCE CASE

In this section we consider the case where the order of magnitude of the time derivative inertia term in (1a) is larger than that of the nonlinear inertia terms in (1a), but the order of magnitude of the time derivative inertia term in (1a) is at most as large as the viscous term in (1a). The results (6)-(7) are valid here are replace (19)-(20) of the previous case. The results (8) and (9a) are replaced by:

$$\frac{1}{r}\frac{\partial}{\partial r}\left(r\frac{\partial \theta_1}{\partial r}\right) = \frac{\partial \theta_0}{\partial t}, \tag{21}$$

$$\theta_1 \sim a^2 \frac{\partial \theta_0}{\partial t}. \tag{22}$$

The results (10)-(12) are valid again, but (13) is now replaced by:

$$w \sim R_m a^2 \frac{\partial \theta_0}{\partial t}\left(1 - A_m z\sin^2\gamma/R_m\right). \tag{23}$$

The results (14)-(15) also follow here, where λ depends on t, z, a, h and P_r.

LOW P_r AND STRONG TIME DEPENDENCE CASE

In this section we consider the last possible case, where the order of magnitude of the time derivative inertia term in (1a) is larger than those of the nonlinear inertia and viscous terms in (1a). The order of magnitude of the velocity components and the stream function in the chimney are now satisfied by:

$$\frac{\partial w}{\partial t} \sim R_1 P_r, u \sim wa, \psi \sim wa^2. \tag{24}$$

From (1) and the condition:

$$1 \ll R_1 P_r w \Big/ \frac{\partial w}{\partial t} \ll 1/a^2, \tag{25}$$

we find $\theta_1 \ll 1$. The results (10)-(12) and (21)-(23) are valid again here. The results (14)-(15) also follow, provided $R_1 a^4$ in the right-hand-side of (14) is replaced by $a^2\sqrt{R_1 P_r}$ and λ now is independent of P_r but depends on a, z, t and h.

CONCLUSIONS

The main results of the present study lead to the following conclusions:

1. For weak rotation, convection in the chimney increases with increasing solutal Rayleigh number but is otherwise unaffected by rotation.

2.   Rotation can be effective only if the rotation axis is inclined with respect to the direction of the high gravity vector.

3.   For moderate rotation, convection in the chimney increases with the solutal Rayleigh number, but decreases with increasing rotational effects.

4.   Freckle formation can be eliminated or at least can be reduced only for moderate rotation.

5.   For strong time-dependent convection, the order of magnitude of the vertical velocity in the chimney depends strongly on the time variation of the temperature.

6.   For strong rotation, convection in the chimney increases with solutal Rayleigh number and with increasing rotational effects.

7.   Some moderate level of high gravity appears to have beneficial effects to eliminate or at least reduce macrosegregation.

8.   The beneficial effects of high gravity on macrosegregation are independent of the type of alloy melt and of the time-dependent features of convection.

REFERENCES

1.   D.N. Riahi and T.L. Sayre, Effect of rotation on the structure of a convecting mushy layer, *Acta Mechanica* (in press).

2.   A.K. Sample and A. Hellawell, The effect of mold precession on channel and macrosegregation in ammonium chloride-water analog castings, *Met. Trans. B* 13:495 (1982).

3.   A.K. Sample and A. Hellawell, The mechanisms of formation and prevention of channel segregation during alloy solidification, *Met. Trans. A* 15:2163 (1984).

4.   S. Kou, D.R. Poirier and M.C. Flemings, Macrosegregation in rotated remelted ingots, *Met. Trans. B* 9:711 (1978).

5.   D.N. Riahi, On the structure of an unsteady convecting mushy layer, submitted for publication.

6.   W.A. Arnold, W.R. Wilcox, F. Carlson, A. Chait, and L.L. Regel, Transport modes during crystal growth in a centrifuge, *J. Crystal Growth* 119:24 (1992).

7.   A.C. Fowler, The formation of freckles in binary alloy, *IMA J Appl. Math.* 35:159 (1985).

8.   M.G. Worster, Natural convection in a mushy layer, *J. Fluid Mech.* 224:335 (1991).

9.   M.J Lighthill, Theoretical considerations on free convection in tubes, *Q. J. Mech. Appl. Math.* 6:398 (1953).

THE EFFECT OF GRAVITY ON THE
WELD POOL SHAPE IN STAINLESS STEEL

Daryush K. Aidun, Scott A. Martin, and Jeffrey J. Domey
Mechanical and Aeronautical Engineering Department
Clarkson University, Potsdam, NY 13699-5725

INTRODUCTION

To obtain a high quality weld we need to understand the mechanical and thermal properties of the material and its melt, and the solidification mode and its morphology. Convective flow during welding is dependent upon buoyancy, surface tension and its temperature coefficient ($d\gamma/dT$), the arc drag and electromagnetic forces, the shielding gas, and the viscosity of the melt. By controlling fluid flow and heat transfer during welding, one should be able to control the weld solidification mode, morphology and segregation, and by that, prevent defects in the fusion zone. Buoyancy and Marangoni flow are major factors in controlling weld solidification.[1,2] There is a lack of knowledge on the effect of gravity, especially macro-gravity ($> 1g$), on weld pool geometry. Thus, the objective of this work is to evaluate the shape and profile of the fusion zone of 304 and 316 gas-tungsten arc welds during centrifugation.

PROCEDURE

Three heats of 304 and two heats of 316 (compositions in Table 1) were subjected to a 7-second stationary GTA (gas-tungsten arc) weld at 1 g and 5 g conditions using the Multi - Gravity Research Welding System (MGRWS) located at Clarkson University (see Figure 1). The MGRWS is the first centrifuge designed and built for the sole purpose of investigating weld solidification under macro-gravity conditions. It has an arm 1.1 meters long and can rotate more than 100 rpm producing a net acceleration of more than 15g acting perpendicular to the weld pool. The welds were sectioned with a diamond-tipped saw. Their depth-to-width ratios were determined using a digital imaging process, with an accuracy of $\pm$ 0.03 mm.

RESULTS AND DISCUSSION

Figures 2A and 2B show macrographs of the fusion zone of 316HS (High Sulfur) GTA welds performed at 1 g and 5 g, respectively. Table 2 summarizes the results of the depth-to width (D/W) ratio measurements for the 304 and 316 heats that were subjected to welds at 1 g and 5 g. Noting the difference in the $d\gamma/dT$ values for the two materials, these results indicate that as the acceleration increases above 1 g, the D/W ratio decreases. This can be explained by the buoyancy driven flow, which is a radially outward flow, becoming much larger than the Marangoni flow despite the sign of $d\gamma/dT$. This result had been predicted for aluminum by our numerical simulations,[3] which showed that as gravity increased above 1 g, the D/W ratio of the fusion zone should decrease.

Table 1. Composition of the austenitic stainless steels in wt.%

I.D.*/Thickness	C	Si	Mn	P	S	Cr	Ni	Mo	Cr_{eq}/Ni_{eq}**
304LS / 3 mm	0.1	0.6	1.1	0.03	0.003	18.09	8.45	-----	1.76
304HS / 3 mm	0.1	0.4	0.9	0.02	0.008	18.29	8.40	-----	1.78
304C / 3 mm	0.1	1.0	2.0	0.05	0.030	18/20	8/12	-----	1.53
316LS / 2 mm	0.1	0.5	0.8	0.02	0.001	17.58	11.74	2.07	1.50
316HS / 2 mm	0.1	0.5	0.8	0.03	0.006	17.55	11.83	2.25	1.46

* LS = low sulfur; HS = high sulfur; C = commercial grade
** Cr_{eq} = %Cr + %Mo + 1.5x%Si + 0.5x%Nb
Ni_{eq} = %Ni + 30x %C + 0.5x%Mn

Figure 1. Multi-Gravity Research Welding System (MGRWS).
From left to right: Steve Zannon, Daryush Aidun, Dave Williams.

Table 2. Average depth to width (D/W) ratio of the weld fusion zone.

Material	D/W @ 1g (# of welds)	D/W @ 5g (# of welds)
304LS	0.312 (3)	0.277 (3)
304HS	0.393 (2)	0.323 (3)
304C	0.297 (2)	0.249 (2)
316LS	0.311 (2)	0.297 (3)
316HS	0.370 (2)	0.221 (2)

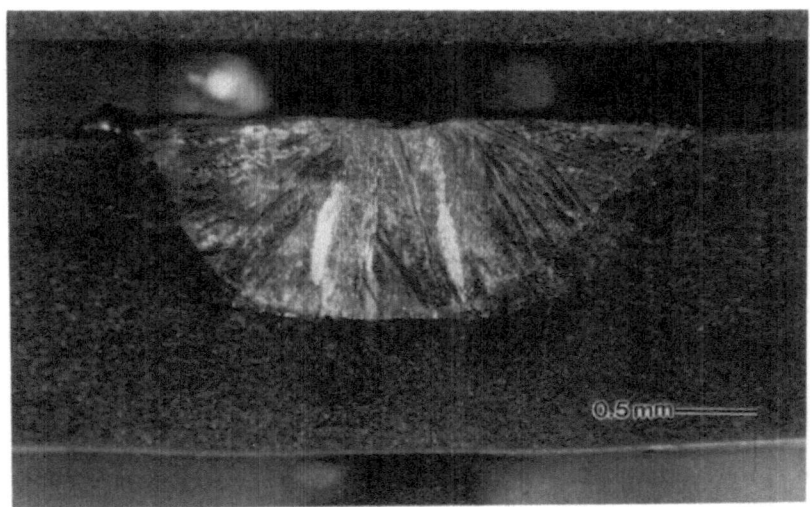

Figure 2A. GTA weld performed at 1 g on 316HS.

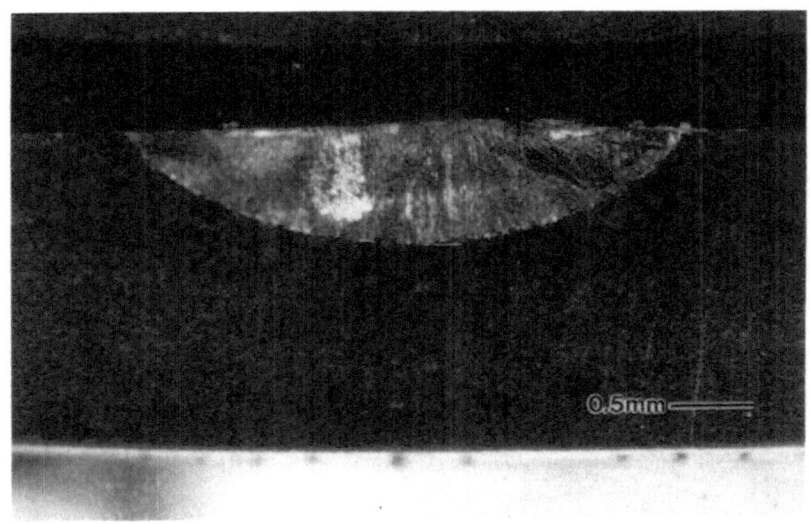

Figure 2B. GTA weld performed at 5 g on 316HS

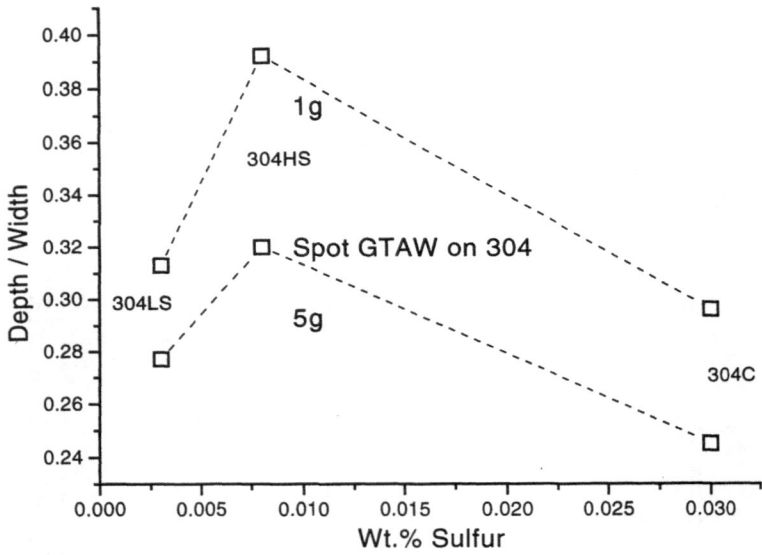

Figure 3. Effect of sulfur on the depth/width ratio of GTA welds performed on 304 steel at 1 g and 5 g.

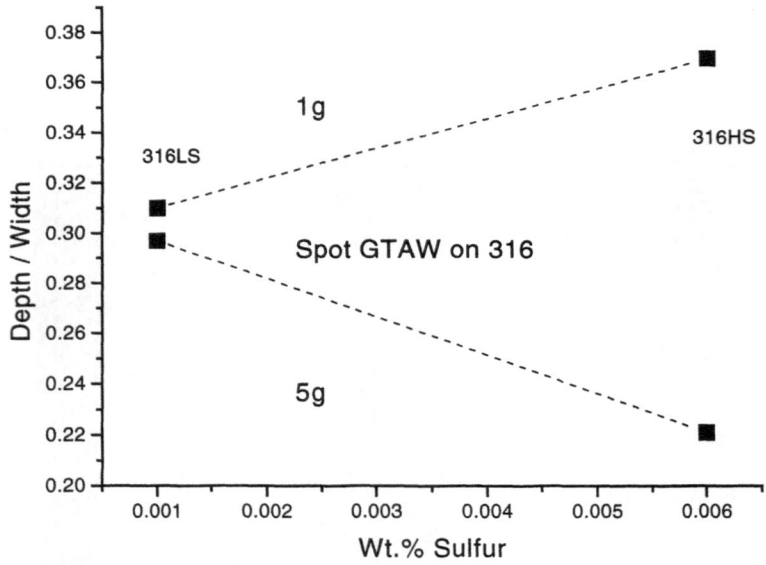

Figure 4. Effect of sulfur on the depth/width of GTA welds on 316 steel performed at 1 g and 5 g.

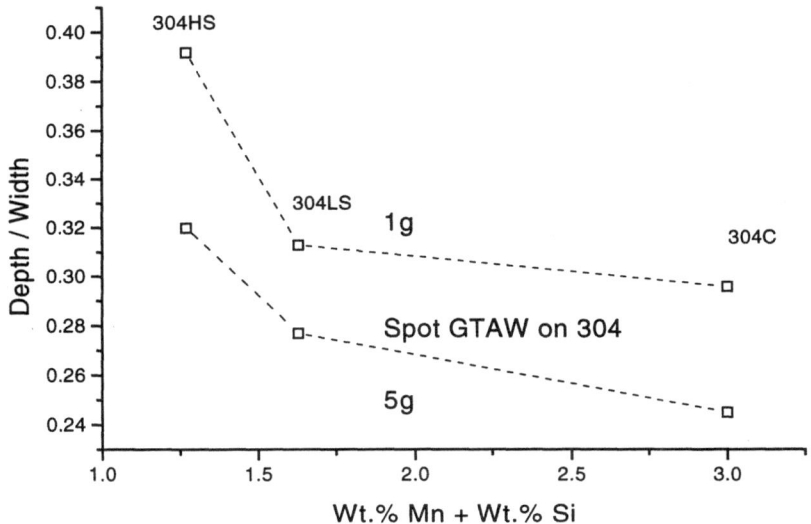

Figure 5. Effect of manganese and silicon on depth/width ratio of GTA welds performed on 304.

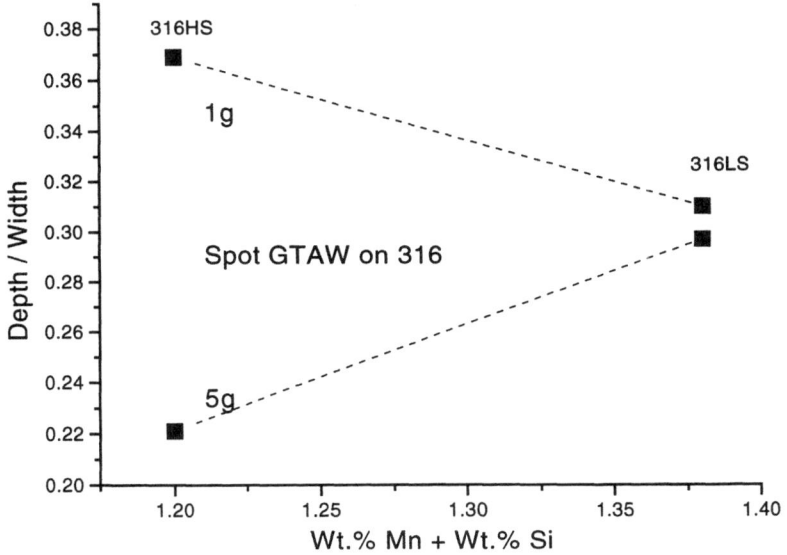

Figure 6. Effect of manganese and silicon on depth/width ratio of GTA welds performed on 304.

Figures 3 and 4 show the effect of sulfur content on the D/W ratio of 304 and 316 GTA welds performed at 1 g and 5 g; respectively. Figures 5 and 6 show the combined effect of manganese and silicon content on the D/W ratio of 304 and 316 welds performed at 1g and 5g, respectively. These figures indicate that active minor elements (surfactants) affect the D/W ratio of the fusion zones created at 5 g, similar to those that have been previously reported for 1g.

SUMMARY

Compared to welds performed at 1 g, the depth/width ratio at 5 g decreased, despite the composition of the 304 and 316 base materials. The 316HS heat displayed a greater difference in the depth/width ratio between welds performed at 1 g and 5 g than was displayed by the 316LS heat. The 304 heats that contained higher amounts of Mn and Si, exhibited lower depth/width ratios. However, this was not true for the 316 welds. Further work will include the effect of centrifugation on the depth/width ratio of materials such as single crystals, metal matrix composites and nickel subjected to both GTA and GMA (gas-metal arc) welding processes.

Acknowledgments

The partial financial assistance of the American Welding Society Fellowship Foundation for S.A. Martin is greatly appreciated. We thank Joseph Beckham and Hobart Welding Systems for donating GTA and GMA power supplies for the MGRWS. We are grateful to Thomson Industries and ALCOA's Massena Operations for their in-kind contributions to the MGRWS.

REFERENCES

1.   T. Zacharia, A.H. Eraslan, and D.K. Aidun, Modeling of autogeneous welding process,
        Welding Journal **67**:18s (1988).
2.   K.C. Mills and B.J. Keene, Factors affecting variable weld penetration,
        International Materials Review **35** (1990).
3.   J. Domey, D.K. Aidun, G. Ahmadi, L.L. Regel, and W.R. Wilcox, Numerical simulation of the
        effect of gravity on weld pool shape, *Welding Journal* **74**:263s (1995).

SEPARATION EFFECTS IN Bi-TYPE
HIGH-Tc SUPERCONDUCTORS
PREPARED DURING CENTRIFUGATION

M.P. Volkov,[1] B.T. Melekh,[1] N.F. Kartenko,[1] and L.L. Regel[2]

[1]Ioffe Physical Technical Institute, Russian Academy of Science
St. Petersburg, Russia
[2]International Center for Gravity Materials Science and Applications
Clarkson University, Potsdam, New York

ABSTRACT

We compared crystal structure, phase distribution and superconducting parameters of Bi-type high-temperature superconductors (HTSC), prepared by melting and recrystallizing during centrifugation, followed by annealing. Results obtained can be explained by the different sensitivities of the 2212 and 2223 phases of Bi-type HTSC to small deviations in stoichiometry. Stable compounds with superconducting critical temperatures up to 10 K higher than the boiling point of liquid nitrogen were prepared by recrystallizing and annealing the 2212 stoichiometric composition with partial substitution of Cu by Li.

INTRODUCTION

Among known high-temperature superconductors, two compounds are most interesting for possible applications: complex cuprates $YBa_2Cu_3O_{7-x}$ with $T_c \sim 91$ K; and Bi-containing complex cuprates Bi-Sr-Ca-Cu-O with $T_{c\ max} \sim 110$ K. Both systems are prepared usually by ceramic methods. Bulk samples of Bi-type cuprates produced by solid state reaction show extremely low Jc values because of a porous microstructure, strong anisotropy, and Josephson weak links at boundaries between superconducting grains. In order to overcome these problems, a melt process has been employed by several authors.[1,2]

There are at least three superconducting phases in the Bi-type system. The composition of these phases can be written as $Bi_2Sr_2Ca_{n-1}Cu_nO_{2n+4}$, where n = 1,2,3. Two of these phases (n = 2, the so called 2212 phase; and n = 3, the 2223 phase) have their critical temperatures T_c higher than the liquid nitrogen temperature (~ 85 K and ~ 110 K, respectively). Bi-type compounds of different initial compositions, 2201 and 2212, can be obtained more easily. The 2223 phase with maximum critical temperature can be obtained

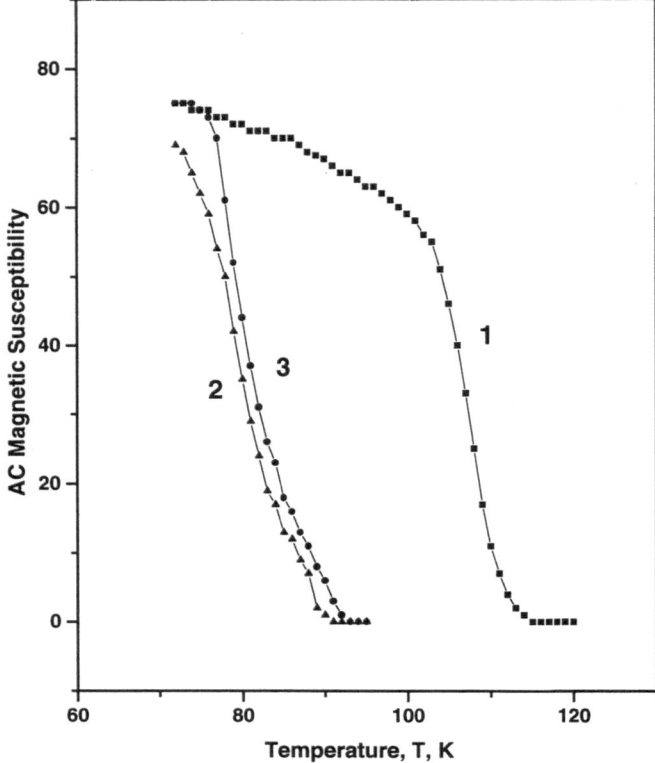

Figure 1. Temperature dependence of AC magnetic susceptibility of Bi-Pb samples of high-temperature superconductors prepared by recrystallization during centrifugation and subsequent annealing. 1 - $1g_0$, 2 - $3g_0$, 3 - $5g_0$.

of the middle of the transition). This value corresponds to well-formed 2223 phase. The T_c values for the 3 g_0 and 5 g_0 samples are approximately equal (~ 80 K) and correspond to the 2212 phase of this compound. X-ray investigation determined that the $1g_0$ sample consisted of a mixture of: the 2223 phase with lattice parameters a = 0.5422 nm, c = 3.718 nm; the 2212 phase with a = 0.5397 nm, b = 0.5431 nm, c = 3.074 nm; 10 wt.% of $(Ca_{1-x}Sr_x)_2PbO_4$; >10 wt.% of Ca_2CuO_3; and a small amount of $CaCu_2O_3$. The $3g_0$ and $5g_0$ samples consisted mainly of the 2212 phase with lattice parameters a=0.5402nm, b=0.542nm and c=3.079nm, and a=0.5404nm, b=0.5426nm and c=3.069nm, respectively. The amount of $(Ca_{1-x}Sr_x)_2PbO_4$ impurity decreased and Ca_2CuO_3 increased as the acceleration was increased.

The Bi-Li samples had different characteristics than the Bi-Pb ones. They did not degrade during 4 month storage, and, as can be seen from Figure 2, consisted mainly of small crystals, 1-2 mm long, oriented normal to the walls of the crucible. Samples from upper and lower parts of the ingots were investigated by AC magnetic susceptibility method and X-ray powder diffraction. Figure 3 shows the temperature dependence of the ac susceptibility for a $1g_0$ Bi-Li sample (curve 1), a $3g_0$ sample from the top (curve 2), and a $3g_0$ sample from the bottom (curve 3). The critical temperatures of the $1g_0$ and $3g_0$-bottom samples were nearly equal (~ 86 K). The T_c of the $3g_0$-top sample was slightly higher (~87.5K). X-ray study showed the co-existence of 2212 and 2201 phases in all Bi-Li samples. The $1g_0$ and $3g_0$-bottom samples contained nearly equal amounts of these phases.

only by long term annealing of stoichiometric composition and partial replacement of Bi by Pb.[3]

The employment of melting for the preparation of these compounds faces some difficulties. There is a temperature range where the compound is partially melted. This range depends on the composition. For instance, for the 2212 composition this range is from 870°C up to ~ 900°C. We performed some experiments on preparation of Bi-type compounds by melting and crystallizing at high gravity, to improve our understanding of the peculiarities of melting and phase transformations in these materials.

In our previous centrifugation experiments,[4] we used the composition $Bi_{1.8}Sr_{1.7}Ca_{1.25}Cu_{2.2}O_x$ for preparation of the 2223 phase. This composition was recrystallized at 950°C under different gravity levels (1 g_o, 8 g_o and 12 g_o). At this temperature, the melt consists of metal oxides with greatly different densities (e.g. 3.4 g/cm^3 for CaO and 8.9 g/cm^3 for Bi_2O_3). Consequently, one might expect a separation of components along the height of a recrystallized ingot. Recrystallized and annealed ingots were cut into slices normal to the gravity direction. A vertical density gradient was observed in these ingots. AC magnetic susceptibility measurements revealed a gradient in superconducting properties - the critical temperature T_c increased from the bottom samples to the top ones. Formation of the 2223 phase was observed only in the top parts of the ingots, where T_c increased with increasing acceleration during recrystallization. To elucidate the peculiarities of recrystallizing Bi-HTSC on a centrifuge, we report here on additional experiments performed with different initial compositions and lower gravity levels.

EXPERIMENTAL

Two initial compositions were used: $Bi_{1.8}Pb_{0.3}Sr_{1.9}Ca_2Cu_3O_x$ (Bi-Pb samples) and $Bi_{2.2}Sr_{1.9}Ca_{1.05}Cu_{1.45}Li_{0.7}O_x$ (Bi-Li samples). The first composition is close to the stoichiometric 2223 composition. Partial replacement of Bi by Pb leads to easier formation of the 2223 phase. The second composition is close to the 2212 phase. Partial replacement of Cu by Li promotes single crystal formation of this phase.[5] Starting compositions were obtained by mixing the appropriate metal oxides and high frequency melting in a cold container.[6] After that, the samples were melted and recrystallized in cylindrical zirconium dioxide crucibles under different gravity levels (1 g_o, 3 g_o and 5 g_o) in a gradient-freeze furnace mounted on the arm of Clarkson's HIRB centrifuge. The temperature at the bottom of the crucible was increased at a rate of 350°C/h up to 970°C, where the centrifuge was switched on. After an hour at 970°C, the temperature was decreased at a rate of 110°C/h down to 750°C. Then both the centrifuge and the furnace were switched off. After this recrystallization, the samples were annealed in air at 780°C (30 h for Bi-Pb samples, and 9 h for Bi-Li samples).

RESULTS AND DISCUSSION

The dense black ingots obtained by recrystallization and annealing were stored for 4 months in dry air. After that, their structure and superconducting properties were measured. These measurements revealed quite different behavior for the two compositions studied. The Bi-Pb samples had degraded greatly. The upper parts of these ingots had become powder-like. Only the lower parts of the ingots were single pieces of the material. AC magnetic susceptibility measurements of these (lower) parts of the Bi-Pb samples (Figure 1) showed that the T_c of the 1g_o-sample is ~ 107 K (determined as the temperature

Figure 2. Longitudinal cut of ZrO_2 crucible containing Bi-HTSC with Cu partially replaced by Li, prepared by recrystallization at $3g_0$ and subsequent annealing.

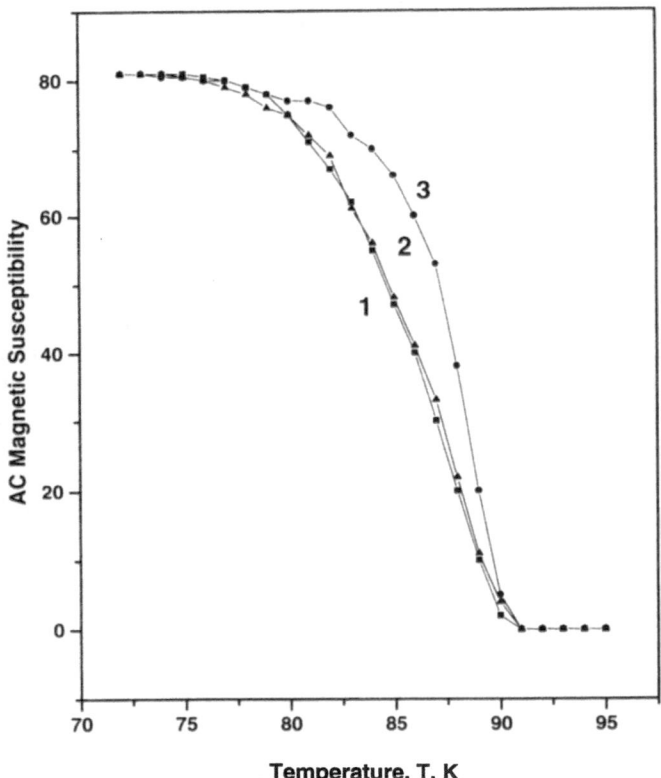

Figure 3. Temperature dependence of AC magnetic susceptibility of Bi-Li samples of high-temperature superconductors prepared by recrystallization during centrifugation and subsequent annealing.
1 - $1g_0$, 2 - $3g_0$ (bottom), 3 - $3g_0$ (top).

The lattice parameters of the 2212 phase were a=0.5418nm, b=0.5449nm and c=3.074nm, and a=0.541nm, b=0.5434nm and c=3.077nm, respectively. For the 2201 phase, a=0.541nm and c=2.466nm, and a=0.536nm and c==2.466nm, respectively. In the $3g_0$-top sample, the 2212 phase was dominant, with lattice parameters a=0.5412nm, b=0.5438nm and c=3.083nm. All Bi-Li samples contained a few percent of $SrCO_3$ and CuO impurities, and an unidentified Li-Ca-Bi phase. In the $3g_0$-top sample the amount of this Li-Ca-Bi phase was nearly the same as the 2212 phase. The Li-Ca-Bi phase could not have been a single phase, but rather a mixture of Li and Ca carbonates with Li_7BiO_6. The carbonates could have been created during storage in air from alkaline metal oxides. Our attempts to prepare Bi-Li samples at $5g_0$ were unsuccessful, probably due to enhanced wetability and/or reactivity of the Li-containing melt with the ZrO_2 crucibles.

At the temperature of recrystallization of Bi-type HTSC compounds, the melt consists of simple metal oxides with greatly different densities. Our results show that separation of the components along the height of the ingot took place during recrystallization. In the Bi-Pb samples, the upper parts probably contained large amounts of Ca and Sr oxides which reacted with moisture and led to the pulverization of the upper parts of the samples. In the lower parts of the Bi-Pb samples there was excess Bi and Pb. Due to the low thermodynamic stability of the 2223 phase, these small deviations from exact stoichiometry led to the creation of other phases. First of these was the 2212 phase, which is much less sensitive to deviations from stoichiometry, and the $(Ca,Sr)_2PbO_4$ and Ca_2CuO_3 impurity phases. The initial composition of the Bi-Li samples was close to the 2212 stoichiometry, and segregation during recrystallization did not lead to drastic changes in structure or superconducting properties. Only the $3g_0$-top sample had a slightly higher critical temperature and a higher value of the lattice constant, c = 3.083 nm.

In our previous investigation,[4] we also observed higher critical temperatures in the upper parts of recrystallized ingots, especially for higher levels of gravity. This observation can be explained by the enrichment of the upper parts of the ingot by Ca and Cu over the exact stoichiometry. Such enrichment usually leads to a more perfect crystal structure of Bi-type superconductors. Note that the critical temperatures of the Bi-Li samples were several K higher than the T_c of Bi-Pb samples (2201 phase). This difference is very important for possible applications of Bi-type superconductors at liquid nitrogen temperature. The more stable and easily obtainable 2201 phase usually has $T_c \sim 80$ K (as had our Bi-Pb samples), which is only ~ 2.5 K higher than the boiling temperature of liquid nitrogen. The Bi-Li samples have the structure of the 2201 phase, but their critical temperatures are about 10 K higher than the boiling point of liquid nitrogen. Thus the partial substitution of Cu by Li in the 2212 phase of Bi-type high-temperature superconductors can be used to produce stable compounds with enhanced T_c by recrystallization and annealing. The enhanced wettability and/or reactivity of Li-containing melts and the nature of the Li-Ca-Bi impurity phase should be investigated in greater depth.

CONCLUSIONS

The low stability of the 2223 phase and the higher stability of the 2212 phase of Bi-type, high-temperature superconductors with respect to small deviations from the exact stoichiometry was confirmed by recrystallization of these compounds during centrifugation. This process led to segregation of components along the height of the ingots. Partial substitution of Cu by Li in the 2212 phase of Bi-type high-temperature superconductors can be used for the preparation of stable compounds by recrystallization and annealing, with critical temperatures up to 10 K higher than the boiling temperature of liquid nitrogen.

Acknowledgments

This research was supported in part by the Russian Basic Research Foundation via grant 96-02-17848 and the United States National Science Foundation via grant DMR-9414304.

REFERENCES

1. E. Yanmaz *et al.*, Preparation of superconducting Bi-Pb-Sr-Ca-Cu-O ceramics by melt casting, *Physica C* 185-189:2415 (1991).
2. J. Bock, E. Preisler, *Solid State Commun.* 72:453 (1989).
3. T. Uzumaki *et al., Appl. Phys. Lett.* 54:2253 (1989).
4. M.P. Volkov, B.T. Melekh, *et al.*, Properties of superconducting Bi-Sr-Ca-Cu-O system remelted under higher gravity conditions, *J. Crystal Growth* 119:122 (1992).
5. T. Horiuchi, K. Kitanava *et al.*, Li substitution to Bi-Sr-Ca-Cu-O superconductors, *Physica C* 185-189:629 (1991).
6. B.T. Melekh, *et al., Izv. Akad. Nauk SSSR, Ser. Neorg. Mater.* 18:1620 (1982).

INFLUENCE OF CENTRIFUGATION ON CLUSTER FORMATION AND NUCLEATION

Alexander F. Izmailov and Allan S. Myerson

Department of Chemical Engineering
Polytechnic University, Brooklyn, New York 11201

ABSTRACT

Phase separation from binary supersaturated solutions is evaluated for the effect of centrifugation. It is demonstrated that the thermodynamics of a centrifuged solution depends on spatial location and on the centrifuge angular velocity ω, when unavoidable random heterogeneities (such as the container walls, dust, impurities) are overshadowed by a heterogeneity due to the centrifugal acceleration g_{centr}. There is a possibility of initiating a deterministic redistribution of subcritical solute clusters. This can lead to *"deterministic"* nucleation, with phase separation occurring exclusively in non-random locations determined by the centrifugal acceleration. When earth's acceleration g is much less than g_{centr} ($g_{centr} >> g$), there is predicted to exist a steady regime where subcritical solute clusters are periodically redistributed in space. In particular, when the length R of the centrifuge arm is much greater than the linear dimensions L_x, L_y and L_z of a cell containing the supersaturated solution, the subcritical solute clusters should be concentrated along a cubic lattice.

INTRODUCTION

The study of supersaturated solutions (liquid metastable states) is of paramount importance for contemporary material processing, since crystallization from solutions is still one of the main industrial technologies for obtaining new materials. However, the experimental study of supersaturated solutions and nucleation is "contaminated" by heterogeneous nucleation, which can be triggered by any heterogeneities like container walls, dust and impurities. On the other hand, theoretical attempts to understand nucleation were developed only for *"pure"* supersaturated solutions and homogeneous nucleation. A combination of *"heterogeneous"* experiments with *"homogeneous"* theories is unlikely to provide conclusive results. At present, we see only two main ways to avoid this obstacle arising from the comparison of results observed experimentally with theoretical predictions. The first way is to eliminate the main source of heterogeneities, the container walls. This has been accomplished by suspending an electrically charged micro droplet of supersaturated solution in an electrodynamic levitator trap in an atmosphere near standard temperature and pressure (this experiment is described in detail[1-3] and its theory given[4-6]). This approach has been used on a number of organic and non-organic solutions, and resulted in relative supersaturations many times larger than those obtained in bulk solutions. These solutions are stable for the periods of time necessary to carry out thermodynamic and spectroscopic studies.[7-10]

The second way to avoid the above obstacle is to overshadow heterogeneities due to container walls, etc. by an artificially created deterministic heterogeneity of known nature. We believe such an opportunity can be provided by the centrifugation of small cells containing supersaturated solutions.[11-17] The centrifugal potential field, directed along the centrifuge arm, determines a Metastable State Relaxation (MSR), followed by nucleation, while suppressing all other heterogeneities. The description of MSR in a strong centrifugal field can be achieved by means of the homogeneous theory developed for the MSR in an external potential field.

The influence of a potential field on crystallization has been investigated in two different sets of experiments. In the first set, vertical columns with supersaturated solutions were subjected to earth's gravity g.[18-23] A fairly good theoretical understanding has been achieved for binary supersaturated solutions.[24-29] Experiments with these columns had demonstrated formation of a solute concentration gradient in gravitational field g. The column experiments were conducted in the following way. The first step was to prepare, inside a vertical column, a homogeneous binary solution in an undersaturated state. Then the solution was gradually cooled, until it became saturated and then supersaturated. If a solution was homogeneous and carefully prepared (little dust), it was possible to obtain and keep a metastable state for a long time t before spontaneous nucleation would take place ($t << t_i$, where t_i is the induction time). For a sufficiently long metastable state lifetime, a vertical solute concentration gradient could be observed. It is important to note that the redistribution of solute under the influence of earth's gravitational field has been observed only in supersaturated solutions.[24-28] No solute redistribution has been observed in undersaturated solutions, even if they were in tall vertical columns. From the experimental point of view, the essence and the most amazing feature of this phenomenon is the unusually rapid formation of the solute concentration gradient. Any attempt to explain this result with the help of conventional approaches based on linear irreversible kinetic theories leads to very long times for redistribution of subcritical solute clusters along the column length. For example, use of linear Fokker-Planck irreversible kinetic formalism gave long times, even for redistribution of large solute clusters.[24] Successful explanation of these experimental results was reached using non-linear approaches based on the Ginzburg-Landau formalism to the MSR in order to appropriately describe the MSR in an applied potential field.[30,31] Such a development has been undertaken.[24-29] One of the most interesting results was the prediction of the possibility to govern and induce nucleation by means of an applied external field.[25]

In other experiments, undersaturated or saturated solutions are centrifuged in order to concentrate solute at the outer edge of the cell and produce supersaturation.[11-15] However, ultracentrifugation ($g_{centr} \geq 13,000g$) is required in order to experimentally observe this effect, e.g., in growth of crystals of energetic materials.[17] These very interesting results for MSR under high gravity have not been analyzed yet.

The goal of the present paper is to develop an adequate theoretical model for MSR in the presence of a heterogeneity produced by a strong external potential field which suppresses all other heterogeneities. Experimentally, such a situation could be created by centrifugation of a cell containing a supersaturated solution, the temperature of which is controlled by a thermal bath. The non-linear process of solute redistribution in a supersaturated solution under the influence of high gravity can be understood as a non-complete (prior to nucleation onset) MSR in the presence of an external field. Such a non-complete MSR implies partial relaxation to an intermediate metastable state corresponding to the redistribution of subcritical solute clusters inside the centrifuged cell. This solute redistribution does not necessary lead to local solute concentrations enough to trigger a phase separation (nucleation). However, the approach suggested in this paper would allow investigation of the case when the solute redistribution induced by high gravity creates local concentrations high enough for nucleation. This investigation is of special interest, since it shows that the locations of critical solute concentrations are deterministically distributed rather than randomly as with conventional nucleation. Creation of a structure of a new solute-rich phase by means of nucleation in a strong external field is predicted to lead to promising new opportunities in materials processing.

THEORY FOR THE MSR SUBJECTED TO HIGH GRAVITY

The general formalism of the MSR is well known.[30,31] Its description requires one to distinguish between slow and fast degrees of freedom. The former determine the dynamics of the MSR, whereas the latter play the role of a thermal bath. The influence of fast degrees is usually taken into account by including random forces into the evolution equation for slow variables. The slow variables are usually identified with the hydrodynamic modes. Thus, the terms depending on the slow "*dynamic*" variables and the fluctuational terms appear in the evolution equation like deterministic and random forces, respectively.

Let us consider the simplest case when a cell containing supersaturated solution is centrifuged, i.e. is subjected to the potential centrifugal field with a constant angular velocity ω. The amount of solute in subcritical solute clusters can be described by a single hydrodynamic mode, i.e. by a scalar field $\varphi(r,t)$. The vector $r = (x, y, z) = (R + \xi_x, \xi_y, \xi_z)$ determines coordinates $(0 \leq \xi_x \leq L_x, 0 \leq \xi_y \leq L_y, 0 \leq \xi_z \leq L_z)$ in the rotating cell, and t is the current time. In order to specify the relaxation equation for the scalar field $\varphi(r,t)$ one has to make an assumption concerning the conservation of the field quantity $\varphi(t)$:

$$\varphi(t) = \frac{1}{V_{cell}} \int_{V_{cell}} d^3r \ \varphi(r,t) \tag{1}$$

in the course of the MSR, where $V_{cell} = L_x L_y L_z$ is the cell volume. Let us consider in this paper only the situation corresponding to a non-conserved scalar field $\varphi(t)$, i.e.:

$$\frac{d\varphi(t)}{dt} \neq const.$$

In this case, $\varphi(r,t)$ can be associated with the amount of solute in the new growing (non-conserved) solute rich phase (subcritical solute clusters). The MSR equation in terms of the non-conserved scalar field $\varphi(r,t)$ has the following Ginzburg-Landau form:[30,31]

$$\frac{\partial \varphi(r,t)}{\partial t} = -\Gamma \frac{\delta F \ (\varphi;t)}{\delta \varphi(r,t)} + f \ (r,t) \tag{2}$$

where Γ is the positive transport coefficient associated with the dissipative effects during the MSR, and $f(r,t)$ is the Gaussian distributed random force modeling the influence of thermal bath. The effective time-dependent free energy functional $F(\varphi;t)$ governing the MSR can be written in the well-known Landau form:[30,31]

$$F(\varphi;t) = \int\limits_{V_{cell}} d^3r \; [\frac{K}{2}[\nabla\varphi(r,t)]^2 + U((\varphi(r,t))], \quad \Gamma K = D, \tag{3}$$

where D is the solute diffusivity. The potential energy of the metastable state given by the term $U(\varphi(r,t))$ is usually interpreted as the coarse-grained free energy density, which is determined by the physics of the problem under consideration. In this paper, we consider a phase transition between a homogeneous (undersaturated) binary solution and its heterogeneous (supersaturated) state with heterogeneities in the form of subcritical solute clusters. As demonstrated by Landau,[32] in this case the potential $U(\varphi(r,t))$ should contain a term proportional to $\varphi^3(r,t)$. For the sake of simplicity, let us also consider the case of small supersaturations, i.e. solutions in the immediate neighborhood of phase equilibrium (binodal or coexistence) line. In addition, we investigate here the behavior of supersaturated solutions subjected to high gravity prior to the onset of nucleation. All these restrictions and conditions allow us to write the following expression for the metastable state potential energy $U(\varphi(r,t))$ in terms of the scalar field $\varphi(r,t)$:

$$U(\varphi(r,t)) = \frac{\mu}{2}\varphi^2(r,t) - \frac{\eta}{3}\varphi^3(r,t) \; - \; \frac{1}{\Gamma VD}[gz + \frac{1}{2}(\omega \times r)^2]\varphi(r,t). \tag{4}$$

where g is the acceleration due to earth's gravity and the Landau coefficients μ and η are determined by the metastable state properties.

ANALYSIS OF THE MSR SUBJECTED TO HIGH GRAVITY

The simplest analysis of metastability can be derived from the fact that the thermodynamic force $J(\varphi(r,t)) = dU(\varphi(r,t))/d\varphi(r,t)$, which drives the system to equilibrium, is equal to zero at thermodynamic equilibrium. This allows us to reach the important intermediate conclusion that the range $0 < \varphi(r,t) < \varphi_s(r)$ of metastable states subjected to centrifugation is dependent on the centrifugal acceleration ω and is inhomogeneous, i.e. it is dependent on the location vector r. The field $\varphi_s(r)$ determines the inhomogeneous boundary (spinodal line) between metastable and unstable states, and can be obtained from the following quadratic equation:

$$J(\varphi_b(r)) = \mu\varphi_b(r) - \eta\varphi_b^2(r) \; - \; \frac{1}{\Gamma VD}[gz + \frac{1}{2}(\omega \times r)] = 0. \tag{5}$$

Thus, the phase diagram, i.e. the thermodynamics, of centrifuged supersaturated solutions is ω-dependent and inhomogeneous. This allows some very interesting applications of centrifugation by governing the thermodynamics of centrifuged solutions. For values of the scalar field $\varphi(r,t)$ greater than $\varphi_s(r)$, the supersaturated solution becomes unstable and instantaneously phase separates. Solution of equation (5) gives:

$$\varphi_b(r) = \frac{\mu}{2\eta}[1 - \sqrt{1 - H(x^2 + y^2, z)}], \qquad H(x^2 + y^2, z) = \frac{4\eta}{\Gamma VD\mu^2}[gz + \frac{\omega^2}{2}(x^2 + y^2)]. \tag{6}$$

Integration by parts of equation (3) allows us to present equation (2) in the form:

$$\frac{\partial \varphi(r,t)}{\partial t} = D\Delta\varphi(r,t) - \Gamma\mu\varphi(r,t) + \Gamma\eta\varphi^2(r,t) + \frac{\Gamma\mu^2}{4\eta}H(x^2+y^2,z) + f(r,t). \qquad (7)$$

It should be mentioned that it has been possible to carry out such an integration only under the following condition:

$$\nabla\varphi(r,t)\big|_{x=R,\ y=0,\ z=0} = \nabla\varphi(z,t)\big|_{x=R+L_x,\ y=L_y,\ z=L_z}$$

The meaning of this boundary condition lies either in the equality of flows of subcritical solute clusters at the locations $(x = R, y = 0, z = 0)$ and $(x = R + L_x, y = L_y, z = L_z)$ at any time t, or in their simultaneous equality to zero. In addition, it is assumed that the process of changing the temperature while obtaining a metastable state, preserves partial thermodynamic equilibrium and is stopped when some supersaturation is achieved. After that, the temperature is held constant. Thus, the following process of solute redistribution under the influence of high gravity is supposed to be isothermal. This means that each point of the cell containing the supersaturated solution is in strong contact with a thermal bath, so that temperature fluctuations may be ignored and equation (7) acquires the form:

$$\frac{\partial \varphi(r,t)}{\partial t} = D\Delta\varphi(r,t) - \Gamma\mu\varphi(r,t) + \Gamma\eta\varphi^2(r,t) + \frac{\Gamma\mu^2}{4\eta}H(x^2+y^2,z). \qquad (8)$$

This form of equation (7) can be also obtained for the situation in which heterogeneities such as container walls and dust are suppressed by the deterministic heterogeneity due to the external field. In this paper we investigate the MSR only in the immediate neighborhood of the coexistence (binodal) line, where the solution of relaxation equation (8) can be presented in the form:

$$\varphi(r,t) = \varphi_b(r) + \psi(r,t). \qquad (9)$$

Equation (9) implies that $sup|\psi(r,t)/\varphi_b(r)| << 1$. This constraint means that the deviation from phase equilibrium at any time t is small. Therefore, the function $\psi(r,t)$ describes the deviation from equilibrium of two coexisting phases: solution (solute-poor phase) and subcritical solute clusters (solute rich phase). In the immediate neighborhood of the coexistence (binodal) line, there exists the following simplifying inequality $H(x^2+y^2,z) << 1$. The next simplifications are due to consideration of the case where $R >> L_x$, L_y, L_z, i.e. the linear dimensions of the centrifuged cell are much less than the length R of the centrifuge arm, and $g_{centr}=\omega^2 R >> g$, i.e. the centrifugal acceleration g_{centr} of the cell is much greater than earth's acceleration g. It is straightforward to demonstrate that under these conditions equation (8) acquires the following simple linearized form:

$$\frac{\partial \psi(r,t)}{\partial t} = D\Delta\psi(r,t) + \alpha\psi(r,t) + \beta, \qquad (10)$$

where $\alpha = (\frac{\omega^2\eta}{VD\mu}R^2 - \Gamma\mu)$; $\beta = \frac{2\omega^2}{\Gamma\mu}(1 + \frac{3\omega^2R^2\eta}{2\Gamma DV\mu^2})$.

In the linearized equation (10) we lose the ability to describe spontaneous homogeneous nucleation, which is essentially a non-linear effect. Therefore, the linearization of the initial non-linear MSR equation (7) implies the consideration of only such redistribution of subcritical solute clusters under high gravity that does not lead to nucleation. Thus, we have obtained the linear inhomogeneous partial differential equation describing early stages of MSR in centrifuged cells containing supersaturated solutions, the metastable states of which are close to the binodal line.

Let us consider the stationary solution $\psi_{st}(r)$ of equation (10) that is conditioned by:

$$\frac{\partial \psi(r,t)}{\partial t}\Big|_{\psi(r,t)=\psi_{st}(r)} = 0. \tag{11}$$

This condition describes the SMR situation in which the amount of solute in the subcritical solute clusters remains unchanged with time. It is understood that the solution $\psi_{st}(r)$ has a limited time applicability. The stationary regime of equation (10) can exist only for times $t << t_i$, since at times comparable with t_i one cannot neglect non-linear effects, which tremendously enhance fluctuations of solute content in the subcritical solute clusters and eventually lead to spontaneous nucleation. The solution $\psi_{st}(r)$ of equation (10) satisfies the following linear inhomogeneous ordinary differential equation:

$$D\Delta\psi_{st}(r) + \alpha\psi_{st}(r) + \beta = 0. \tag{12}$$

The simplest solution of this equation under the boundary conditions,

$$\psi_{st}(r)\Big|_{(x,y,z)=(R,0,0)} = \delta, \quad \frac{\partial \psi_{st}(r)}{\partial x}\Big|_{(x,y,z)=(R,0,0)} = 0 \text{ , can be found in the form:}$$

$$\psi_{st}(x,y,z) = \delta + (\delta+\frac{\beta}{\alpha})(\cos[\sqrt{\frac{\alpha}{3}}(\xi_x+\xi_y+\xi_z)]-1). \tag{13}$$

The first of these boundary conditions implies that the supersaturation (excess over saturation) of a centrifuged solution is equal to δ at the cell location $(R,0,0)$. When $\delta = 0$, the solution at location $(R,0,0)$ is in thermodynamic equilibrium, i.e. saturated. The second boundary condition implies that there is no flux of subcritical solute clusters at location $(R,0,0)$. We find that the total stationary solution of equation (8) describing the redistribution of subcritical solute clusters in the centrifuged cell with supersaturated solution has the form:

$$\varphi_{st}(x,y,z)=\varphi_b(r)+\psi_{st}(x,y,z)=\varphi_{st}(\xi_x,\xi_y,\xi_z) =$$

$$= \frac{\omega^2 R^2}{2\Gamma DV\mu} + \delta + (\delta+\frac{\beta}{\alpha})(\cos[\sqrt{\frac{\alpha}{3}}(\xi_x+\xi_y+\xi_z)]-1). \tag{14}$$

In deriving the above expression we have taken into account that in the limiting case, where $R >> L_x, L_y, L_z$ and $\omega^2 R >> g$, there exists the approximation:

$$\varphi_b\,(r)\;\approx\;\frac{\mu}{4\eta}H(x^2+y^2,z)\;\approx\;\frac{\omega^2R^2}{2\Gamma DV\mu}$$

Let us investigate the solution $\varphi_s(x,y,z)$. In the preceding derivation, we assumed that the characteristic length l of the distribution of subcritical solute clusters is much less than the cell macro-dimensions L_x, L_y, L_z. This assumption allowed us to ignore boundary conditions at the outer edge of the centrifuged cell. However, with large solute molecules and ultracentrifugation ($g_{centr} \geq 10,000g$), the centripetal acceleration can be strong enough to induce a non-zero flux of solute molecules directed to the cell outer edge. In this case, solutions of equation (12) are required to satisfy boundary conditions at the cell boundaries as well. Such a situation has been investigated xperimentally.[17]

For a one-dimensional cell (a long ampule directed along the centrifuge arm) solution (14) describes the periodic redistribution of subcritical solute clusters along the ξ_x- axis. Points of maximum concentration of these clusters are: $\xi_x(n)=2(3/\alpha)^{\frac12}\,\pi n$ $(n = 0,1,2...)$. Therefore, nucleation will take place most likely at the periodically located points $\xi_x(n)$, forming an analogue of a one-dimensional crystal. For a two-dimensional (Hele-Shaw) cell attached horizontally to the centrifuge arm, equation (14) describes the periodic redistribution of subcritical solute clusters with positions of maximum concentration given by:

$$\xi_x(n) + \xi_y(n) = 2(3/\alpha)^{\frac12}\,\pi n\ (n=0,1,2...).$$

Therefore, nucleation is most likely along periodically distributed lines forming an analogue of two-dimensional crystals. For a three-dimensional cell, equation (14) describes the periodic redistribution of subcritical solute clusters with surfaces of maximum concentration given by:

$$\xi_x(n) + \xi_y(n) + \xi_z(n) = 2(3/\alpha)^{1/2}\,\pi n\ (n = 0,1,2...).$$

The most likely nucleation in this case is along three-dimensional surfaces, forming embedded crystalline shells. To the best of our knowledge, such a crystallization process has never been observed and could be created only by means of centrifugation.

CONCLUSIONS

We have examined theoretically the situation when centrifugation of a supersaturated solution suppresses the effect of heterogeneities such as container walls and dust, thereby providing a regime of homogeneous nucleation in an external potential field. Such an opportunity distinguishes centrifugation, since it allows creation of conditions for homogeneous nucleation in bulk supersaturated solutions. Experimental study needs to be carried out in order to determine the acceleration required to suppress the effects of heterogeneities under various physical conditions, such as perfection of container walls, solution purity, temperature, etc.

Our theoretical approach, based on the Ginzburg-Landau formalism, is suggested as the most appropriate one to describe the MSR subjected to high gravity. In this paper we studied only the stationary MSR regime, corresponding to times less than the induction time t_i. For these times, we predict a redistribution of subcritical solute clusters and formation of a deterministic spatial structure induced by centrifugation . This structure would be periodic in space and serve as the seed structure for onset of nucleation. That is, nucleation would most likely occur at the points of maximum concentration of subcritical solute clusters. As follows from equation (14), the most interesting feature of the periodic nucleation structure induced by centrifugation would be the three-dimensional case, where the structure is predicted to be embedded shells. To the best of our knowledge, such a structure has not yet been observed.

REFERENCES

1.   R.F. Weurker, H. Shelton, and R.V. Langmuir, *J. Appl. Phys.* 30: 349 (1959).
2.   E.J. Davis, P. Ravindran, and A.K. Ray, *Adv. Colloid Interface Sci.* 15:1 (1981).
3.   H. Winter and H.W. Ortjohann, *Am. J. Phys.* 59:807 (1991).
4.   S. Arnold, L.M. Folan, and A. Korn, *J. Appl. Phys.* 74:4291 (1993).
5.   A.F. Izmailov, S. Arnold, and A.S. Myerson, *Phys. Rev. E* 50:702 (1994).
6.   A.F. Izmailov, S. Arnold, S. Holler, and A.S. Myerson, *Phys. Rev. E* 52:1 (1995).
7.   A.S. Myerson, H.S. Na, A.F. Izmailov, and S. Arnold, *in:* "Proceedings of the 12th Symposium on Industrial Crystallization," 1:3-013, Warsaw, Poland (1993).
8.   H.S. Na, S. Arnold, and A.S. Myerson, *J. Crystal Growth* 139:104 (1994).
9.   H.S. Na, S. Arnold, and A.S. Myerson, *J. Crystal Growth* 149:229 (1995).
10.   A.F. Izmailov, A.S. Myerson, and H.S. Na, *Phys. Rev. E* (in press).
11.   P.J. Shlichta and R.E. Knox, *J. Crystal Growth* 3: 808 (1968).
12.   L.L. Regel, International Workshop on Materials Processing in High Gravity, Moscow, USSR (1991).
13.   H. Rodot, L.L. Regel, and A.M. Turchaninov, *J. Crystal Growth* 104:280 (1990).
14.   I. Amato, *Science* 253:30 (1991).
15.   P.J. Shlichta, *J. Crystal Growth* 119:1 (1992).
16.   L.L. Regel and W.R. Wilcox, *in:* "Materials Processing in High Gravity," L.L. Regel and W.R. Wilcox, eds., Plenum Press, New York (1994).
17.   M.Y.D. Lanzerotti, J. Autera, J. Pinto, and J. Sharma, *ibid.*
18.   J.W. Mullin and C.L. Leci, *Philos. Mag.* 19:1075 (1969).
19.   Y.C. Chang and A.S. Myerson, *in:* "Industrial Crystallization," Elsevier, Amsterdam (1984).
20.   Y.C. Chang and A.S. Myerson, *AIChE J.* 28:28 (1984).
21.   M.A. Larson and J. Garside, *Chem. Eng. Sci.* 41:1285 (1986).
22.   A.S. Myerson and P.Y. Lo, *J. Crystal Growth* 99:1048 (1990).
23.   R.M. Ginde and A.S. Myerson, *J. Crystal Growth* 116:41 (1991).
24.   A.F. Izmailov and A.S. Myerson, *J. Crystal Growth* 121:723 (1992).
25.   A.F. Izmailov and A.S. Myerson, *Physica A* 183:549 (1992).
26.   A.F. Izmailov and A.S. Myerson, *Physica A* 192:85 (1992).
27.   A.F. Izmailov and A.S. Myerson, *J. Crystal Growth* 128:139 (1993).
28.   A.F. Izmailov and A.S. Myerson, *J. Phys. A: Math. Gen.* 26:2709 (1993).
29.   A.S. Myerson and A.F. Izmailov, *J. Phys. D: Appl. Phys.* 26:B123 (1993).
30.   L.D. Landau and E.M. Lifshitz, "Physical Kinetics," v. 10 in "Course of Theoretical Physics," Pergamon Press, London (1980).
31.   J.D. Gunton, M. San Miguel, and P.S. Sahni, *in:* "Phase Transitions and Critical Phenomena," C. Domb and J.L. Lebowitz, eds., v. VIII, Academic Press, New York (1983).
32.   L.D. Landau, *Soviet Physics: JETP* 7:627 (1937).

INFLUENCE OF GRAVITY ON THE HABIT OF SINGLE CRYSTALS
OF COMPOUNDS GROWN FROM SOLUTIONS

V.N. Gurin,[1] S.P. Nikanorov,[1] L.L. Regel,[2] and L.I. Derkachenko[1]

[1]Ioffe Physical-Technical Institute, Russian Academy of Sciences
St. Petersburg 194021, Russia
[2]International Center for Gravity Materials Science and Applications
Clarkson University, Potsdam, NY 13699-5814

ABSTRACT

The effect of different gravity levels on the growth habit of single crystals of some compounds has been determined. Isometric crystals were obtained both at microgravity and at 10-11g. Accelerations of 2-6g promoted the formation of elongated and flattened crystals. The microhardness of KCl and KBr crystals increased with increasing acceleration at which they were obtained).

INTRODUCTION

The growth habit of single crystals plays an important part, both in studying the properties of compounds and in their practical applications.[1] The growth habit describes the usual external morphology of single crystals, by their dominant crystal form or forms as determined by goniometric analysis.[2]

Preparation of single crystals from solutions gives rise to the largest diversity of crystal habits.[1,3,4] (Solvents typically consist of water for inorganic salts, organic liquids for organic compounds, and molten salts and metals for complex oxides and refractory compounds.) Special features of solution growth methods, such as the possibility of using different solvents and of varying the parameters of the crystallization process within wide limits, enable one to study crystallization processes under the most uncommon conditions, e.g., under microgravity (in space) and macrogravity (on a centrifuge). It seems interesting to compare for a given compound the results of crystallization experiments carried out at these conditions with those at normal (terrestrial) conditions. However, for many systems such comparison is not feasible because of technical difficulties.

For solution growth, the factors influencing crystal habit have been classified as chemical, physical-chemical and physical.[1] Each of these three factors exerts its own peculiar influence on the mechanism and kinetics of crystallization, which, in turn, influence the growth habit and many properties of the resulting crystals.

Chemical factors concern the solvent's composition, the nature, concentration and stoichiometry of the initial components and impurities, and their combination. Alteration of these factors changes the chemical composition of the solution, which, in turn, affects the solution structure, the scheme of chemical transformations and the surface composition of the growing crystals.

Physical-chemical factors concern the crystallization regime, soaking temperature and cooling rate. Alteration of these factors influences the solution's structure and the kinetics of phase and structural transitions.

Physical factors include gravity, pressure, magnetic and electrostatic fields, ultrasound, irradiation. These factors affect the solution's structure, convection, and the chemistry and kinetics of all processes proceeding in a system. The influence of these factors on the growth habit is practically unexplored; studies for some of them being only recently begun.

In the present work, we studied the influence of micro- and macrogravity on the growth habit of single crystals of some compounds in comparison with those grown under normal terrestrial conditions.

EXPERIMENTAL

Influence of Microgravity on the Habit of Chromium Disilicide Single Crystals

Experiments on the crystallization of chromium disilicide from a solution in molten zinc were carried out in space.[5] The results showed substantial alteration of the growth habit of $CrSi_2$ single crystals under microgravity (~0.02 g). A new crystal form - pinacoid - appeared in the habit of $CrSi_2$ single crystals. This form had never been observed in terrestrial experiments. Moreover, among the crystals grown in space neither needles nor plates were found, i.e. a general isometrisation occurred.

An explanation of the above phenomenon can be given, as follows. Figure 1 shows schematically that the convective flows near the surface of crystals growing in space differ from the flows occurring in terrestrial experiments. The existence of "top" and "bottom" directions in terrestrial conditions causes a higher concentration of solute in such flows near the bottom of the growing crystal and a lower concentration near its top. It follows that a higher growth rate would be observed at the crystal's bottom and a lower growth rate at the crystal's top. In space, where there is neither top nor bottom, the solute concentration will be relatively uniform along the crystal surface and the crystal should grow more uniformly in all directions. In the limit, such a crystal should grow isometric, exactly as observed for $CrSi_2$ single crystals grown at microgravity conditions. When the acceleration deviates from zero, one would expect to obtain different results. However, the main conclusions one can make from the experiments on crystallization of $CrSi_2$ from a molten zinc solution under microgravity are the isometrisation of crystals, occurrence of new forms in the crystal habit, and a 1.5-2 fold increase in the crystals' size. Of course, in the present paper we do not touch upon the alteration of crystals' properties and characteristics, which also takes place.[1]

Growth of Alkali Halide Crystals from Aqueous Solutions at High Gravity

It was not feasible for us to study the $CrSi_2$-Zn system described above under high

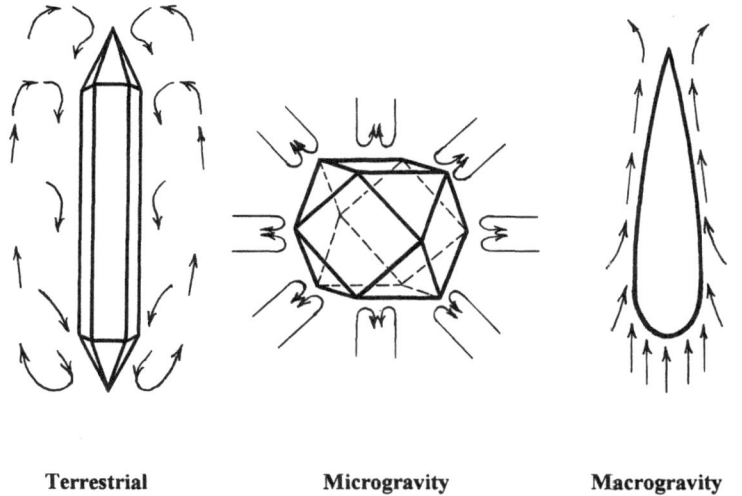

| Terrestrial | Microgravity | Macrogravity |

Figure 1. Schematic diagram of convective flows near the surface of a crystal growing under different accelerations.

gravity. Consequently, for the present studies we employed aqueous solutions of KCl and KBr as model systems.

One can suppose that at high gravity conditions, as compared to the normal terrestrial conditions, a decrease in crystal size would be observed. This follows from the observation that under microgravity the crystal size increases. As for the growth habit, the possibility of its alteration does not directly follow from the above considerations. For this case, experimental facts are indispensable.

In the present work, experiments were carried out with the systems KCl-H_2O and KBr-H_2O at T≈50-60°C. Alkali halide salt, in excess of its solubility in water at 50°C, was put into a small glass container with ~40-50 ml of distilled H_2O. This mixture was heated on an electric hot plate up to 70-90°C, poured into 2 rotary test-tubes which were balanced and then placed into the centrifuge K24D. The temperature controller was set to cool the tubes from room temperature down to 5°C in 30 min. The centrifuge was switched on.

Crystallization of KCl and KBr was studied at three values for acceleration. The results obtained are presented in Table 1. They demonstrate that macrogravity exerts an essential influence on the crystallization processes studied. Crystallization under high gravity (~11g) did not produce a morphology change in the crystals. Only isometric crystals (sometimes slightly elongated) of KCl and KBr were formed, their size being about 3 to 5 times smaller than those obtained at normal conditions.

A decrease of acceleration by about 2 times to 6g produced noticeable effects. In the crystal habit, in addition to cubes, another crystal form, rhombododecahedron, appeared, as shown in Figure 2. The number of such crystals was small, less than 1% of the total number of crystals. Besides, more than 80% of the cubic crystals were somewhat flattened. A few elongated crystals, thickening towards one end, were found. The thick ends of such crystals were aligned "downwards," i.e. in the direction of the centrifugal force.

Table 1. Influence of acceleration on habit of alkali halide crystals grown from aqueous solutions.

Gravity	Compound	Crystal forms	Morphological variety and crystal size, mm		Knoop microhardness, H_K (P=5g)
1g (normal conditions)	KCl	cube {100}	needles	≈ 8	17.5 kg/mm²
			plates	≈ 5	
			isometric	≈ 3	
	KBr	cube {100}	needles	≈ 10	12.5 kg/mm²
			plates	≈ 7	
			isometric	≈ 5	
2g	KCl	cube {100}	needles	≈ 2	18.5 kg/mm²
			isometric	≈ 1	
	KBr	cube {100}	needles	≈ 3	18.1 kg/mm²
			isometric	≈ 3	
6g	KCl	cube {100}	needles	≈ 1	21.8 kg/mm²
			isometric	≈ 1	
	KBr	cube {100}, rhombododecahedron {110}	needles	≈ 2	27.1 kg/mm²
			isometric	≈ 1	
11g	KCl	cube {100}	isometric	≈ 1	22.3 kg/mm²
	KBr	cube {100}	isometric	≈ 1	30.3 kg/mm²

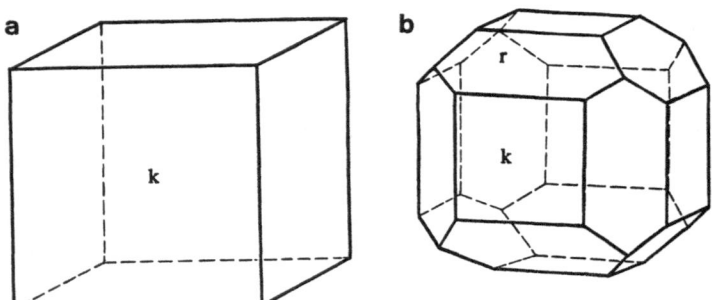

Figure 2. Habit of KCl and KBr crystals: (a) grown at normal 1g conditions; (b) grown at high gravity, 6g. Here k is a cube face and r is a rhombododecahedal face.

A further decrease of gravity to 2g resulted in an increased portion of elongated crystals, up to ~10 % of the total number of crystals. Very often these elongated crystals were covered with small badly shaped crystals. The size of crystals slightly increased, up to 3 mm.

At all macrogravity values the material sedimented at the bottom of the rotary test-tubes in the form of large aggregates of single crystals intergrown in different orientations.

Most interesting is the alteration of properties and characteristics of crystals prepared under various accelerations as compared with normal conditions. In this field remarkable achievements have been reported.[6] The model compounds of the present study provide a vast subject for future research.

DISCUSSION

Changing the acceleration, as expected, modified the growth habit of crystals of various compounds prepared by crystallization from solutions by cooling. We arrive at this general conclusion by comparing the results of experiments with various compounds performed under considerably different gravity values: from 0.02 to ~11g. The more concrete conclusions that one can make from scarce data available so far are as follows. Compared with normal (terrestrial) conditions:

1) microgravity (~0.02g) leads to isometrisation of single crystals, to an alteration of their growth habit, and to an increase of their size;
2) macrogravity (~10-11g) leads to isometrisation of single crystals and to a decrease of their size;
3) moderate gravity (2-6g) leads to an alteration of the growth habit of single crystals and to a decrease of their size;
4) with increasing macrogravity, a noticeable microhardness increase (by more than two fold).

The last conclusion is difficult to explain. These findings call for more detailed studies.

For the analysis of the results obtained under macrogravity, one can use a scheme of convective flows near the surface of the growing crystal similar to that realised at terrestrial conditions. Inside a centrifuge, the buoyancy-driven convection should be higher than at normal conditions (see Figure 1). Therefore, growth will be more rapid on the crystal's "head" from the flow of the solution. The lateral surfaces of the crystal will be streamlined by the impoverished flows. The growth of the lateral faces and of the crystal's "tail" will be slowed down. This would cause the crystal to take the shape of a tadpole. Crystals of this shape have been observed by growth of KCl from H_2O under ~2g.

At normal conditions, the crystal slowly settles down in the solution. Therefore, the solute-impoverished upward flow has time to mix back into the solution's volume, and than come back to the lateral surfaces of the crystal. Therefore the crystal grows more or less uniformly. Sometimes crystals may elongate and thicken downwards, according to the above scheme. This effect will be not as pronounced as in a centrifuge.

Similar results have been obtained in all experiments on growing single crystals of refractory compounds from solutions in molten metals.[7]

The proposed scheme is rather general and tentative. These processes are difficult to investigate. In space as well as in centrifuge many physical-chemical parameters of these processes need to be measured.

In conclusion, one should briefly discuss one of the most important challenges of centrifugal materials processing. This is the possibility of formation of new solid solutions and to an alteration of the width of the homogeneity range of compounds under the action of macrogravity. One can imagine that such promising materials as mixed crystals of rare earth hexaborides (e.g., $Nd_xLa_{1-x}B_6$[8]), which are of importance for practical applications in emissive electronics, under macrogravity might form low defect single crystals with better emissive characteristics. Materials for cold emission such as NbC and TaC, which have wide homogeneity range,[9] under macrogravity might display different limits of this range, which, in turn, would modify the properties of these materials. This might open an opportunity for optimising the known properties and for searching for new properties of prospective functional materials. These assumptions call for experimental verification. The experiments performed with the model compounds KCl and KBr, which practically do not have homogeneity ranges (they are very narrow), cannot clear up the situation so far. In any case, the action of macrogravity on crystallization from solution of the above-mentioned systems should bring about essential alterations in both the crystals' habit and in their functional properties and characteristics.

Acknowledgments

We thank Z.I. Uspenskaya, Institute of Cytology of the Russian Academy of Sciences, for her technical assistance with centrifuging. The help of M.M. Korsukova, Ioffe Physical-Technical Institute of the Russian Academy of Sciences, with preparation of the present publication is gratefully acknowledged.

REFERENCES

1. V.N. Gurin and L.I. Derkachenko, Growth habit of crystals of refractory compounds prepared from high temperature solutions, *Prog. Crystal Growth Charact. Mat.* 27:163(1993).
2. L.G. Berry, B. Mason and R.V. Dietrich, "Mineralogy. Concepts, Descriptions, Determinations," W.H. Freeman and Company, San Francisco (1983).
3. B. Honigman, Rost i forma kristallov, "Inostrannaya Literatura," Moscow (1961).
4. "Modern Crystallography," B.K. Vainstein, ed., Nauka, Moscow (1980) vol. 1-3 (in Russian).
5. V.N. Gurin, L.I. Derkachenko and S.P. Nikanorov, Solutions in metallic melts under microgravitation conditions, *in:* "Proceedings AIAA/IKI Microgravity Science Symposium," AIAA, Washington, (1991) pp. 134-137.
6. "Materials Processing in High Gravity," L.L. Regel and W.R. Wilcox, eds., Plenum Press, NY (1994).
7. V.N. Gurin and M.M. Korsukova, *Prog. Crystal Growth Charact.* 6:59(1983).
8. T.I. Serebryakova, V.A. Neronov and P.D. Peshev, Refractory borides, *in:* Metallurgya, Moscow (1991)(in Russian).
9. G.V. Samsonov, G.S. Upadhaya and V.S. Neshpor, "Refractory carbides," Naukova Dumka, Kiev, (1974)(in Russian).

CRYSTALLIZATION OF URIDINE PHOSPHORYLASE FROM *E. COLI* ON EARTH, MACROGRAVITY AND MICROGRAVITY

E. Blagova,[1] E. Morgunova,[1] E.Smirnova,[1] A.Mikhailov,[1] S. Armstrong,[2] C. Mao,[2] and S. Ealick[2]

[1]Institute of Crystallography, Russian Academy of Sciences
 Moscow 117333, Russia
[2]Section of Biochemistry, Molecular and Cell Biology
 Cornell University
 Ithaca, New York 14853

ABSTRACT

Uridine phosphorylase from *E. coli* (Uph) has been crystallized on Earth by a vapour diffusion technique, at zero gravity in space (space station "Mir"), and in macrogravity (ultracentrifuge G-62). We compare results obtained under these three conditions. We show the advantages of macrogravity growth of Uph crystals, i.e. high stability, rate and reproducibility. Crystallization of this protein in the ultracentrifuge allowed us to choose the optimal conditions for the production of large crystals suitable for X-ray analysis. The protein structure was determined by the molecular replacement method, using crystals obtained in macrogravity and on Earth.

INTRODUCTION

Uridine phosphorylase from *E. coli* (Uph) belongs to the class of nucleoside phosphorylases. It catalyzes the reversible phosphorolysis of the product of enzyme destruction of nucleic acids - uridine - with the formation of riboso-1-phosphate and uracil (Figure 1), realizing the transfer of ribose residue from the nucleoside to the phosphoric acid. This protein is involved in the degradation of pyrimidine nucleosides and their utilization as carbon and energy sources in *E. coli* cells. Uph has been identified as the enzyme that is also responsible for the cleavage of some pyrimidine nucleoside analogs, possessing antitumor activity. For example, 5-fluorouridine and 5-fluorodeoxyuridine are converted by Uph to 5-fluorouracil that gives rise to strong toxicity. Thus the intensive investigation of structure-function dependence of Uph to clarify the inhibitor action

Centrifugal Materials Processing
Edited by Regel and Wilcox, Plenum Press. New York. 1997

203

mechanism is very important for the selective suppression of enzyme activity. It is connected with the search for new selective inhibitors that might be used as drugs for therapy of some solid tumors. Uph is also of interest as a tool in the laboratory synthesis of nucleoside analogs for certain drugs. So to understand the action mechanism of the enzyme it is very important to know its structure.

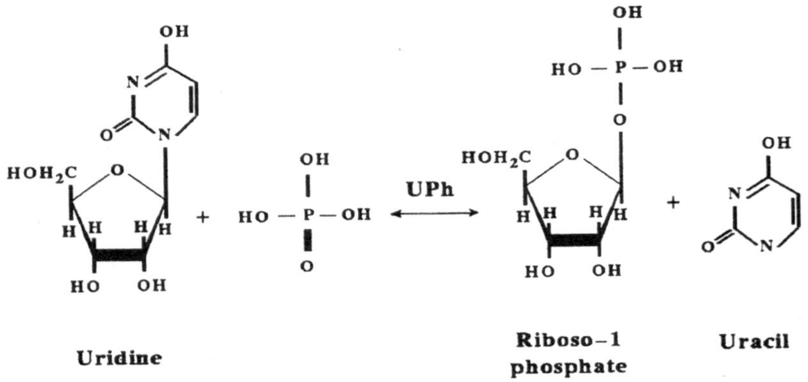

Figure 1. The reaction catalyzed by uridine-phosphorylase.

The success of an investigation of protein structure by X-ray crystallography depends on a reliable supply of sufficiently large, well-ordered single crystals. The production of such crystals is a critical rate-limiting step in three-dimensional X-ray analysis.

Many parameters influence nucleation and crystal growth of biological macromolecules, namely the concentration of the protein and precipitants, temperature, pH, time, ionic strength and purity of the chemicals, diffusion and convection, volume and geometry of the samples and set-ups, solid particles, wall and interface effects, density and viscosity effects, pressure, electric and magnetic fields, biochemical parameters, conformational heterogeneities, and batch effects. Crystal growth and nucleation are also affected by the method used.

It is known that every protein demands an individual approach for its crystallization. No method can be considered universal. Thus to reach the result it may be good to try different methods: crystallization at normal pressure and gravity (on the Earth), experiments at microgravity in space, and at high gravity in a centrifuge. Whatever the crystallization method used, it requires a high concentration of biological macromolecules as compared to normal biochemistry conditions. The common feature for all these methods is the production of a supersaturated solution.

Among the crystallization micromethods which are usually used at normal pressure and gravity, the vapour diffusion technique is probably the most widely used all over the world.[1] It provides an easy way to practical crystallization. Proteins have traditionally been crystallized by inducing supersaturation through the addition of precipitant (salts, alcohols), altering the pH towards the isoelectric point, slowly changing the temperature of the system, or increasing the solute concentration by evaporation of the solvent.

The principle of vapour diffusion crystallization is shown in Figure 2. A droplet containing the protein, buffer, crystallizing agent and additives is equilibrated against the reservoir containing a solution of crystallizing agent at a higher concentration than the droplet. The cell for crystallization is hermetically sealed. Diffusion of the volatile species (water or organic solvent) into the vapor leads to the gradual equilibration of its vapour pressure over the droplet and the reservoir. Consequently, the concentration of all constituents in the drop will change. If supersaturated conditions are reached, precipitation of crystals is observed. Such crystallization experiments are performed over a period of time that can be days, weeks, months or even years.

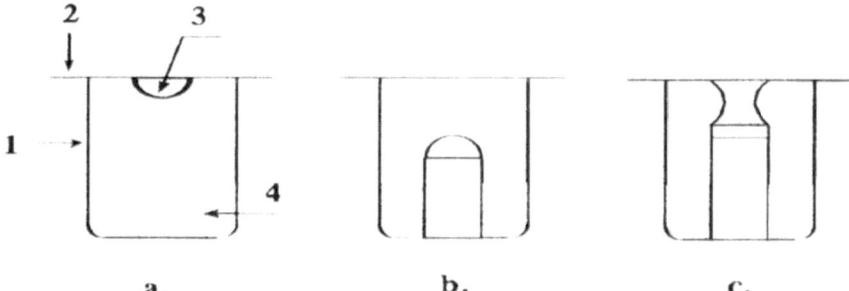

Figure 2. Schematic representation of three varieties of the vapour diffusion technique.
a - pendant drop; b - sessile drop; c - sandwich drop;
1 - reservoir; 2 - glass with hydrophobic surface; 3 - drop of protein solution;
4 - buffer solution with the precipitant.

EXPERIMENTS

Figure 3 shows Uph crystallized by the vapor diffusion method at room temperature. The trigonal crystals were obtained in sessile drops of 0.05M TRIS-HCl buffer, pH 7.0, containing 10-12 mg/ml of the protein and 4-6% of PEG 4000. The reservoir solution consisted of 0.1M TRIS-HCl buffer, pH 6.2 , 20% PEG and 0.04% sodium azide. The crystals reached the size of 0.4x0.2x0.3 mm after 4-6 weeks, and diffracted only to 4.5 Å resolution.

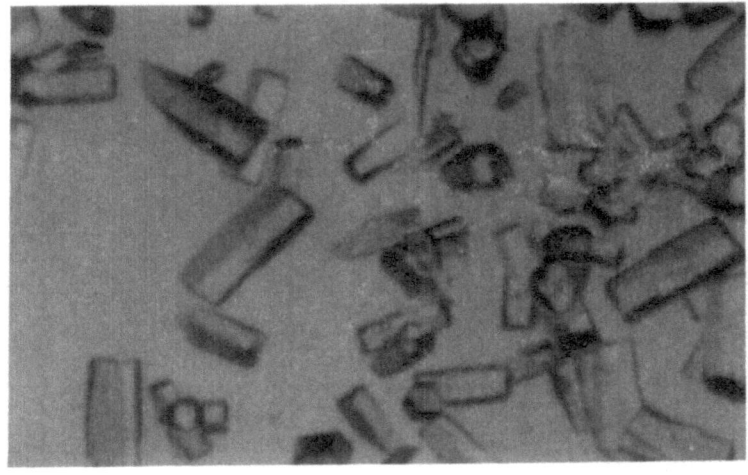

Figure 3. Crystals obtained on the Earth (magnification 50X).

Thus, this well-known procedure did not allow us to obtain good crystals of Uph for X-ray studies. They had rather high mosaicity and small dimensions. We decided to use less standard crystallization methods in this particular case. By using an ultracentrifuge, it is possible to concentrate the protein, to prepare a supersaturated protein solution, and to carry out rapid crystallization of the protein.

The sedimentation rate of spherical particles in a centrifuge can be estimated from their mass and the rotation speed.[2,3] When a protein solution is centrifuged, the large protein molecules sediment to the bottom of the tube more rapidly than the smaller salt molecules. The time needed to precipitate a molecule with sedimentation coefficient S is on the order of:

$$t = \frac{(\ln x_1 - \ln x_0)}{\omega^2 S} \tag{1}$$

where ω is the angular speed of rotor rotation (rad/sec), and x_0 and x_1 are the distances between the rotation axis and the top and the bottom of the tube, respectively. For an ideal solution, the sedimentation coefficient S is connected with the molecular weight of the precipitating molecules by:

$$S = \frac{M(1 - \bar{\upsilon}\rho_0)}{f} \tag{2}$$

where $\bar{\upsilon}$ is the partial specific volume of the dissolved substance, ρ_0 is the density of the solution, and f is the molar friction, which is proportional to $M^{1/3}$. So, the sedimentation coefficient increases with increasing molecular weight. Modern ultracentrifuges can give rates from 40,000 to 70,000 revolutions per minute, so it is possible to precipitate any protein this way. Proteins with a molecular weight of 100 kDa or more are concentrated the most rapidly. During centrifugation of protein solutions, a supersaturated solution is formed at the tube bottom and crystallization occurs there. The initial protein concentration in the solution can be rather low in comparison with other methods of crystallization. It should be noted that the requirement for enzyme preparation homogeneity may not be so strict.

Uridine phosphorylase has a molecular weight of 165 kDa, and so it is suitable for carrying out crystallization experiments in the centrifuge. Its crystallization was performed in centrifuge tubes containing 30 ml of the solution in ultracentrifuge G-62 with magnetic rotor suspension (RB 30 rotor, R_{max} = 121 mm; R_{min} = 43 mm, 30000 rev/min). The acceleration was 50,000 to 85,000 g. The experiments were carried out varying the parameters shown in Table 1.

Table 1. Conditions of uridine phosphorylase crystallization in the centrifuge.

N°	Altering parameter	Range	Step
1	PEG 6000 concentration	4 - 8 %	1 %
2	0.05 M TRIS-mal/NaOH buffer with different pH values	5.5 - 7.2	0.2
3	Protein concentration	0.04 - 0.2 mg/ml	0.01 mg/ml
4	Temperature	6 - 20°	1°
5	Centrifugation time	50 - 72 hours	2 hours
6	Speed	25000 - 30000 rev/min	1000 rev/min

Crystallization occurred under all conditions and did not depend considerably on the speed, time of centrifugation or initial protein concentration. Under the action of the centrifugal force, crystalline precipitant formed at the bottom of the tube, as shown in Figure 4a. It was a crystalline druse which was easily split into separate sectors - monocrystals - along the junction of monocrystal blocks, by light tapping on the tube bottom. The form and dimensions of these fragments differed from each other (Figure 4b). In crossed nicols of a polarizing microscope, we could see that the crystalline fragments homogeneously became dim. This provides evidence for their monocrystalline nature.

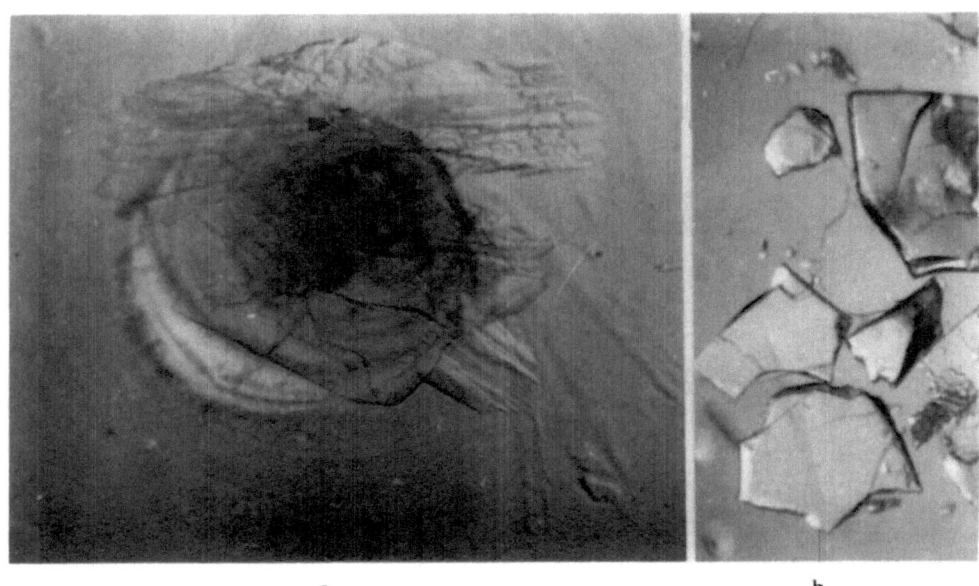

a.   b.

Figure 4. Crystals of UPh obtained in the centrifuge.
a - crystalline druse at the bottom of the tube (magnification 4X).
b - separate crystalline fragments (magnification 30X).

The best crystals were obtained under the following conditions: 50mM TRIS-mal/NaOH buffer, pH 6.3; 6% PEG 6000; 7°C; t=72 hours; ω=25000 rev/min. The crystal dimensions were about 1.5x0.6x0.3 mm^3. According to precession photographs, Uph crystals belong to the $P2_12_12_1$ space group with the following unit cell dimensions: a = 90.4; b = 128,8; c = 136.8 Å.

So, macrogravity conditions producing protein sedimentation into the growth zone allowed us to crystallize micro-amounts of proteins with low solubility. The method is characterized by rapid production of high-quality crystals from rather low initial concentrations of protein solutions. The growth time of the crystals is often decreased. In the case of Uph, the period required for growth was 7-15 times shorter. Centrifugation also promotes an improvement of the crystals' size. Moreover, protein crystallization in the centrifuge has very high reproducibility and allows to obtain big crystals, suitable for X-ray investigation.

Uph also was crystallized under microgravity conditions on the space station "Mir" in 1993. The results may be considered to be quite hopeful for further possible experiments. Equipment for protein crystallization in space was worked out in a special design bureau connected with the creation of space apparatus. Figure 5 shows a schematic diagram.

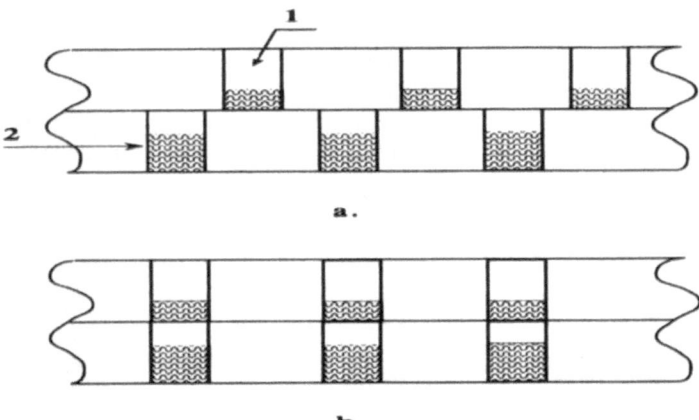

Figure 5. Equipment for protein crystallization in space.
a - the position before and after crystallization; b - the position during crystallization;
1 - chamber for the protein solution; ; 2 - chamber for the precipitating agent.

This device provides an opportunity to use the vapour diffusion technique in the pendant drop variant. It has 48 chambers for protein solutions (each volume equal to 30 μl), and 48 chambers for the precipitating agent (each volume is approximately 1ml). The chambers were loaded with protein and precipitating agent on the Earth. The upper part of the diagram corresponds to the position before and after crystallization, the lower part to the position during crystallization. The device was put on space station "Mir" in zone protected against vibration, at temperature 22 ± 3°C. The crystallization lasted for 57 days. A control experiment was carried out in the same chambers on Earth. The results of the experiment were examined and analyzed on Earth within 24 hours after landing.

The protein solution was in 0.1M TRIS-mal/NaOH buffer, pH 6.0. The initial concentration of the protein was 8.5 mg/ml. The conditions used for Uph crystallization in space are shown in Table 2.

Comparison of crystals grown in space and on Earth under the same initial conditions showed that the crystals were practically commensurate. But the crystals from the microgravity experiments were more isometric and better shaped. Various forms (cubes, plates, sticks, pyramids) were obtained using PEG 4000. Crystals obtained with PEG 2000

Table 2. Conditions used for uridine phosphorylase crystallization in space.

18% PEG X	18% PEG X with the addition of PEG 300	21% PEG X	21% PEG X with the addition of PEG 300
PEG 4000	PEG 4000	PEG 4000	PEG 4000
PEG 2000	PEG 2000	PEG 2000	PEG 2000
PEG 1500	PEG 1500	PEG 1500	PEG 1500

and PEG 1500 were smaller and less perfect than those grown with PEG 4000. The crystals from the Earth experiments had grown together. All of the well-shaped crystals (e.g., Figure 6) obtained in space were too small to use in X-ray analysis (their dimensions were approximately 0.2 - 0.3 mm), though they diffracted to 1.7 - 1.9 Å resolution. Thus the crystals grown in microgravity were of good quality, but they were rather small. Evidently the crystallization conditions were not optimal for this device. A drawback of this method is the unpredictable conditions for transport of the crystals from space to the x-ray laboratory on earth.

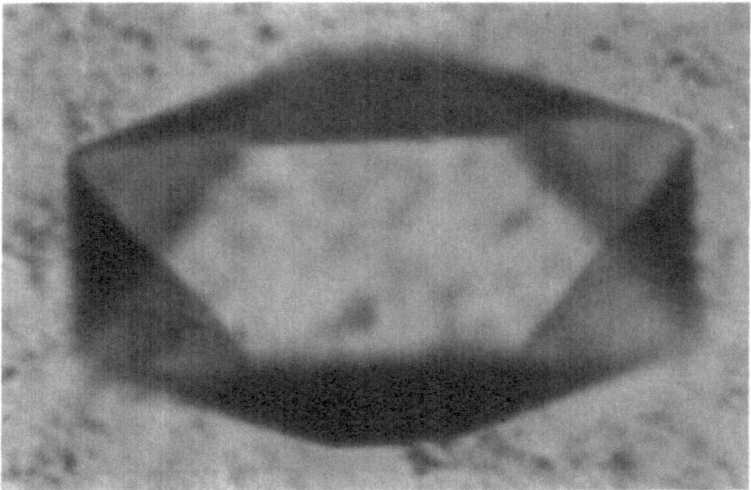

Figure 6. Uph crystal grown in space (magnification 400X).

Table 3 compares growth conditions and crystals of uridine phosphorylase obtained using different methods of crystallization. One can see the advantages of the macrogravity method of crystallization. The only disadvantage of the crystals obtained in macrogravity conditions is their low lifetime in the x-ray beam. But this method allowed us to choose the optimal conditions for crystal production and to use them in crystallization experiments on Earth. After optimization of the growth conditions, we managed to obtain monoclinic crystals on Earth that diffracted to 2.5 Å resolution.

The uridine phosphorylase structure was solved using crystals grown under macrogravity and Earth's gravity at 2.5 Å resolution by the molecular replacement method.[4,5] It was shown that the enzyme molecule is a hexamer (Figure 7). Each subunit in the hexamer has close contact with the identical one related by a 2-fold non-crystallographic axis, and 3 dimers are packed in a hexamer by a 3-fold non-crystallographic axis which is perpendicular to the 2-fold axes. Thus the hexamer belongs to a 32 point group symmetry. Each monomer is an α/ß-protein which contains 6 alpha-helixes and 2 twisted beta-sheets forming the main core of the monomer. There are two flexible loops playing an important role in enzyme catalysis. The intrinsic part of the molecule is formed by the hydrophobic residues. The active site is supposedly located near the interface of two subunits. The structure of uridine phosphorylase was carefully analyzed, but further investigation of the enzyme is needed to explain its action mechanism.

Table 3. Growth conditions and crystals of uridine phosphorylase.

N	Growing and crystal characteristics	Earth	Macrogravity (centrifuge)	Microgravity (space)
1	Initial protein concentration	10-12 mg/ml	0.04-0.2 mg/ml	8.5 mg/ml
2	Crystallization time	4-6 weeks	50-70 hours	8 weeks
3	Crystal forms	cubes, plates sticks, pyramids tetrahedrons	crystalline fragments	cubes, plates sticks, pyramids, tetrahedrons
4	Crystal dimensions	0.4x0.2x0.3 mm	1.5x0.6x0.3 mm	0.2x0.2x0.2 mm
5	Quality of crystals	high mosaicity, diffract to 4.5Å, growing together	quite perfect, absence of faceting, diffract to 2.5Å, low living time in the X-ray beam	high quality, more isometric, well shaped, diffract to 1.5-1.7Å
6	Reproduction of the experiment	partial	high	unpredictable conditions of crystals delivery

Figure 7. The Uph molecule solved by the molecular replacement method at 2.5 Å resolution.

REFERENCES

1. A. McPherson. "Preparation and Analysis of Protein Crystals," J. Wiley and Sons, New York, p. 372 (1982)
2. S.Ya. Karpuhina, V.V. Barynin, and G.M. Lobanova , Crystallization of catalase in ultracentrifuge, *Kristallographia* 20:680 (1975).
3. V.V. Barynin and V.R. Melik-Adamyan, Mechanism of protein crystallization in ultracentrifuge, *Kristallographia* 27:981 (1982).
4. E.Yu. Morgunova, A.M. Mikhailov, A.A. Komissarov, Ch. Mao, E.V. Linkova, A.S. Mironov, A.N. Popov, S. Armstrong, A.A. Burlakova, D.V. Romanova, E.V. Blagova, E.A. Smirnova, V.G. Debabov, and S.E. Ealick, Structure-functional aspect of the structural investigation of Uph from E. coli., *Kristallographia* 41:672 (1995)
5. E.Yu. Morgunova, A.M. Mikhailov, A.N. Popov, E.V. Blagova, E.A. Smirnova, B.K. Vainshtein, *et al.,* Atomic structure at 2.5 Å resolution of uridine phosphorylase from E. coli as refined in monoclinic crystal lattice, *FEBS Letters* 367:183 (1995)

CRYSTAL GROWTH OF ENERGETIC MATERIALS
DURING HIGH ACCELERATION

M.Y.D. Lanzerotti,[1] J. Autera,[1] L. Borne,[2] and J. Sharma[3]

[1]U.S. ARMY ARDEC
Picatinny Arsenal, NJ 07806 5000
[2]French-German Research Institute of Saint-Louis (ISL)
France
[3]Naval Surface Warfare Center
Carderock Dividions
Silver Spring, MD 20903

ABSTRACT

Studies of the growth of crystals of energetic materials under conditions of high acceleration in an ultracentrifuge are reported. When a saturated solution is accelerated in an ultracentrifuge, the solute molecules move individually through the solvent molecules to form a crystal at the outer edge of the tube if the solute is more dense than the solvent. Since there is no evaporation or temperature variation, convection currents caused by simultaneous movement of solvent and solute are minimized and crystal defects are potentially minimized. Crystal growth is controlled by the g-level of the acceleration. We present results of TNT, RDX, and TNAZ grown at high g from various solutions. Crystals grown at high g were free of solution inclusions and gas bubbles.

INTRODUCTION

Crystal growth from a solution can be considered a heterogeneous chemical reaction of the type in which a portion of the liquid goes into crystal form.[1,2] In the laboratory at 1 g, crystal growth methods include solvent evaporation at constant temperature and slow cooling. Crystal growth occurs when the solution becomes supersaturated. The crystal growth is controlled by simultaneous movement of solute and solvent in convection currents.

Supersaturation can also occur in an initially saturated solution during high acceleration.[3-6] If the solution is initially saturated, then under acceleration the solution at the outer edge of the accelerating tube becomes supersaturated. A density gradient is established. Thus at high g (above 1000 g), the crystal growth mechanism is different. The solute molecules individually move through the solvent molecules to form a crystal. Crystal growth is controlled by the acceleration. In this new method, crystal defects caused by temperature variation or evaporation are potentially minimized. Two international conferences on crystal growth at high g addressed these issues.[7,8] although only a few of the reports were for accelerations above 1000 g.[4,9]

We believe that voids and solution inclusions are less likely to form in a crystal grown under high acceleration. Since the mechanical sensitivity of an explosive is significantly influenced by defects in the crystal,[10-14] this feature is important for numerous applications utilizing energetic materials[15-20] The long term objective of this program[6,9,21] is to understand the fundamentals of the crystal growth process and thereby to reduce the formation of defects in crystals of energetic materials so that they will be less sensitive to mechanical shock.

TECHNIQUE

The experiments on crystal growth were performed using saturated solutions. The solutions were filtered prior to insertion into the centrifuge tube in order to remove seed crystals. A Beckman preparative ultracentrifuge model L8-80 with a swinging bucket rotor model SW 60-Ti accelerated the saturated solution up to 500,000 g. Polyallomer centrifuge tubes with hemispherical ends were used. After an experimental run, the centrifuge tube with saturated solution sample was removed from the bucket and the saturated solution was poured off if a crystal had formed. If necessary, the polyallomer tube was cut lengthwise with a razor blade to remove the crystals so we study their physical features and habit without damaging them.

RESULTS

Polycrystalline materials were found on the curved interior surface of the polyallomer centrifuge tube. In the experiments performed to date, this curved surface appeared to inhibit single crystal formation. A hemispherical Teflon insert with a flat surface interfacing with the saturated solution was inserted into the tube to provide a flat surface that yielded single crystal growth.

A number of experimental runs were made on TNT (trinitrotoluene), RDX (cyclotrimethylene - trinitramine), and TNAZ (1,3,3-trinitroazetidine). These runs were performed for various values of temperature, time and acceleration. The results are shown in Table 1 and Table 2 for TNT and RDX, respectively. The results are shown in Tables 3 through 5 for TNAZ.*

The results of Table 1 show that 2-5-mm size TNT crystals were grown from TNT-saturated ethyl acetate solutions at 50,000 g at approximately 24 x 10^6 Pa and 25°C for 16 hours. The pressure at the growth surface depends on the density and the height of the saturated solution and the acceleration. The density of the TNT saturated ethyl acetate solution (~ 1.44 g/cc) was estimated from the solubility of TNT in ethyl acetate (59.8 g/100 g ethyl acetate at 21°C,[23] the density of TNT (1.65 g/cc),[24] and the density of ethyl acetate (0.9 g/cc).[25] Acceleration at 50,000 g at 25°C for 92 hours resulted in polycrystalline TNT. The crystal structure of the 5-mm size TNT crystal was determined to be monoclinic by x-ray diffraction.[23,26]

Pressure in psi is found by dividing Pascals by 6890.

Table 1. TNT crystal growth experiments at high g at 25°C using saturated ethyl acetate solutions.

Acceleration (x 10^3 g)	Pressure (x 10^6 Pa)	Growth surface	Time (hr)	Solution Filtered?	Results
13	6	Curved	16	No	No crystals
29	13	Flat	64	Yes	No crystals
50	24	Curved	17	No	Polycrystalline
50	24	Flat	15	Yes	Individual crystals aligned parallel to acceleration, 2 mm long, coffin-like habit.[22]
50	24	Flat	92	Yes	Polycrystalline
50	24	Flat	16	Yes	One crystal aligned perpendicular to acceleration, 5 mm long, coffin habit.[22]
50	24	Flat	21	No	Many individual crystals, 2 mm long, coffin habit.[22]

Table 2. RDX crystal growth experiments at high g at 25°C using saturated acetone solutions.

Acceleration (x 10^3 g)	Pressure (x 10^6 Pa)	Growth surface	Time (hr)	Solution Filtered?	Results
50	14	Curved	17	No	No crystals.
200	56	Flat	17	No	Individual crystals, 2 mm long, orthorhombic.[27]
200	56	Flat	23	No	Individual crystals, 3 mm long, orthorhombic.[27]
200	56	Flat	118	No	Polycrystalline

The results of Table 2 show that 2-mm size RDX crystals were grown from RDX-saturated acetone solutions at 200,000 g at approximately 56 x 10^6 Pa and 25°C for 17 hours. The density of the RDX-saturated acetone solution (0.85 g/cc) was estimated from the solubility of RDX in acetone (7.3 g RDX/100 g acetone at 20°C),[23] the density of RDX (1.81 g/cc),[24] and the density of acetone (0.79 g/cc).[25]

The RDX crystals grown at 200,000 g and at 1 g were studied in liquids of matching refractive index to reveal gas bubbles, particles and solution inclusions inside the crystals. Figure 1 shows optical micrographs of RDX crystals grown at 200,000 g and at 1 g, using 1-bromonaphthalene (R.I. = 1.600). The initial growth surface of the RDX crystal grown at 200,000 g is indicated by the dark spots across the lower edge of the crystal in Figure 1. Apparently insoluble impurities present in the saturated solution reached the bottom of the centrifuge tube before sufficient solute molecules reached the growth surface to begin crystal growth. A comparison of the two crystals shows clearly that individual RDX crystals grown at 200,000 g had far fewer voids than did crystals grown at 1 g.

Table 3 shows that 2-mm x 3-mm TNAZ[28] crystals were grown from TNAZ-saturated acetone solutions at 50,000 g at approximately 20 x 10^6 Pa during a 19- hour run. The density of the TNAZ saturated acetone solution (1.23 g/cc) was estimated from the solubility of TNAZ in acetone (0.44 g/cc),[29] the density of TNAZ (1.84 g/cc)[30] and the density of acetone.[25]

Table 4 shows that TNAZ crystals up to 2 mm x 5 mm in size grew at 50,000 g at approximately 15 x 10^6 Pa during a 64 hour run. Table 4 also shows that 2 mm x 6 mm TNAZ crystals grew at 200,000 g at approximately 59 x 10^6 Pa during 19 hours. The density of TNAZ-saturated methyl alcohol solution (0.88 g/cc) was estimated from the solubility of TNAZ in methyl alcohol (0.09 g/cc), the density of TNAZ, and the density of methyl alcohol (0.79 g/cc).[26]

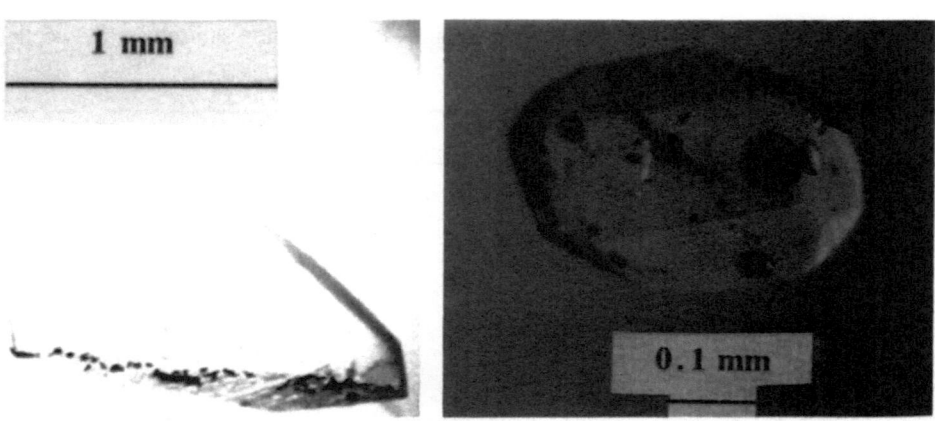

Figure 1. Optical micrographs of RDX crystals grown at 200,000 g (left) and at 1 g (right).

Table 3. TNAZ crystal growth experiments at high g on a flat growth surface and 25°C in saturated acetone solutions.

Acceleration (x 10^3 g)	Pressure (x 10^6 Pa)	Time (hr)	Solution Filtered?	Results
13	5	15	Yes	No crystals
50	20	19	Yes	Crystals, 2 mm x 3 mm
200	82	17	No	Polycrystalline

Table 4. TNAZ crystal growth experiments at high g on a flat growth surface at 25°C in saturated methyl alcohol solutions.

Acceleration (x 10^3 g)	Pressure (x 10^6 Pa)	Time (hr)	Solution Filtered?	Results
50	15	64	No	Crystals, 2 mm x 5 mm
200	59	93	No	Polycrystalline
200	59	19	No	Crystals, 2mm x 6 mm

Table 5 shows that 0.5 mm x 3 mm TNAZ needles grew at 200,000 g at approximately 58 x 10^6 Pa during 23 hours. Table 5 also shows that 2 mm x 6 mm and 1 mm x 4 mm TNAZ crystals grew at 50,000 g at approximately 14 x 10^6 Pa during 96 hours and 64 hours, respectively. The density of the TNAZ-saturated ethyl alcohol solution (0.85 g/cc) was estimated from the solubility of TNAZ in ethyl alcohol (0.06 g/cc)[29] the density of TNAZ, and the density of ethyl alcohol (0.79 g/cc).[26]

Table 5. TNAZ crystal growth experiments at high g on a flat growth surface at 25°C in saturated ethyl alcohol solutions.

Acceleration ($\times 10^3$ g)	Pressure ($\times 10^6$ Pa)	Time (hr)	Solution Filtered?	Results
200	58	23	No	Needles, 0.5 mm x 3 mm
50	14	96	No	Crystal, 2 mm x 6 mm
50	14	64	No	Crystal, 1 mm x 4 mm

DISCUSSION

The objectives of this investigation are to understand the fundamental chemistry and physics of crystal growth during high acceleration and to make explosives more insensitive to mechanical shock by reducing the formation of defects in the crystals.[16-21] For the first time TNT, RDX, and TNAZ crystals have been grown by this new method, in which crystal growth is controlled by acceleration in an ultracentrifuge. TNT crystals as large as 5 mm in size were grown at 50,000 g from saturated ethyl acetate solutions. RDX crystals as large as 3 mm in size were grown at 200,000 g from saturated acetone solutions. TNAZ crystals as large as 3 mm in size were grown at 50,000 g from saturated acetone solutions. TNAZ crystals as large as 6 mm in size were grown at 200,000 g from saturated methyl alcohol solutions. TNAZ crystals as large as 6 mm in size were grown at 50,000 g from saturated ethyl alcohol solutions.

The crystals grown at high g were compared with crystals grown at 1 g in the laboratory by slow evaporation. Under optical microscopy, the crystals grown at high g appeared to be free of the voids and the solution inclusions that are found in crystals grown at 1 g. The RDX crystals grown at high g were observed in a fluid of matching refractive index to reveal defects and solution inclusions inside the crystals. Individual RDX crystals grown at 200,000 g had far fewer voids than RDX crystals grown at 1 g.

Since this is the first investigation to explore the possibility of producing improved crystals of energetic materials, a wide range of parameter space was explored in these initial experiments. A wide range of acceleration was used. It was found that no crystals were obtained from TNT-saturated ethyl acetate solutions accelerated at 13,000 g for 16 hours or 29,000 g for 64 hours. When the acceleration was increased to 50,000 g, individual crystals were obtained. These results show that the acceleration must be sufficient for a sufficient number of molecules to reach the end of the tube during the run to form a crystal. On the other hand, polycrystalline material formed when the duration at 50,000 g was increased to 92 hours. Unfiltered solutions at approximately the same conditions producing a few crystals yielded many small individual crystals. Suspended particles may have provided nuclei.

No crystals were obtained from a RDX-saturated acetone solution at 50,000 g for 17 hours. When the acceleration was increased approximately four times to 200,000 g individual crystals were obtained. Polycrystalline material formed when the duration at 200,000 g was increased to 118 hours.

No crystals were obtained from a TNAZ-saturated acetone solution accelerated at 13,000 g. When the acceleration was increased approximately four times to 50,000 g, crystals are obtained. Polycrystalline material resulted when the acceleration was increased an additional four times to 200,000 g. The run times were approximately the same for each of these three experiments.

Solubility and solvent molecular structure can also influence crystal growth. TNAZ is approximately equally soluble in methyl alcohol and ethyl alcohol. TNAZ crystals grew at 200,000 g from methyl alcohol, whereas only very small TNAZ needles grew at 200,000 g from ethyl alcohol. The run times were approximately the same for these two experiments. The difference in the molecular structure of methyl alcohol and ethyl alcohol may have been the reason. For example, one would expect TNAZ molecules to sediment faster in methyl alcohol molecules than in ethyl alcohol, because of the differences in solvent density and molecular weight.

This initial investigation has only begun to elucidate the crystal physics and chemistry of TNT, RDX, and TNAZ. Further work in these directions will provide more understanding and will likely produce important results that can be used in applications for energetic materials.

Acknowledgments

We thank Drs. J. Lannon, R. Surapaneni, C. Choi, S. Iyer, Messrs. B. Travers, and M Joyce, all at U. S. Army ARDEC, Mr. W. Lukasavage at GEO-CENTERS, Inc., and Dr. S. A. Mogren of Columbia, MD for their helpful comments.

REFERENCES

1. R.A. Laudise, "The Growth of Single Crystals," Prentice-Hall, Inc., Englewood Cliffs (1970) p. 39.
2. A. Holden and P. Singer, "Crystals and Crystal Growing," Doubleday & Company, Inc., New York, (1960).
3. P.J. Shlichta and R.E. Knox, Growth of crystals by centrifugation, *J. Crystal Growth* 3:808 (1968).
4. P.J. Shlichta, Crystal growth and materials processing above 1000 g, *J. Crystal Growth* 119:1 (1992).
5. I. Amato, The high side of gravity, *Science* 253:30 (1991).
6. M.Y.D. Lanzerotti, J. Autera, J. Pinto, and J. Sharma, Crystal growth of energetic materials during high acceleration using an ultracentrifuge, *in*: "High Pressure Science and Technology - 1993," S.C. Schmidt, J.W. Shaner, G.A. Samara, and M. Ross, eds. , American Institute of Physics, New York, NY (1994) pp. 489-491.
7. L.L. Regel, M. Rodot and W.R. Wilcox, eds. *J. Crystal Growth* 119:1-175 (1992).
8. L.L. Regel and W.R. Wilcox, eds., "Materials Processing In High Gravity," Plenum Press, New York (1994).
9. M.Y.D. Lanzerotti, J. Autera, J. Pinto, and J. Sharma, Crystal growth of energetic materials during high acceleration using an ultracentrifuge, *ibid*, pp. 181-184.
10. F. Baillou, J.M. Dartyge, C. Spyckerelle, and J. Mala, Influence of crystal defects on sensitivity of explosives, *in*: "Proceedings of the Tenth Symposium (International) on Detonation," (1993) pp. 816-823.
11. L. Borne, Influence of intragranular cavities of RDX particle batches on the sensitivity of cast wzx bonded explosives, *ibid*, pp. 286-293.
12. A. Van Der Steen, H.J. Verbeek, and J.J. Meulenbrugge, Influence of RDX crystal shape on the shock sensitivity of PBXs, *in*: "Proceedings of the Ninth Symposium (International) on Detonation," (1989) pp. 83-88.
13. I.B. Mishra and L.J. Van de Kieft, Novel approach to insensitive explosives, *in*: "Proceedings of the 19th International Annual Conference of ICT," Karlsruhe (1988) pp. 25-1 to 25-21.
14. L. Borne, Microstructure effects on the shock sensitivity of cast plastic bonded explosives, *in*: "Proceedings of Europyro 95, 6th Congress Internationale de Pyro," Tours-France (1995) pp. 125-131.
15. J.J. Dick, Plane shock initiation of detonation in gamma-irradiated pentaerythritol tetranitrate, *J. App. Phys.* 53:6161 (1982).
16. J.J. Dick, Effect of crystal orientation on shock initiation sensitivity of pentaerythritol tetranitrate explosive *J. App. Phys. Lett.* 44:859 (1984).

17. J.J. Dick, POP plot and Arrhenius parameters for <110> pentaerythritol tetranitrate single crystals, *in*: "Shock Waves In Condensed Matter," Y.M. Gupta, ed., Plenum Press, NY (1986) pp. 903-907.

18. J.J. Dick, R.N. Mulford, W.J. Spencer, D.R. Petit, E. Garcia, and D.C. Shaw, Shock response of pentaerythritol tetranitrate single crystals, *J. App. Phys.* 70:3572 (1991).

19. J.J. Dick, E. Garcia, and D.C. Shaw, Shock initiation of pentaerythritol tetranitrate crystals: steric effects due to plastic flow, *in*: "Shock Compression of Condensed Matter - 1991," S.C. Schmidt, R.D. Dick, J.W. Forbes, D.G. Tasker, eds., Elsevier Science Publishers, The Netherlands (1992) pp. 349-352.

20. J.J. Dick, E. Garcia, and D.C. Shaw, Crystal orientation dependence of elastic precursor strength in pentaerythritol tetranitrate, *in:* "Shock Compression Of Condensed Matter - 1993," S.C. Schmidt, J.W. Shaner, G.A. Samara, and M. Ross, eds., American Institute of Physics, NY (1994) pp. 1373-1376.

21. M.Y.D. Lanzerotti, J. Autera, J. Pinto, and J. Sharma, Crystal growth of explosives during high acceleration using an ultracentrifuge, *in:* "Army Science Conference Proceedings," Vol. I (1994) pp. 69-75.

22. H.G. Gallagher and J.N. Sherwood, The growth and perfection of single crystals of TNT, *in*: "Structure and Properties of Energetic Materials," D.H. Liebenberg, R.W. Armstrong, and J.J. Gilman, eds., Material Research Society Proceedings 296, Pittsburgh, PA 1993) pp. 215-219.

23. S. Morrow, U. S. Army ARDEC (1989).

24. B.M. Dobratz and P.C. Crawford, "Properties of Chemical Explosives and Explosive Simulants," UCRL-52997, Lawrence Livermore National Laboratory, University of California, Livermore (1985).

25. R.C. Weast, "Handbook of Chemistry and Physics," CRC Press, Cleveland, OH (1975-1976).

26. S.M. Kaye, "Encyclopedia of Explosives and Related Items," Picatinny Arsenal Technical Report 2700, 9:T263 (1980).

27. J.T. Rogers, "Physical and Chemical Properties of TDX and HMX," Control No. 20-P-26 (1962).

28. T.G. Archibald, R. Gilardi, K. Baum, and C. George, Synthesis and X-ray crystal structure of 1,1,1-trinitroazetidine, *J. Org. Chem.* 55:29920 (1990).

29. D. Stec, Geo-Centers, Inc. (1992).

30. S. Iyer, S. Eng, M. Joyce, R. Perez, J. Alster, and D. Stec, Scaled-up preparation of 1,3,3-trinitroazetidine (TNAZ), *in:* "Proceedings of the Joint International Symposium on Compatibility of Plastics and Other Materials with Explosives, Propellants, Pyrotechnics and Processing of Explosives, Propellants, and Ingredients," (1991) pp. 80-84.

CENTRIFUGAL DIAMOND FILM PROCESSING

Liya L. Regel,[1] Yoshiki Takagi,[2] and William R. Wilcox[1]

[1]International Center for Gravity Materials Science and Applications
Clarkson University
Potsdam, NY 13699-5814
[2]Teikyo University of Science and Technology
Yamanashi Prefecture, Japan

ABSTRACT

Three diamond deposition techniques were tested for their potential for centrifuge studies. It was decided to use a sealed chamber containing hydrogen gas in which a graphite rod is heated by passing electric current through it. Diamond particles grow on a nearby silicon substrate. The apparatus was constructed at Teikyo University of Science and Technology and mounted on the centrifuge at Clarkson University. Early experiments show that centrifugation significantly influences the deposition pattern and growth rate.

INTRODUCTION

Diamond films have a unique combination of physical and electronic properties. Many studies are currently underway to investigate the effect of various processing parameters on the quality of the resulting diamond films. No study, however, has so far considered the possible influence of centrifugation on the quality and the growth rate of diamond films by chemical vapor deposition (CVD). It is known that buoyancy-driven thermal convection influences diamond synthesis using CVD methods. Since thermal convection is affected by centrifugation, CVD and the resulting diamond films should be altered at high gravity. We hope to be able to synthesize diamond films with fewer defects or unique structures, as well as to obtain new information about growth mechanisms and the relationship between growth conditions and properties.

Diamond has superior properties, for example, high hardness, high thermal conductivity, high transparency, high electron and hole mobilities, and a wide bandgap. These unique properties suggest various potential applications, such as high-temperature semiconductor devices, high-power microwave devices, hard coatings, heat sinks, and blue wavelength optoelectronics. To synthesize diamond semiconductor devices, we have to be able to grow single crystal diamond films. The epitaxial growth of diamond films on diamond[1] and c-BN[2] substrates has been reported. But on other substrates, for example, Si and Mo, only particles or polycrystalline films have been obtained thus far.

The usual techniques for depositing diamond at low pressure utilize a flow of a hydrocarbon gas combined with a method for generating atomic hydrogen, such as a plasma, hot tungsten filament, or flame. A flow system is not considered practical for mounting on a centrifuge. Similarly, a flow system is impractical for use in space, and preliminary work had been performed on development of a closed system for that application.[3] Here, we report on successful development of a closed system suitable for mounting on large centrifuges. This system could also find application in microgravity experiments, for a thorough investigation of the influence of convection on diamond deposition.

EXPERIMENTS

As a preliminary study for centrifuge experiments, we synthesized diamond using three different carbon sources. As a gaseous carbon source we used vaporized ethyl alcohol; as a liquid source, methyl alcohol; and as a solid source, graphite.

Gaseous Carbon Source

Diamond particles were deposited on Si substrates using a hot-filament CVD method.[4] We were able to control the deposition rate with the DC-bias electric field between the tungsten filament and the Si substrate (this method is usually called EACVD; electron assisted chemical vapor deposition[5]). The apparatus is shown in Figure 1. Ethyl alcohol was vaporized by bubbling hydrogen through it. This gas mixture was introduced into the reaction chamber. Exhaust gas flowed out through outlet tubing in this flow system. We also tried a closed system. Once the gas mixture was introduced into the chamber, the inlet and outlet valves were closed.

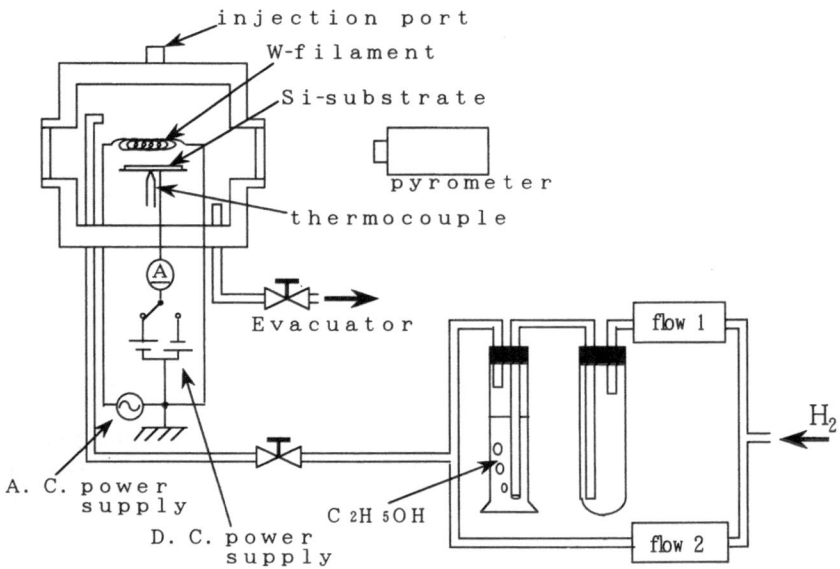

Figure 1. Schematic diagram of EACVD apparatus.

The experimental conditions of the flow system are listed in Table 1.

Table 1. Experimental conditions
used with the flow system.

Filament temperature (°C)*	2550 - 2700
Substrate temperature (°C)	530 - 600
Distance between filament	
and substrate (mm)	1.0 - 2.0
Flow rate 1 (SCCM)	10
Flow rate 2 (SCCM)	40
Reaction pressure (Torr)	730 - 740
Reaction time (min)	30
Bias voltage (V)	0, +100, +150
Bias current (A)**	0.01

* The filament temperature was controlled so
as to keep the substrate temperature constant.
** The bias current listed is the initial setting
of the DC power supply.

Liquid Carbon Source

In this method, methyl alcohol was used alone for diamond synthesis, without hydrogen gas. Methyl alcohol was heated with a tungsten filament, and presumably decomposed to carbon monoxide and hydrogen. Diamond particles formed on a Si substrate. No other gas was introduced to the reaction chamber, while reactant gases flowed out from the chamber. This method might be called "quasi-closed." On the other hand, since carbon monoxide and hydrogen gas were continuously supplied from liquid methyl alcohol, so from another point of view, this method could be called "quasi-flow." Gas chromatographic analysis showed that the fractions of gaseous species, for example, carbon monoxide, methane, carbon dioxide, acetylene and ethylene, were very stable.

Solid Carbon Source

Spitsyn[6] reported chemical crystallization of diamond from an activated vapor phase using a chemical transport reaction (CTR) in a graphite-hydrogen-diamond sandwich system. We synthesized diamond on Si substrates using CTR with graphite as the sole carbon source. A graphite rod 1 mm in diameter and 20 mm in length was inserted into a tungsten filament in a regular hot filament CVD chamber. Hydrogen gas was introduced to 10 to 50 Torr. Then the graphite rod was heated with the tungsten filament. Diamond particles grew on a nearby Si substrate.

In this new method, no source gas is introduced and no reactant gas is evacuated. So, this method is completely closed.

RESULTS AND DISCUSSION

EACVD Results

With EACVD, in a flow system, the diamond growth rate was enhanced by increasing the bias voltage. The scanning electron micrographs in Figure 2 show the influence of bias voltage. But at the same time, deposition of amorphous carbon was also enhanced.

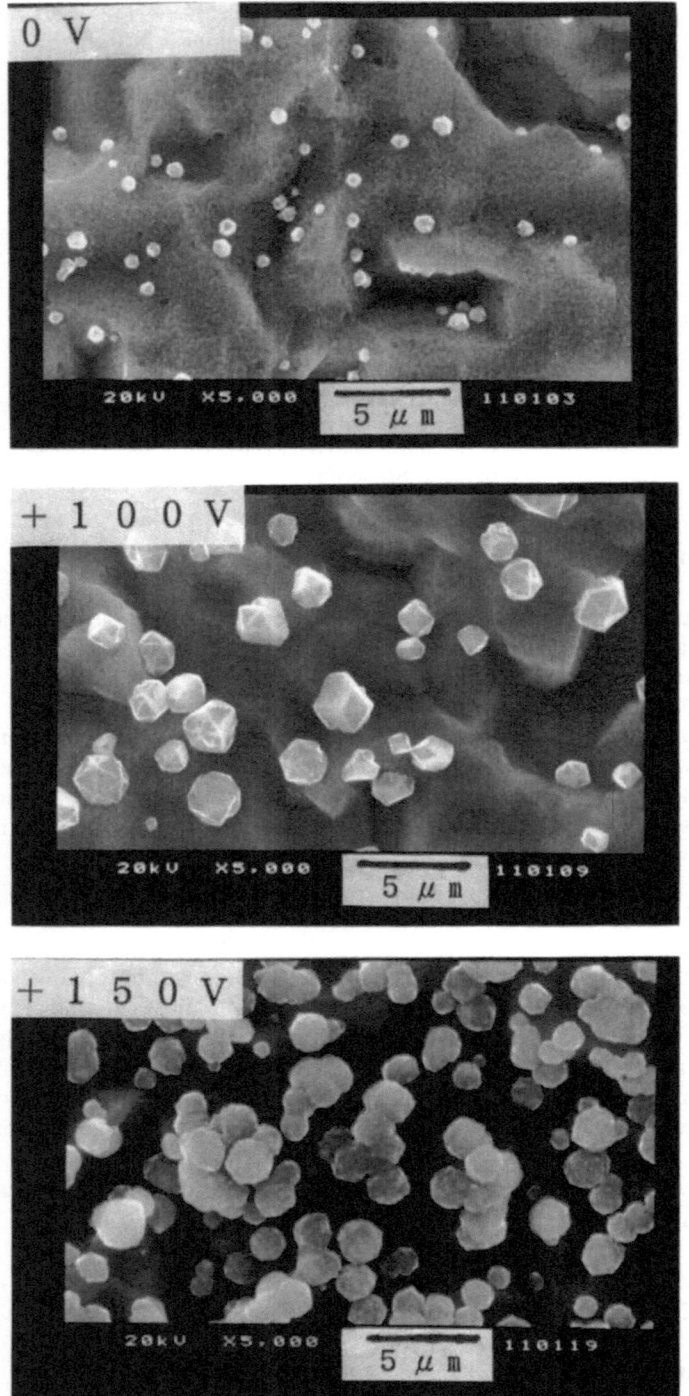

Figure 2. Scanning electron micrographs of diamond crystals resulting from EACVD experiments. From top to bottom, the bias voltage was 0, 100 and 150V.

In a closed system, with normal bias voltage applied, the diamond growth rate was enhanced. With an inverse bias voltage, diamond particles also were produced.

Liquid Carbon Source Results

With this method, the presence of diamond particles was confirmed by Raman and SEM. One hour reaction time was not enough to form a continuous film, and longer reaction times lowered the growth rate.

Solid Carbon Source Results

Diamond was synthesized, and confirmed with Raman spectroscopy (Figure 3). X-ray diffraction analysis suggested tungsten contamination from the filament.

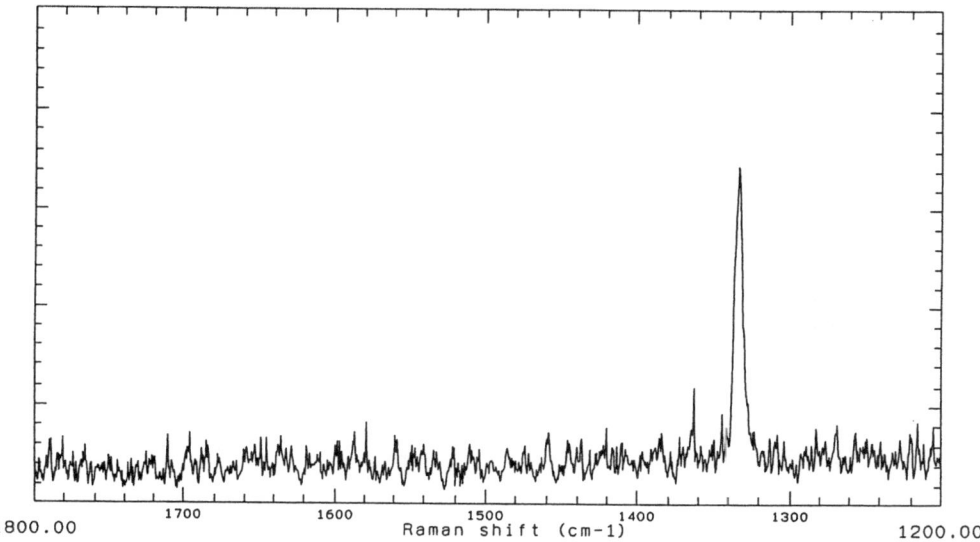

Figure 3. Raman spectra for diamond synthesized from graphite.

CENTRIFUGE EXPERIMENTS

From among the three different carbon sources, we selected the solid carbon source for high gravity experiments. The reasons for that choice are as follows:
1) With a graphite source, no source gas is introduced into the reaction chamber and no reactant gas comes from the chamber. We do not need gas inlet-outlet tubing, gas cylinders, liquid sources, or flow control systems mounted on the centrifuge.
2) We can control and monitor the entire reaction system with electrical contacts, i.e. slip rings, on the centrifuge.
3) Hydrogen is the only gas charged into the chamber, prior to sealing the chamber.
4) The graphite rod is the only heat source. (We have successfully dispensed with the tungsten filament.) Since resistance heating is used, the power supply and its control system are simple.
5) Theoretical analysis should be simplified, since there are no electrical discharges, plasmas or high-speed flows. The chemical species should be near their equilibrium concentrations.

We are now synthesizing diamond in an apparatus mounted on the centrifuge. As shown in Figure 4, faceted diamond particles are deposited on silicon substrates held a few mm away from a Joule-heated graphite rod in a hydrogen atmosphere. Centrifugation can increase the growth rate, the nucleation density, and the area of deposition.

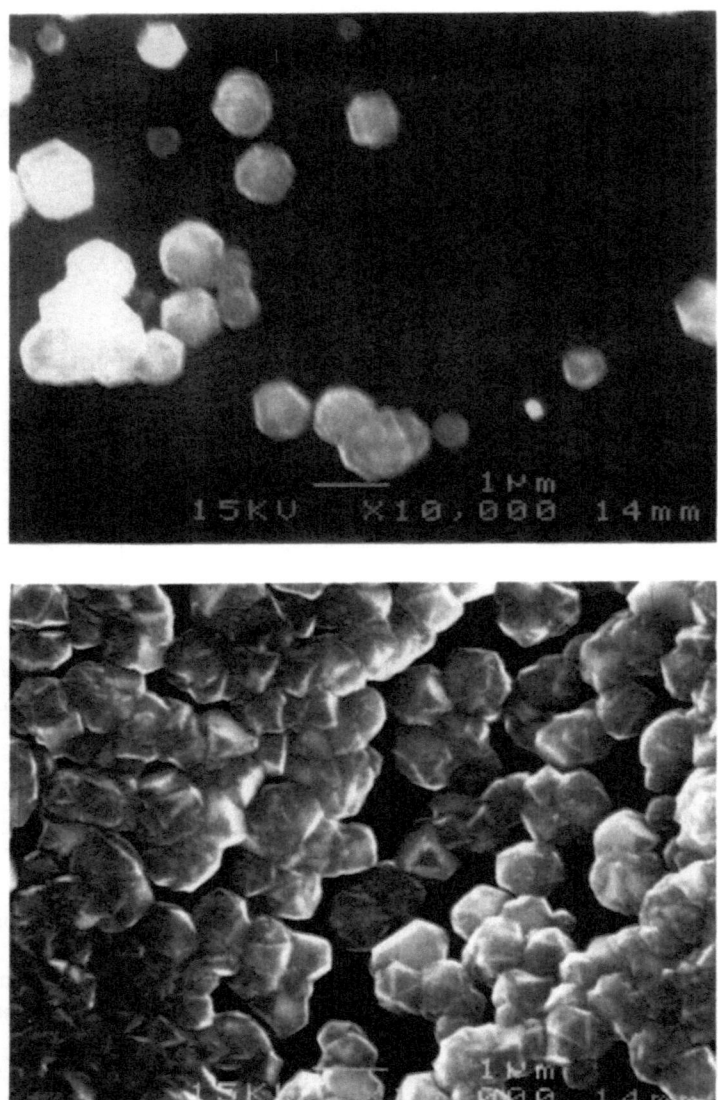

Figure 4. Scanning electron micrographs of diamond particles deposited with 30 Torr hydrogen pressure. Top: without centrifugation. Bottom: with centrifugation at 2.5g.

Acknowledgments

This research is supported by the National Science Foundation under grant DMR-9414304. We are grateful to Peter Skudarnov and Yu Yoshizaki for their assistance.

REFERENCES

1. H. Shiomi, K. Takabe, Y. Nishibayashi, and N.Fujimori, *Jpn. J. Appl. Phys.* 28:34 (1990).
2. S. Koizumi, T. Murakami, T. Inuzuka, and K. Suzuki, *Appl. Phys. Lett.* 57:563 (1990).
3. Y. Takagi, S. Sato, K. Kaigawa, A.B. Sawaoka, and L.L. Regel, *Micrograv. Q.* 2:39 (1992).
4. S. Matsumoto, Y. Sato, M. Kamo, and N. Setaka, *Jpn. J. Appl. Phys.* 21:L183 (1982).
5. Q. Lichang, X. Zhenwu, Y. Peichun, P. Xin, and H.Li, *J. Cryst. Growth* 112:580 (1991).
6. B.V. Spitsyn, The state of the art in studies of diamond synthesis from the gaseous phase and some unsolved problems, *in:* "Applications of diamond films and related materials," Y. Tzeng, M. Yoshikawa, M. Murakami, and A. Feldman, eds., Elsevier, Amsterdam (1991).

THIN FILM PRODUCTION UNDER ELEVATED GRAVITY CONDITIONS

Yoshiyuki Abe,[1] Giovanni Maizza,[1] Hervé Rouch,[1] Noboru Sone,[2] and Yuji Nagasaka[2]

[1]Electrotechnical Laboratory, Tsukuba 305, Japan
[2]Keio University, Yokohama 223, Japan

ABSTRACT

The effect of elevated gravity has been reported for crystallization of various semiconductor materials and alloys. The results have shown many unique properties never attained under normal gravity or microgravity conditions. Although no attempt has, so far, been made to produce thin films under elevated gravity, it is expected that a higher gravity condition would yield a higher nucleation site density, which should increase the deposition rate. The grain size and therefore crystal morphology should also be under the strong influence of gravity, so that the thermal, physical, electrical and mechanical properties of the thin films produced under elevated gravity condition may show unexpected characteristics suitable for practical applications.

From this standpoint, we plan to produce thin films at moderate pressures under elevated gravity conditions at the Electrotechnical Laboratory. A centrifuge facility for chemical vapour deposition (CVD) up to 100 g was designed and constructed. For the first step of a demonstrative study with the facility, we plan diamond thin film production by DC-plasma CVD.

The present paper is divided into two parts. The first part briefly reviews the main features of DC-plasma CVD processes and examines the potential advantages in using DC-plasma CVD in conjunction with high gravity. The second part introduces the mathematical model that describes the macroscopic heat and mass transport phenomena with chemical reactions occurring in a DC-plasma CVD (HGCVD) reactor under high gravity. Numerical evaluations were accomplished aimed at investigating the fluid dynamic performances of several reactor configurations under normal gravity and high gravity conditions, with and without the influence of the plasma source. Selected results are presented that show the criteria for optimal hardware configurations of DC-plasma CVD reactor chambers under high gravity.

INTRODUCTION

External fields play important roles in many materials processes, such as a magnetic field in solidification, an electromagnetic field in sputtering, an electric field in condensation, electrostatic, electromagnetic and acoustic forces for levitation, etc. This is also the case with centrifugation -- high gravity has already been utilized for separation, casting and many other industrial processes for many years. In the era of space utilization, the microgravity environment has drawn particular attention from materials scientists. Recently, new attention has been focused on more sophisticated utilization of high gravity to produce unique and high quality materials, with morphology, purity and homogeneity that cannot be obtained in either a normal gravity environment or in microgravity.[1]

In contrast to the numerous experiments on crystal growth from the liquid state at high gravity, high gravity crystallization from the gas phase has been extremely limited, although one might expect the effect of gravity to be even more pronounced in the gas phase than in the liquid phase. Wiedemeier et al.[2] reported very unexpected experimental results for vapour transport of GeSe in a Xe atmosphere up to 10 g. In a destabilized configuration where strong natural convection exists, the mass flux at 10 g was one order of magnitude larger than at 1 g, and the size of the GeSe platelet crystals deposited at 10 g was more than three orders of magnitude larger than for those obtained at 1 g. Furthermore, the overall surface morphology and crystallographic quality of the larger GeSe platelets were rather good. We do not believe that these experimental results can be explained by the Coriolis effect, which has been shown to stabilize, in a particular geometrical configuration, oscillatory flows in melts.[3,4] The effect of gravity on the characteristics of those materials produced from gas phase is totally veiled, and gravity, as a new parameter of materials processing, may provide the possibility to produce a variety of advanced materials with new functions and characteristics which have never realized in present processes.

One of the present authors (YA) initiated a study on materials processing from the gas phase under elevated gravity. In particular, interest is focused on thin film production due to its industrial significance. Thin films -- metallic, ceramic, amorphous, polymer, etc. -- enjoy unique characteristics in physical, chemical, optical, thermal, magnetic, mechanical and electrical aspects, which have so far brought a large amount of industrial applications. The market for thin films is expected to further spread in the future.

It is not possible at present to state the influence of high gravity on the characteristics of thin films, but we may expect the following to be affected:
- nucleation site density,
- deposition rate,
- size of particles deposited,
- homogeneity,
- morphology,
- adhesiveness,
- compactness.

All of these items are important parameters in most current thin film production processes.

The production processes for thin film may be categorized into either physical vapour deposition (PVD) or chemical vapour deposition (CVD). The CVD method allows for a much wider variety of materials to be deposited on a large variety of substrates compared to the PVD method. The deposition pressure is normally higher than for PVD, which means that a more significant gravity effect should be expected in CVD processes.

Although a comprehensive work to try to produce many different thin films under elevated gravity conditions is beyond the scope of the present study, the purpose and the framework of the present study are as follows:
(1) To develop an apparatus for thin film production under elevated gravity.

(2) To demonstrate high gravity thin film production.

(3) To characterize thin films grown at high gravity and to identify the gravity effects on their characteristics and properties.

(4) To develop *in situ* process diagnostics methods in order to determine the influence of gravity on thin film growth processes.

As a first demonstration of thin film production in high gravity, we decided to make an attempt to produce diamond thin film by CVD. Diamond is regarded as an important material of interest for many industrial applications due to its features, *i.e.*:

- wide band gap,
- large breakdown field,
- short carrier lifetime,
- small dielectric constant,
- high electron velocity,
- small lattice constant,
- high refractive index,
- high chemical stability,
- high electrical resistivity,
- high thermal conductivity,
- low thermal expansion coefficient.

These unique characteristics yield a variety of potential applications for diamond, such as:

- high temperature electronics devices,
- heat sinks,
- coatings for machine tools,
- optics (coatings, windows),
- optical switches, etc.[5]

Production of diamond thin film via CVD has been developed in the last decade, and a huge number of papers have been published. The synthesis of diamond film by CVD is believed to occur in the presence of a precursor gas, usually a few percent of methane diluted in hydrogen, in combination with the achievement of an extreme non-equilibrium condition on the substrate surface. Normally a suitable activation technique is used to fragment the precursor in order to produce high concentrations of active species such as CH_3, C_2H_2 and atomic H. Such atoms and free radicals produced from hydrocarbon decomposition are supersaturated and highly reactive at the non-equilibrium deposition surface, thus contributing to the deposition of diamond films. However, the role of atomic hydrogen in the deposition process is not yet clear. Practical evidence suggests that it enhances the rate of diamond growth and reduces the undesirable graphite co-deposition.

Table 1 lists and compares typical CVD methods currently employed for the production of diamond thin films.[6] From this comparison, we selected the DC-plasma CVD method as a potential activation technique to be used in conjunction with high gravity for diamond thin film production. Following are the main reasons we selected the DC-plasma HGCVD (high gravity CVD) method:[7]

- moderately high operation pressure,
- unnecessary to use an artificial nucleation procedure (such as scratching, imposition of a bias voltage, etc.),
- rather simple experimental components,
- good reproducibility,
- relatively high deposition rate.

It is noteworthy that, under contract from the National Space Development Agency (NASDA), a Japanese research group produced a diamond thin film by means of the DC-plasma CVD method in microgravity on the Japanese

Table 1. Comparison of CVD methods for diamond film growth

Method	dep. rate μm/h	press. mbar	quality Raman	temperature K substr.	gas	advantage	drawback
COMBUSTION FLAME	30-100	1000	+++	700-1400	~3500	simple	area, stability
HOT FILAMENT	0.3-2	20-40	+++	1100-1300	~2300-2700	simple, large area	contaminations, stability
DC PLASMA (low pressure)	< 0.1	20-40	+	900-1300	~1500	simple, large area	rate, quality
DC PLASMA (moderate pressure)	20-250	200	+++	900-1200	~4000-6000	rate, quality	area
DC PLASMA JET (atmospheric press.)	930	0.1-1000	+++	1100-1400	>5000	highest rate, quality	area, stability homogeneity
RF PLASMA (low pres.)	< 0.1	0.1-40	+/-	1000-1300	~1500	scale-up	quality, rate contaminations
RF PLASMA (atmosph.)	180	1000	+++	1000-1500	>5000	rate	area, stability homogeneity
MW PLASMA (0.9- 2.45 GHz)	1 (<<p) 30 (>>p)	20-100 1000	+++	600-1300	~3000	quality, stability (reasonable rate & area)	rate, quality
MW PLASMA (ECR 2.45 GHz)	0.1	20-100 1000	+/-	600-1300	>5000	area (?), low P	quality, rate contaminations

un-manned space free flyer unit (SFU), which flew about six months in a low earth orbit. The experimental module was returned to the PIs in April of 1996, and analysis of the specimen is now underway. Before the space experiment, a series of parabolic aircraft flight tests were performed using He gas instead of CH_4-H_2, for safety reasons. They found changes in the substrate temperature, the shape of the plasma and the intensity of plasma emission spectrum during the low gravity period,[8] although the pressure was relatively low (2.7 kPa) due to the limited electric power available on the aircraft. This experimental evidence suggests a possible gravity effect on plasma characteristics in the DC-plasma CVD process. Comparison of the results from the space experiment with those from our future high gravity experiments will enable us to develop a comprehensive understanding of the role of gravity on plasma CVD.

EXPERIMENTAL

High Gravity Facility

The construction of the high gravity facility in the present study is different from the rotating arm normally employed in high gravity materials processing. As shown in figure 1, the present facility is a rotating drum structure, in which the CVD chamber and the counterweights are fixed. Although the fluctuation of total acceleration in the chamber is appreciable, especially for a low rotational speed, this rotating drum structure was preferred in the present study in order to guarantee:
  - relatively high gravity conditions (10's to 100 g),
  - gravity gradient,
  - more practical structure for large scale production under high gravity.
  A cut-away schematic view is depicted in figure 2. The maximum rotation speed is designed as 600 rpm, although 100 g acceleration at the midpoint of the anode and cathode in the CVD chamber is attained at about 500 rpm. The rotating drum has 1,060 mm inner

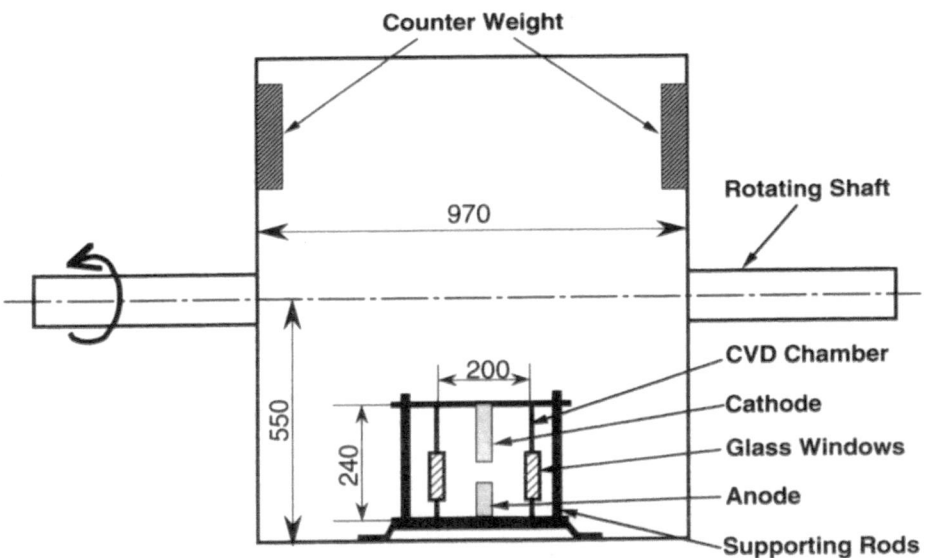

Figure 1. Photograph of high gravity CVD facility.

Figure 2. Structure inside the rotating drum.

diameter and 970 mm length. The total length of the high gravity facility is 4,220 mm. It is installed in a pit in the laboratory.

Four independent gas lines, one vacuum line of ½" and three feed-gas lines of ¼", are connected between the chamber and the outside of the rotating drum with the aid of a magnetic seal. The chamber could be evacuated to 10^{-3} Pa in a vacuum test at the maximum rotational speed. Twelve pairs of slip rings allow transmission of experimental data from inside the chamber to a data acquisition system, including four special leads for R-type (platinum and platinum/rhodium) thermocouples,. In addition, electric power through three independent lines can be supplied to the chamber via other slip rings. A cooling water line is also available.

CVD Chamber

Figure 3 shows the CVD chamber, which is made of 316 stainless steel. The chamber has two sight glass windows 140 mm in diameter, a gas inlet and an outlet, and several feed-throughs for electric power, thermocouples and the probe of a pressure transducer. The basic orientation of the chamber is as shown in the figure -- the direction of the acceleration vector is from the cathode to the anode. Other orientations, such as opposite or vertical configurations, also will be examined.

Before performing a CVD experiment, the inside of the chamber is evacuated to the order of 10^{-3} Pa with the aid of a rotary vacuum pump and a turbo molecular pump. When a high vacuum is established, a mixed gas of hydrogen and methane is fed into the chamber through the gas inlet at a constant flow rate under the control of a flow-control valve. The pressure in the chamber is also kept constant by adjusting the vacuum valve located between the chamber and the rotary pump. The pressure in the chamber is monitored with a diaphragm pressure gauge.

A molybdenum substrate is placed on the surface of the grounded molybdenum anode. A DC voltage, nearly 1 kV, is imposed at the molybdenum cathode. The distance between the cathode and the substrate is adjusted by a screw that allows cathode displacement in the axial direction. R-type sheathed thermocouples are embedded in the anode, and the substrate temperature is estimated by the temperature gradient in the anode. Other R-type sheathed thermocouples are located throughout the chamber to monitor the temperature profile. Since the substrate surface is spontaneously heated up by electron bombardment in the DC-plasma CVD process, no auxiliary heating device for the substrate is necessary. The surface temperature may be controlled either by the pressure in the chamber or by the DC voltage, and therefore no cooling water is introduced.

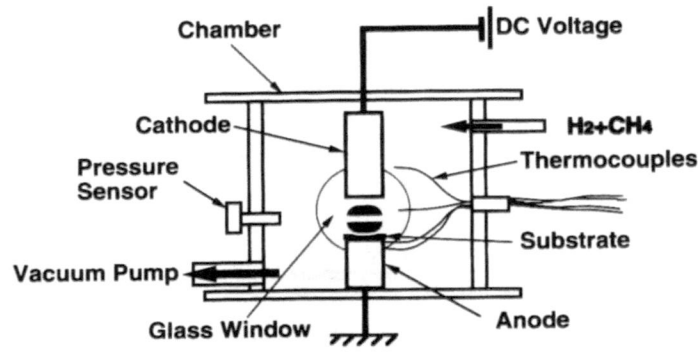

Figure 3. DC plasma CVD chamber.

Plasma-Gas Interactions in DC-plasma CVD

The purpose of the DC-glow discharge is to produce a fragmentation (pyrolysis) of the precursor (1-2% CH_4), which is normally diluted in a carrier gas (H_2). Very high fragmentation is desirable, since this assures the high concentration of active species that is important, although not critical, for diamond deposition.[9-10] The temperature of the gas in the plasma is directly related to the operating pressure of the CVD reactor. For a typical pressure range of 13 to 27 kPa, as assumed in our reactor, the gas temperature is roughly equal to that of the electrons and considerably higher than that of the substrate (≈ 1000 K). For this reason, these type of plasmas are usually called 'thermal plasmas,' since the local thermal equilibrium condition is normally satisfied. In a similar DC-plasma reactor, under the same operating conditions assumed here, Suzuki[7,11-13] estimated a gas temperature range between 4000 K (with a density equal to 0.5 A/cm²) and 6000 K (with a current density equal to 10 A/cm²). Such a large temperature difference, together with the relatively short distance between the plasma source and the substrate surface, ensures a large temperature gradient, which is considered essential to grow diamond films.[9,10,21,22] It is felt that in DC-plasma CVD, such a large temperature gradient is mainly responsible for the high diamond nucleation rate and growth rate.[16-19] The high nucleation rate is explained as a direct consequence of such a large temperature gradient, whereas the higher growth rate follows from the consequent manifestation of complex non-equilibrium thermodynamic and chemical processes near the surface. The most important of these processes being: a) the achievement of a quite severe metastability condition; and b) the occurrence of rapid 'quenching' of the chemical reactions, with the subsequent formation of a supersaturation region of active species (mainly atomic hydrogen) near the surface. Moreover, Suzuki observed that the plasma consisted of a relevant abundance of carbon and hydrogen atoms, indicating that most of the free radicals were completely dissociated at these temperatures.

According to such experimental observations,[7,11-13] we suggest a possible overall reaction for diamond growth in a DC-plasma CVD process:

$$C_{(g)} + H_{(g)} + H_{(s)} \rightarrow C_{(s)} + H_{2(g)} \tag{1}$$

This reaction is possible thanks to the relatively high reactivity of monatomic C. In addition, the identification of atomic carbon as the main growth species has also been proved workable in atmospheric-thermal plasma CVD[23,24] and confirmed by semi-empirical calculations.[25] However, several authors[22] claim that under metastable conditions diamond deposition could be controlled by kinetic factors, i.e. by the faster rate of diamond deposition compared to other competing processes such as the co-deposition of graphite. A common way to avoid the co-deposition of graphite in diamond CVD processes, is to ensure a suitable flux of atomic hydrogen to the surface.[21,22] A super-equilibrium concentration of atomic hydrogen near the surface is desired since it stabilizes the deposited diamond film, avoids the co-deposition of graphite, and enhances the rate of diamond growth.[9,10,21] Below we argue that these conditions should be better fulfilled in a centrifugal DC-plasma CVD reactor.

Effects of High Gravity on DC-plasma CVD

In this section we describe how high gravity may be effective for controlling DC-plasma CVD processes and how it may enhance the deposition kinetics.

In general, centrifugation entails several independent forces, including acceleration, the Coriolis force and pressure gradients. All these forces, individually or simultaneously, may significantly affect macroscopic heat and mass transfer transport phenomena, microscopic phenomena related to nucleation, and homogeneous/heterogeneous chemical reactions.

In a centrifugal DC-plasma CVD reactor, heavier species (mainly C atoms) should reach the substrate surface in a shorter time and with a higher temperature. A decreased time would reduce the probability of undesirable intermediate reactions, such that no other chemical species other than C and H atoms will take part in the deposition process. A higher temperature would enhance the metastability conditions, with all the consequent advantages on the nucleation rate and growth rate. Accordingly, we believe that equation 1 may effectively describe our HGCVD process, since the plasma source with centrifugal effects will provide the necessary C atoms with high atomic hydrogen flux expected on the surface for successful deposition of diamond.

Another factor that arises from a HGCVD process is enhanced mass transport. In standard CVD processes, diffusion is recognized to be one of the main mechanisms in thin film growth. Under high gravity conditions, mass transport will involve two additional contributions, buoyancy-driven convection and pressure diffusion. These two new terms may contribute favorably to increasing the probability for species to react with the surface, such that growth rate and the deposition area can be increased. Convection induced by centrifugation is a further factor that can be exploited in HGCVD processes by varying the angular frequency.

Another interesting property induced by centrifugation is gas stratification. This feature may be utilized, for instance, for improving film quality and uniformity by optimal selection of the angular speed ϖ and the inter-electrode distance L. We may also take advantage of gas stratification, by optimizing the parameter L, to promote diamond film production under undersaturated conditions. This feature has scarcely been investigated by conventional CVD methods. Indeed, thermodynamic studies[9,10] demonstrate that co-deposition of graphite can be safely avoided even with undersaturated gas mixtures, provided that the transport distance of the active species between the carbon source and the substrate surface is very short.

MACROSCOPIC MODELING OF DIAMOND CVD UNDER HIGH GRAVITY

Heat and Mass Transfer

In this section, we write the general fundamental equations that describe the dominant physics and chemistry of diamond CVD processes in a DC-plasma reactor under high gravity conditions. The use of a moderate operating pressure in our facility justifies the applicability of a fluid-mechanical approach. In this way, macroscopic quantities are reasonably averaged over an appreciable number of mean free paths of the molecules in the gas.

Most of the reported heat and mass transfer models developed for simulating diamond film deposition neglect the interaction of the plasma with the gas phase or, in other cases, assume one- or two-dimensional simplifications. This is mostly due to the inherent difficulties encountered when a plasma is modeled as a continuum fluid. The corresponding numerical solutions are often very costly in terms of overall computer time and memory usage. There are even cases where unstable solutions are observed.[14]

In the following, attention is mainly focused on a three-dimensional model governing gas flow and heat transfer in a centrifugal DC-plasma CVD reactor. Heat and mass transfer processes are modeled by the corresponding continuum conservation equations by assuming the gas to be an ideal, compressible, and non-dissipative continuum. The gas flow is assumed to be laminar. Most of these assumptions are fairly well satisfied in practice at the prescribed experimental conditions.

Due to design requirements, a 3D analysis of the underlying phenomena in the reactor chamber is demanded. In figure 4, the rotation distance Ro -- i.e., the distance between the substrate and the rotation axis -- is a functional parameter. The reactor chamber rotates around a horizontal axis (x-axis in figure 4) at a constant velocity ϖ. We make Ro relatively large with respect to the substrate size in order to avoid possible non-uniform film coverage during the deposition process. As shown in figure 4, heat and mass transfer phenomena are studied with respect to a reference system fixed with the chamber.

If no rotation is applied, the set of conservation equations in a 3D Cartesian coordinate system reads:

$$\frac{\partial}{\partial t}(\rho\phi) + \nabla \cdot \left(\rho v \ \phi \ - \eta_\phi \nabla\phi \ \right) = S_\phi \qquad (2)$$

where ϕ is a general (scalar or vectorial) field variable (i.e., density, velocity, enthalpy, mass fraction) as shown in Table 2.

The meaning of the various parameters is indicated in the table of nomenclature at the end. All transport properties (viscosity, thermal conductivity and diffusion coefficient) are mixture-averaged. Most of these are available from the literature, or can be evaluated from standard gas kinetic theory using the LJ intermolecular potential.[15] The thermal diffusion coefficient is computed by a simplified equation suggested by Jones.[15]

High gravity effects can be included in the transport model given by equation 2, by adding the inertial terms to the left hand side of the momentum equation, as shown in the third column of Table 2. Since species conservation is also affected by high gravity, a new contribution has to be added to the mass flux in order to account for separation induced by the pressure gradient. For the sake of clarity, the pressure diffusion term $F(\nabla p)$ in Table 2 will be treated separately in the following. It is worth to noting that, from Table 2, the

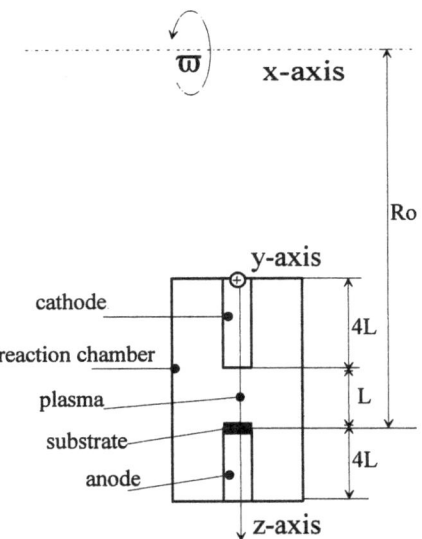

Figure 4. Schematic diagram of the rotating HGCVD reactor system.

Table 2. Individual terms appearing in the general conservation equation 2.

Equation	Field ϕ	Property η_ϕ	Source term S_ϕ	High Gravity
Continuity	1	0	0	0
x-mom.	u	μ	$-\partial P/\partial x + C_x$	0
y-mom.	v	μ	$-\partial P/\partial y + C_y$	$-2\varpi w - \varpi^2 y - \varpi' z + g\cdot\sin(\varpi t - \varphi_0)$
z-mom.	w	μ	$-\partial P/\partial z + C_z$	$2\varpi v - \varpi^2 z - \varpi' y + g\cdot\cos(\varpi t - \varphi_0)$
Energy	H	k/Cp	S_p	0
Species	m_i	$\rho\,D_i^{eff}$	$D_i^T \nabla(\ln T) + Q_{Bi}$	$F(\nabla p)$

energy and continuity equations are not affected by high gravity due to the particular choice of the reference frame. In the momentum equation, the inertial effects consist of a centrifugal acceleration term, which is the product of the squared angular frequency ϖ and the distance from the center of rotation, a Coriolis term, which is the product of the angular frequency ϖ and the velocity, and a periodic (earth) gravitational term which is function of the rotational speed ϖ and time. The Coriolis term appears as a secondary flow effect that drives the flow along the negative y-direction and the positive z-direction. In equation 2, the conservation of chemical species includes transport due to convection and diffusion for each chemical species. Besides ordinary molecular diffusion, terms corresponding to pressure and thermal diffusion also appear in the equation. Thus, for conservation of chemical species i, equation 2 can be rewritten as:

$$\frac{\partial}{\partial t}(\rho m_i) + \nabla\cdot\left(\rho v\, m_i - j_i\right) = Q_{Bi} \tag{2'}$$

where, the overall diffusion mass flux j_i in a multi-component system consists of three terms, specifically, molecular diffusion j_i^C, thermal diffusion j_i^T, and pressure diffusion j_i^P, such that:

$$j_i = j_i^C + j_i^T + j_i^P \tag{3}$$

In a dilute multi-component mixture, the molecular diffusion term may be computed by using the Fick-Wilke approximation:[15]

$$j_i^C = -\rho D_i^{eff} \nabla m_i \tag{4}$$

where $D_i^{eff} = (1 - x_i)\left(\sum_{j\neq i} x_j / D_{ij}\right)^{-1}$ (5)

is the effective diffusion coefficient for the ith species.

The thermal diffusion term can be computed by:

$$j_i^T = -D_i^T \nabla(\ln T) \tag{6}$$

where the thermal diffusion coefficient D_i^T is a function of gas composition and temperature. The thermal diffusion term describes the tendency of species to separate under the influence of a temperature gradient. Due to the presence of a plasma source, this term can play a significant role in diamond DC-plasma CVD.

The pressure diffusion flux is given by:

$$j_i^P = -\frac{\rho^2 M_i m_i}{P}\left(\frac{\overline{V}_i}{M_i} - \frac{1}{\rho}\right)\nabla p \tag{7}$$

which describes the tendency of species to separate under the influence of a pressure gradient. Therefore, equations 6 and 7 account for gas separation in a centrifugal DC-plasma CVD reactor, with heavy active species (i.e., atomic carbon, radicals, etc.) tending to separate from light species (i.e., atomic hydrogen).

Finally, the density of the mixture is evaluated by the equation of state for an ideal gas:

$$P = \rho RT/M, \quad \text{with} \quad M = \sum_i x_i M_i \qquad (i = 1,2,\ldots n \text{ species}) \tag{8}$$

Model Results and Discussion

Numerical investigations were performed in order to explore the performance of the centrifuge DC-plasma CVD reactor geometry sketched in figure 4. The model reactor consists of a cylindrical chamber of 0.2 m diameter and 0.24 m height and two cylindrical electrodes of variable gap. The rotation axis is located at $R_0 = 0.363$ m with respect to the substrate surface. The chamber is made of 316 stainless steel and electrodes are made of molybdenum. The inlet cavity is located on the top left hand side of the cylindrical chamber, and the outlet is located at the bottom right hand side. Due to the asymmetric design of such cavities, a full integration of the 3D partial differential equations, summarized in Table 2, is demanded. For simplicity, the chamber is assumed to be filled by pure hydrogen gas; therefore only one set of governing equations needs to be solved. This approximation is justified by the relatively low concentration of the reactant CH_4 which, in this kind of application, is below ≈ 5 %. Details of the chemistry are not included in the present calculations. The inlet flow rate is 2.67×10^{-8} kg s^{-1}, which corresponds to an initial velocity of 0.04 m s^{-1} along the y-direction. The gas flow is assumed to be laminar and the ideal gas law is applied everywhere in the chamber, including the plasma region. In the energy equation, the viscous dissipation and compressibility terms are neglected. An operating pressure of 27 kPa is introduced as a fixed-pressure boundary condition at the outlet.

The temperature values in the plasma core, at the reactor walls, and at the surface of both electrodes are taken from DC-plasma CVD experiments.[7,12,13] These are about 6000 K, 330K and 1020 K, respectively. The temperature at the reactor walls is introduced as a fixed-temperature boundary condition. The temperature conditions at the electrode surfaces and in the plasma region can be fulfilled implicitly, through estimation of an *effective* heat source distribution giving the same temperature values at the given points . This estimation was accomplished for a stationary (non-rotating) reactor. Then, the same distribution was used, as a first guess, for simulating the heat and mass transfer in a centrifugal CVD reactor. The use of such *effective* heat source is quite convenient for achieving our design purposes, since it attempts to reproduce the actual energetics in the stationary reactor without solving additional equations with respect to current density, flow of electrons, etc. It takes into account all the thermal effects due to the plasma source and radiation heat losses. The accuracy of this estimation can be improved if temperatures in other selected monitoring points are available. Experimental measurements are underway to validate this procedure.

For comparison, all of the input data above are used for normal gravity as well as for high gravity simulations. All the thermal properties (except density) are assumed to be constant during the calculations.

The overall solution consists of the velocity, temperature, pressure and density fields, along with the estimated heat source distribution. For this purpose a suitable off-line iterative procedure is demanded. The governing equations and the related boundary conditions are solved in a body-fitted coordinate by a control-volume formulation.[29,27] The computations are accomplished on a non-uniform grid in order to resolve the boundary layer behaviour near the wall and substrate regions. The grid mesh consists of 30x16x56 elements.

Several rotational regimes, from zero to high gravity conditions, were investigated. However, the highest gravity condition permissible by our experimental device (i.e. $\varpi = 50$ rad s^{-1}) could not be examined faithfully, since we could not reach a fully converged solution. Due to space limitation, only two cases are presented here. These results are presented in figures 5, 6, 7, and 8. They show the effects due to the absence/presence of a plasma at a rotational speed of $\varpi = 2$ rad s^{-1} in a selected longitudinal section passing through the inlet and outlet regions. In both cases, Earth's gravity was omitted in order to appreciate the individual effects generated by the chamber's rotation on the gas flow. These results may also give an initial insight on the heat and mass transfer that would occur at higher rotational regimes where Earth's gravity is reasonably negligible.

In figure 5 and 6 we show the pressure distribution without and with the plasma heat source, respectively. Note how the pressure profiles are tilted to the left with respect to the cylinder axis due to the Coriolis effect. The highest pressure region is located at the left bottom hand side of the chamber, whereas the lowest pressure region is located at the opposite corner, say, the top right hand side of the reactor. The presence of the plasma source alters the pressure distribution significantly and reduces the pressure limits caused by rotation.

The corresponding velocity fields are shown in figures 7 and 8. The flow patterns appear quite asymmetric, proving the need for a full 3D analysis. In figure 7, the gas enters the reactor, a small amount impinges on the cathode, and a larger amount undergoes a helical motion around the electrodes, due to the chamber rotation. Such helical motion is very important because part of the gas returns to the first half of the chamber through the electrode gap, while part of it undergoes a recirculation in the lower part of the reactor, before exiting. In figure 7, the highest gas velocities appear to be concentrated at the inlet and the outlet, indicating that the pressure difference between the inlet and the outlet is the main driving force. In figure 8 the situation changes dramatically due the presence of a plasma.

The highest velocities are found inside the plasma region, with the two largest values near the edge of the anode. Except for these, the flow seems to be mainly driven by the pressure gradient induced by rotation, which drives the low density gas towards the lowest pressure region of the reactor.

Figures 6 and 8 suggest a criterion for optimizing the design of HGCVD reactors. Indeed, it appears more reasonable that gas inlet and outlet would function better if placed in the highest and lowest pressure regions, respectively, in order to facilitate the entry of new gas and the exit of the exhausted gas from the chamber.

In order to be able to test this new possible hardware configuration and to provide the largest flexibility to our experimental system, we constructed a chamber with four cavities, reproducing the present configuration and the one suggested by computer simulations.

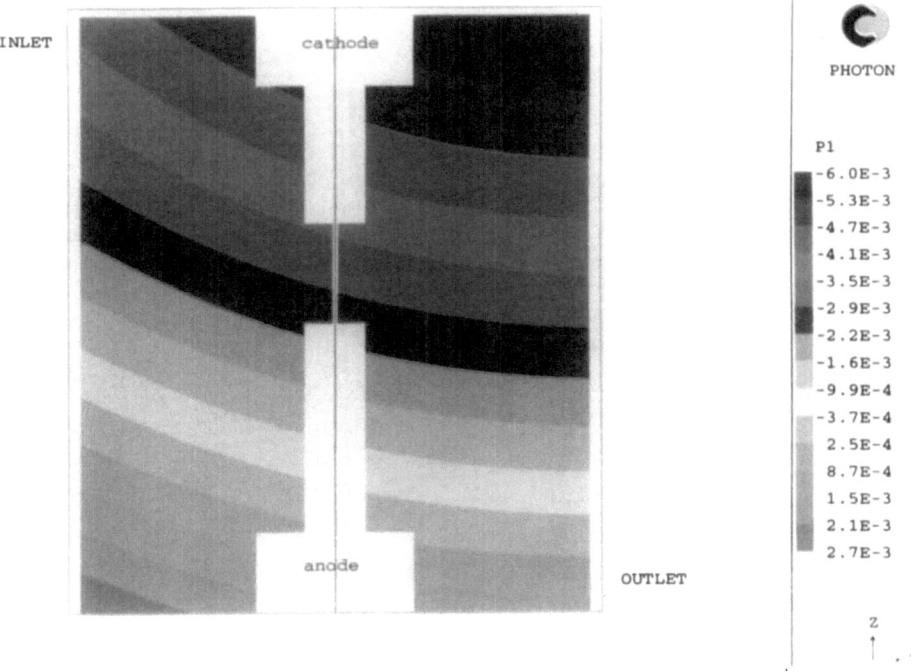

Figure 5. Computed iso-pressure map on a longitudinal section passing through the inlet and outlet regions of the HGCVD reactor rotating at 2 rad/s, without a plasma heat source or Earth's gravity.

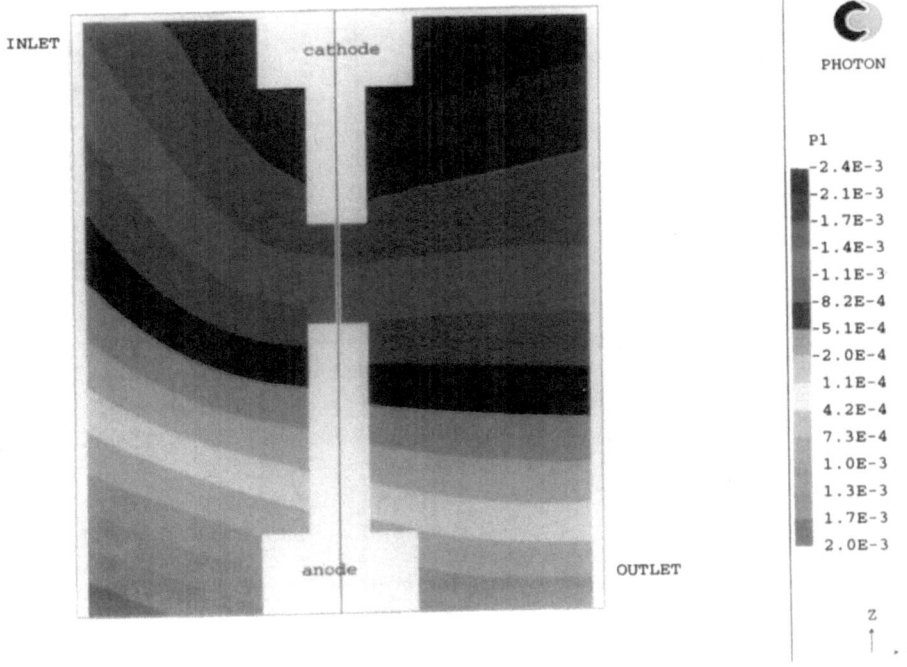

Figure 6. Computed iso-pressure map on a longitudinal section passing through the inlet and outlet regions of the HGCVD reactor rotating at 2 rad/s with a plasma heat source, without Earth's gravity.

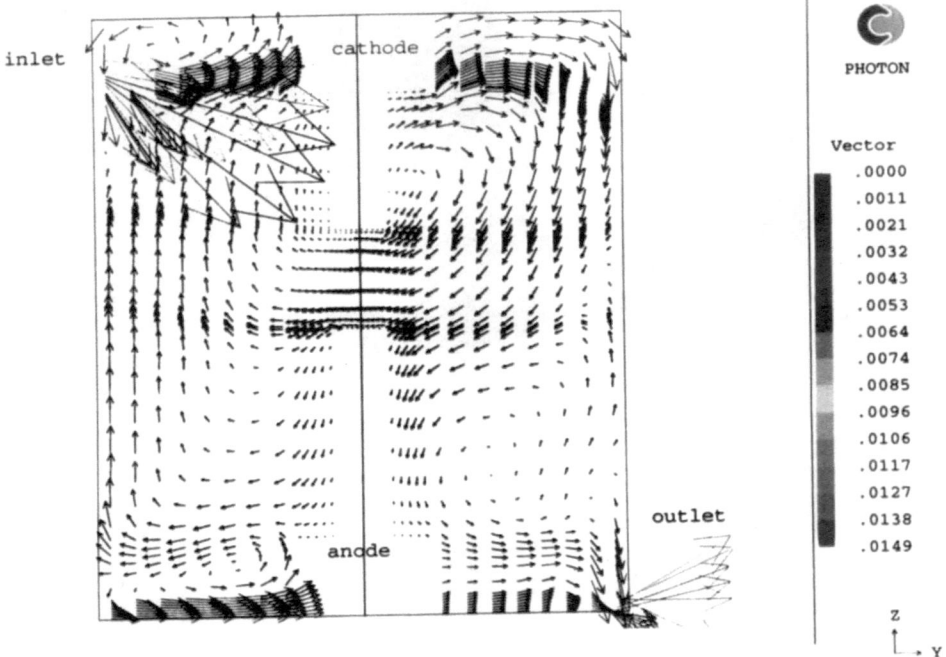

Figure 7. Computed velocity vectors on a longitudinal section passing through the inlet and outlet regions of the HGCVD reactor rotating at 2 rad/s, without a plasma source or Earth's gravity.

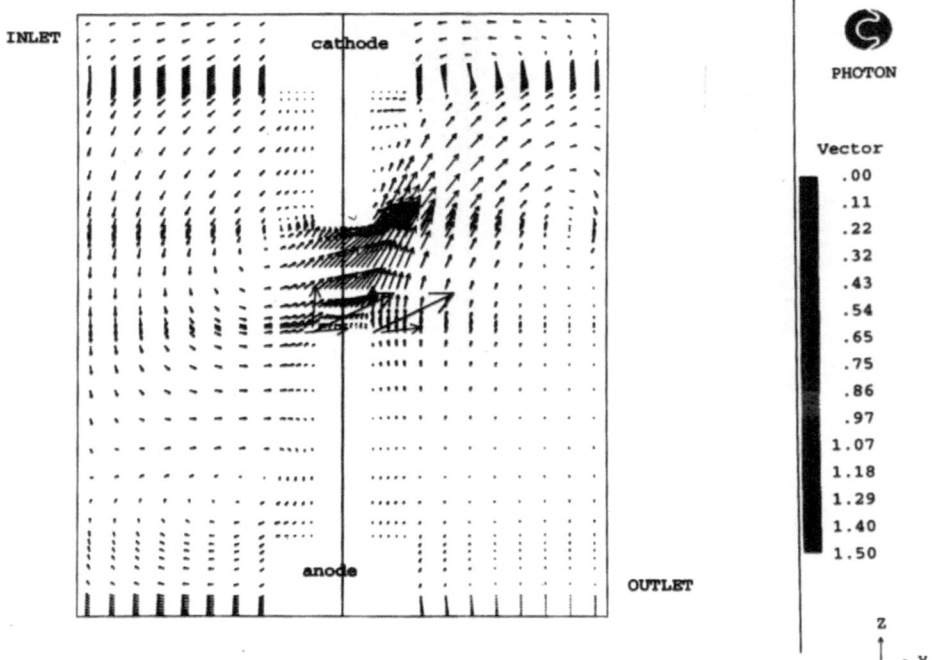

Figure 8. Computed velocity vectors on a longitudinal section passing through the inlet and outlet regions of the HGCVD reactor rotating at 2 rad/s, with a plasma source, and without Earth's gravity.

CONCLUSIONS

Since high gravity is quite a new experimental technique in thin film growth technology, there are neither experimental nor theoretical studies that fully support all the expectations mentioned in this paper. What can be stated at this stage is that high gravity will at least alter the kinetics of the deposition processes and probably the corresponding morphology of the deposited diamond thin film. Therefore, it is of fundamental and practical interest to study the influence of film production under high gravity.

Combined experimental and theoretical work, through macroscopic flow modeling and microscopic computer simulation, will provide rapid and effective answers to most questions and will substantially help to establish a systematic link between the variables controlling the deposition kinetics and the input parameters.

In our future experimental program, we will evaluate the properties, especially thermophysical properties,[28] of thin films produced under high gravity, in addition to the usual characterization of thin films (e.g., SEM observation, Raman spectroscopy). We will identify the effects of high gravity on the characteristics of thin films. We will also utilize *in-situ* diagnosis of the reaction taking place in high gravity, in order to understand how gravity influences the CVD process. As a first approach, an optical diagnostic method is under consideration for monitoring qualitative characteristics of the plasma, such as relative density and size of radicals.

ACKNOWLEDGMENT

The authors express sincere gratitude and deep condolence to the late Prof. T. Inuzuka of Aoyama-gakuin University who gave us many helpful suggestions when this study was started. We also would like to thank Hiroji Suzuki of CRC-Japan for many helpful discussions about the implementation of the computational model into the Phoenics environment.

TABLE OF NOMENCLATURE

C	additional dissipation terms, other than those accounted for by viscosity μ	η_ϕ	generic mixture property related to ϕ
D_i^{eff}	effective diffusion coefficient of a multi-component mixture	ϕ	generic field variable (**v**, T, H, m)
D_i^T	thermal diffusion coefficient	φ_0	angular location of the chamber at t=0
g	acceleration of gravity	ω'	angular acceleration
j	mass flux of species	ϖ	angular frequency
k	thermal conductivity	μ	absolute viscosity
H	enthalpy	ρ	mixture density
m_i	mass fraction of the i-th species		
M	molar mean molecular weight	**Subscripts**	
P	thermodynamic pressure	(g)	gas phase
Q_{Bi}	generation term in the bulk	(s)	surface
S_p	heat source due to the plasma	i	i-th species

S_ϕ	source per unit volume related to the field ϕ
T	absolute temperature
u,v,w	velocity components along x, y, z
v	average mixture velocity vector
$\overline{V}$	partial molal volume
x	mole fraction

REFERENCES

1. L.L. Regel and W.R. Wilcox, Introduction to materials processing in large centrifuges, *in*: "Materials Processing in High Gravity," L.L. Regel, M. Rodot, and W.R. Wilcox, eds., Plenum Press, NY (1994) p 1.
2. H. Wiedemeier, L.L. Regel, and W. Palosz, Vapour transport and crystal growth of GeSe under normal and high acceleration, *J. Crystal Growth* 119:79 (1992).
3. G. Muller, G. Neumann, and W. Weber, *J. Crystal Growth* 119:8 (1992).
4. F. Tao, Y. Zheng, W.J. Ma and M.L. Xue, Unsteady thermal convection on melts in a 2-D horizontal boat in a centrifugal field with consideration of the Coriolis effect, *in:* "Materials Processing in High Gravity", Plenum Press, NY (1994) p 67.
5. S. Albin, "Study of diamond film growth and properties," NASA-CR-187709 (1990).
6. P.K. Bachmann and H. Lydtin, High rate versus low rate diamond CVD methods, *in*: "Diamond and Diamond-Like Films and Coatings," Plenum Press, NY (1991) p 829.
7. K. Suzuki, A. Sawabe, H. Yasuda, and T. Inuzuka, Growth of diamond thin films by DC-plasma chemical vapour deposition, *Appl. Phys. Lett.* 50:728 (1987).
8. K. Shibukawa, K. Murakami, S. Kamei, T. Hanyu, M. Ishikawa, Y. Sato, T. Inuzuka, and N. Fujimori, Preliminary experiments of SFU/EFFU/GDEF utilizing airplane, *in:* Reports of Microgravity Experiments by Aircraft (NASDA) vol. 2, no. 3 (1992).
9. W. Piekarczyk and W.A. Yarbrough, Application of thermodynamics to the examination of the diamond CVD process, *J. Crystal Growth*, 108:583 (1991).
10. W. Piekarczyk, R. Messier, R. Roy, and C. Engdahl, Diamond deposition by chemical vapor transport with hydrogen in a closed system, *J. Crystal Growth*, 106:279 (1990).
11. K. Suzuki, Growth of diamond thin films in a DC-discharge plasma, *Appl. Surf. Science* 33/34:539 (1988).
12. L. Suzuki, A. Sawabe, and T. Inuzuka, Growth of diamond thin film by DC-plasma CVD and characteristics of the plasma, *Jap. J. Appl. Phys.* 29:153 (1990).
13. L. Suzuki, Characterization of the DC-discharge plasma during CVD for diamond growth, *Appl. Phys. Lett.* 53:1818 (1988).
14. S.R. Vamas, Gas flow and heat transfer in DC-plasma heated reactors, *J. Comp. Fluid Dyn. Appl.* 4:362 (1991).
15. R.B. Bird, W.E. Stewart, and E.N. Lightfoot, "Transport Phenomena," J. Wiley & Sons, NY (1960).
16. N.V. Samokhvalov, V.E. Strel'nitskij, V.A. Bolous, V.P. Zubar, A.G. Timchuk, E.D. Obraztsova, and V.G. Ral'chenko, Production of polycrystalline diamond films by DC-glow discharge CVD, *Diamond & Related Mater.* 4:964 (1995).
17. H. Liu and D.S. Dandy, Studies on nucleation process in diamond CVD: an overview of recent developments, *Diamond & Related Mater.* 4:1173 (1994).
18. E. Molinari, R. Polini, V. Sessa, M.L. Terranova, and M. Tomellini, Diamond nucleation from the gas-phase: a kinetic approach, *J. Mater. Res.* 8:785 (1993).
19. M. Tomellini, R. Polini, E. Molinari, and V. Sessa, A model kinetics for nucleation at solid surface with application to diamond deposition from the gas phase, *J. Appl. Phys.* 70:7573 (1991).
20. F.C. Eversteyn, P.J. Severin, C.H.J. van der Brekel, and H.L. Peek, A stagnant layer model for the epitaxial growth of silicon from silane in a horizontal reactor, *J. Electrochem. Soc.* 117:929 (1970).
21. K.E. Spear, Diamond-ceramic coating of the future, *J. Am. Ceram. Soc.* 72:171 (1989).
22. J.C. Angus and C.C. Hayman, Low-pressure, metastable growth of diamond and diamond like phases, *Science*, 241:913(1988). See references herein.
23. S.L. Girshick, C. Li, B.W. Yu, and H. Han, Fluid boundary layer effects in atmospheric-pressure plasma diamond film deposition, *Plasma Chem. and Plasma Process.* 13:169 (1993).

24. B.W. Yu and S.L. Girshick, Atomic carbon vapour as a diamond growth precursor in thermal plasmas, *J. Appl. Phys*. 75:3914(1994).
25. D. Huang and M. Frenklach, *J. Phys. Chem*, 96:1868 (1992).
26. D.S. Knight and W.B., *J. Mat. Res*. 4:385 (1989).
27. J.R. Heritage, Phoenics-CVD - A customized CFD code for the simulation of CVD processes, *in*: "Proceedings of 6th International Phoenics Conference," Waseda University, Tokyo (1996).
28. M. Akabori, Y. Nagasaka and A. Nagashima, Measurement of the thermal diffusivity of thin films on substrate by photoacoustic, *Int. J. Thermophys*. 13:499 (1992).
29. D.B. Spalding, *Math. Comp. Sim.* XXXIII:267 (1981).

COMPARATIVE INVESTIGATION OF THERMAL AND PHOTOPOLYMERIZATION UNDER THE ACTION OF CENTRIFUGAL FORCES: BASIC MECHANISMS OF HEAT AND MASS TRANSFER

V. Briskman, K. Kostarev, and T. Yudina

Institute of Continuous Media Mechanics, Russian Academy of Science
1, Korolev Street, Perm, Russia 614061

INTRODUCTION

It may seem strange that we began studying heat and mass transfer involved in polymerization under high gravity conditions while being engaged with polymerization problems in microgravity.[1-12] Such transformation of our interests from micro- to high gravity was due to the following reasons.

Frequently, transferring material processing into microgravity results in an improvement in properties. Such was the case with electrophoresis matrices based on polyacrylamide gels (PAG) formed by photoinitiation. The electrophoretic resolution of the orbital PAG specimens was higher than that of their analogs prepared on earth. A series of experiments was carried out in the laboratory aimed at gaining insight into the gravitational sensitivity mechanisms of gelation. These experiments showed that at normal gravity, photopolymerization is accompanied by free-convection in the reaction mixture[3-6]. Convection stirs liquid with different degrees of conversion, and thus changes the character of the reaction and the final polymer structure. We have identified the causal relations between the intensity of convective motion in the reaction mixture and the reaction evolution in the bulk of the material. However, the question arises of whether convection is the single mechanism for the influence of gravity or if it conceals the action of some other gravitational mechanisms. One might expect experiments in high gravity to give an answer to this question.

MATERIALS, EXPERIMENTAL SETUP, METHODS OF MEASUREMENT

The simplest example of a nonuniform high gravity field is that formed by rotation. In this case, the force gradient has a constant value and is aligned with a radius of rotation. For the object of investigation, we used aqueous solutions of acrylamide transformed by polymerization into three-dimensional net-sponges whose pores are filled with water. Figure 1 is an electron micrograph of such a gel. [13]

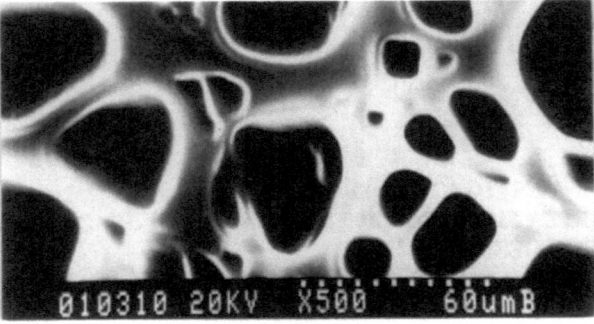

Figure 1. Electron microscope photograph of gel structure.

The reaction mixture consisted of the monomer (15% aqueous solution of acrylamide), the cross-linking agent (0.5% methylenebisacrylamide), and the catalyst (0.01% tetramethylenediamine). The reaction was initiated either by ammonium persulfate (thermal initiator, 0.02%) or riboflavin (photoinitiator, 5×10^{-4} %). We need to discuss these two particular means of reaction initiation in more detail. If we assume that there are several mechanisms of polymerization sensitivity to gravity, then we have great interest in their proper identification and a deep understanding of the role they play and the manner by which they interact during gel formation. The latter is essentially determined by the type of reaction initiation. During photopolymerization there are actually three zones in the reactor over the whole period of the process: gel, liquid monomer, and a fairly narrow transition zone. Each of these regions has its own characteristic features, which determine the type of gravitational action within its boundaries. By contrast, during chemical initiation the entire reaction mass is at the same level of progress of reaction and, consequently, the mechanisms operating in the bulk of the reaction mixture are similar. Here the intensity of the influence and interaction of the gravitational mechanisms is defined only by the force, and in the case of a centrifugal force field – by its gradient. When comparing the results of polymerization for one or the other type of initiation we may define each mechanism and give a reasonably accurate estimation of the role played by it.

For an experimental model we used a horizontal plane disk, 5.5 *mm* thick and 60 *mm* in diameter filled with the reaction mixture (Figure 2a). Transparent acrylic plastic plates formed top and bottom surfaces (1).

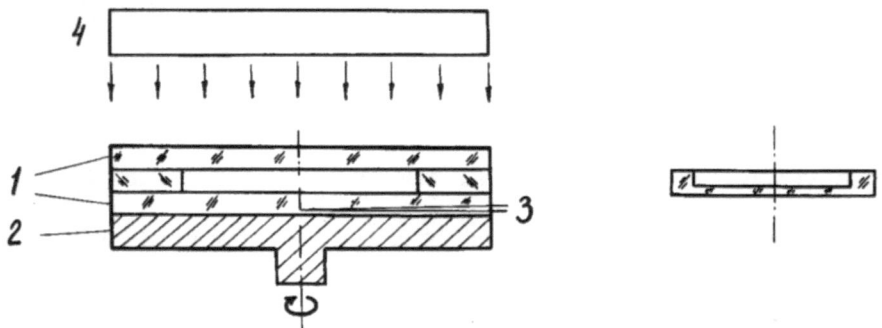

Figure 2. Experimental model. (a): 1 – transparent plates, 2 – steel platform, 3 – differential thermocouple, 4 – photoinitiating light source. (b): specially designed mold for gel samples.

The model was fixed on a heavy steel platform (2) fitted on the shaft of the motor. The rotation axis of the model coincided with that of the shaft. The rotation axis was 30 arc sec out of the vertical. The rotational speed of the motor was 3000 revolutions per minute.

The tests were performed at room temperature (293 – 295K). PAG polymerization is an exothermic reaction. The temperature of the reaction mixture at the central point of the model cavity was measured by differential copper-constantan thermocouple (3). Free convection of air and heat transfer to the steel platform through the bottom surface allowed heat removal from the model. A special light source (4) with a maximum in radiation intensity at 445 *nm* was used for photoinitiation. The light source was placed above the model at a height of 15 *cm*.

To gain information on the spatial distribution of properties in the final gel structure, we chose a procedure of measuring the local elasticity modulus at different points of the gel specimen (the elastic modulus of the gel is known to depend on the pore dimensions). To this end, two methods were used. The first consisted in measuring gel samples in the form of cylinders of 4.5 *mm* radius and 5.5 *mm* height, which were cut from the gel matrix. The samples for each set were taken at increasing distances from the disk center so as to allow estimation of the elastic modulus distribution along the radius.

In the other method, the elastic modulus was determined by measuring deformation of disk sections subject to local loads. This method involved preparation of gel in a specially designed mold with an inner diameter of 50 *mm* and a height of 4 *mm* (Figure 2b). This mold was placed inside the model. A mathematical treatment of the measurements obtained by the two methods gives results which agree within experimental error.

All experiments were run according to a rather simple scheme. The reaction mixture was bubbled with argon for 30 minutes in order to withdraw oxygen, which is a reaction inhibitor. Then, in the case of thermal initiation, an initiator was incorporated in the reaction stock and the mixture was poured in the mold, which was put in rotation at the prescribed time. In order to choose the instant for starting rotation, we began each set of experiments with a preliminary test on a motionless model to analyze the behaviour of the heat emission curve during polymerization in relation to the initiation mode. Then we successively determined the times of reaction initiation (interval *OA* in Figure 3), gel-points (gel formation in the bulk of mixture – *OB*), maximum heat emission (*OC*), and completion of the reaction (*OD*). With the centrifuge switched off or on for the specified time intervals, we could control (strengthen or weaken) the action of one or another gravitational mechanism. The total time of each experiment was 1 hour.

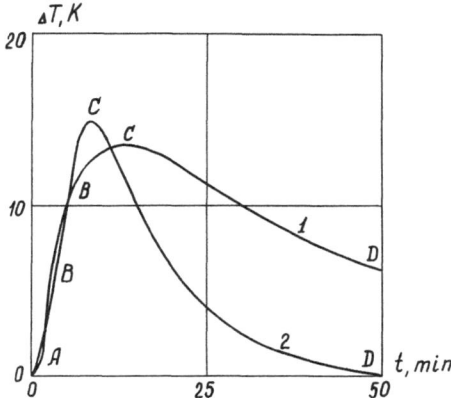

Figure 3. Heat emission in a reaction mixture during polymerization.
1 – photoinitiator; 2 – thermal initiator

After completion of an experiment, the model was disassembled and the extracted polymeric disk was used to measure the variation of the elastic modulus with radius.

EXPERIMENTAL RESULTS AND DISCUSSION

In what follows, we shall discuss the experimental data for the effect of high gravity on the thermally initiated polymerization in detail.   A low thermal conductivity at the boundaries of the model cavity (acrylic plastic) allowed us to avoid high temperature gradients in the near-wall region, so that the usual thermal convection was absent. Due to a uniform distribution of the initiator, the reaction in the bulk of the monomer proceeded at a constant rate. This provided a constant value of the elastic modulus in the whole body of the gel disk obtained without rotation, with the exception of two regions (Figure 4): a near-wall region, 2-3 *mm* wide, where the increased modulus resulted from an instrumental error, and a region close to the center where an air cavity formed due to shrinkage of the reaction mixture during polymerization.

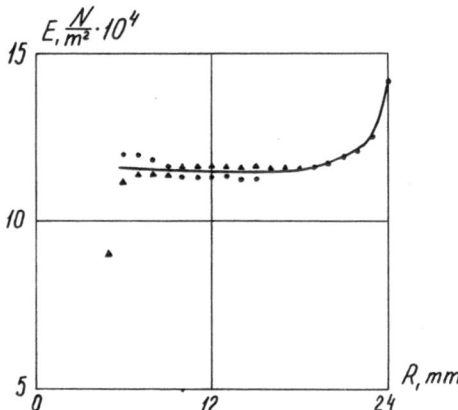

Figure 4. Elastic modulus distributions over the disk radius for the gel matrices polymerized at rest with thermal initiation.

The situation changed if the cuvette was rotated during the liquid-phase stage of polymerization, and was stopped as soon as the reaction mixture passed through the gel-point (the gel, in this case, maintained the current distribution of the monomer conversion). Figure 5 shows that the properties of the final gel structure varied linearly along the disk radius. This behaviour can be attributed to the action of sedimentation in the centrifugal force field. (As was evident from additional experiments with light scattering particles, thermal convection was absent even with rotation.)

It should be noted that in aqueous solutions, sedimentation was also observed in the normal gravitational field, although several days later.[9] In the case under consideration, the sedimentation time was several minutes, implying that, in the course of polymerization, polymer particles were formed that were far in excess of the monomer molecule size. The hypothesis has been that these polymeric particles (globules) are not linked with each other and are randomly distributed in the buffer solution along with monomer particles. While new particles are generated at each instant of time, the old ones continue to grow.

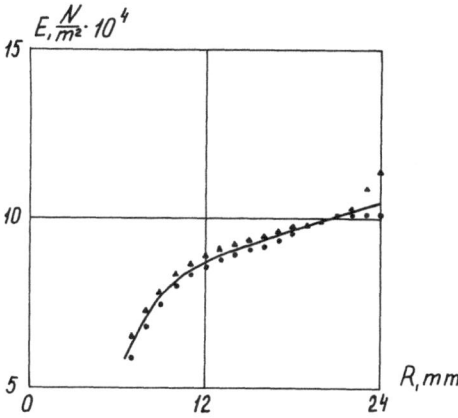

Figure 5. Elastic modulus distribution over a disk radius for matrices polymerized in the presence of centrifugal forces with thermal initiation. Rotation was stopped when the reaction mixture passed through the gel-point.

At a later time, as the number of particles essentially increases, they are gradually involved in formation of the three-dimensional polymer network. This hypothesis was advanced several years ago,[14] but only recently has it received true support. Because of sedimentation, the concentration of polymeric globules close to the side wall increases and gelation evidently begins in this part of the cavity (Figure 6).

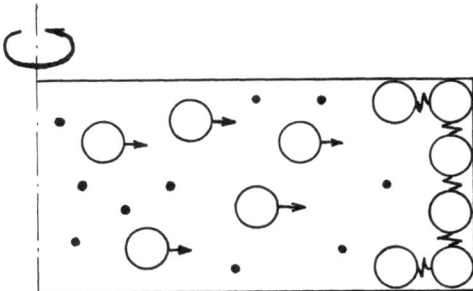

Figure 6. Schematic illustration of the influence of sedimentation on gel formation.

Figure 7 shows the variation of elastic modulus with radius for experiments in which the rotation was switched on at the instant the reaction mixture passed through the gel-point.

These experimental results can be explained by the action of another gravitation-sensitive mechanism. It is known that in order for the three-dimensional network to be generated, the mixture needs to reach a degree of conversion of 35-50%. The remainder of the chemical links are formed in the next stage of gel formation. If only the generating polymeric network (Figure 8a) is subject to loading (in our case, by placing it in the centrifugal force field), it deforms and displaces the buffer solution to the center of the polymeric disk (Figure 8b). As the deformed gel polymerizes further, new links are continuously generated, this time, however, free of deformation (Figure 8c). After unloading, these links remain partially deformed (Figure 8d).

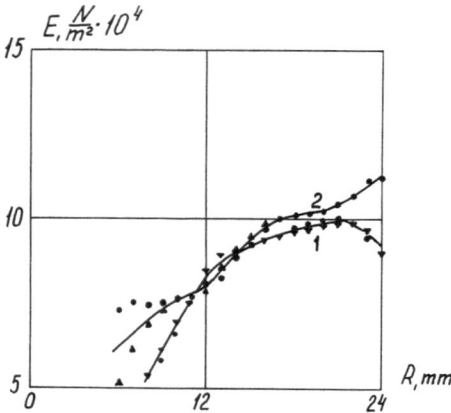

Figure 7. Radial variation of elastic modulus for matrices polymerized in the presence of centrifugal forces with thermal initiation. The two curves are for different times between the beginning of the reaction and the start of rotation: curve 1 – 5 min, curve 2 – 15 min.

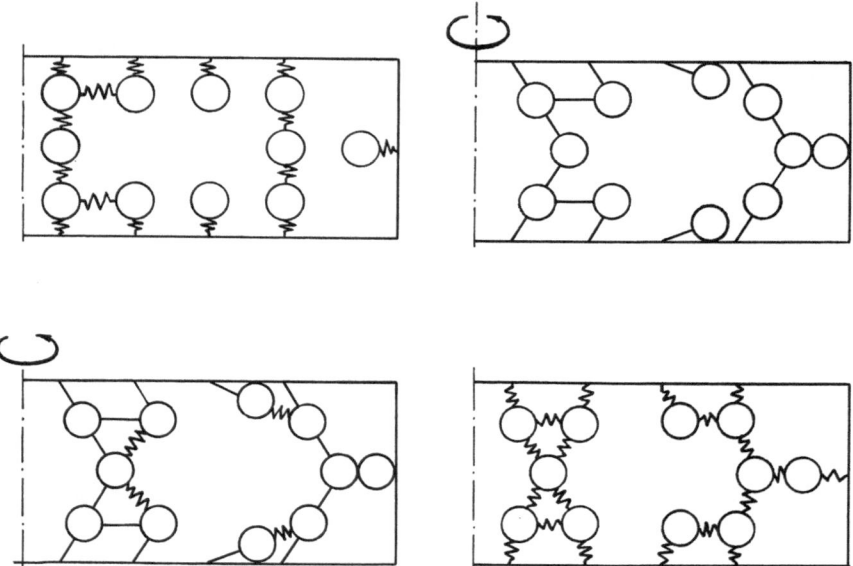

Figure 8. Schematic illustration of gel network deformation caused by force field.
a – gel structure at rest, b – gel structure deformation in force field,
c – link generation in force field, d – new gel structure at rest

(By analogy, we may consider a sponge which is compressed, stitched at some places and then released. The new sponge shape will differ from the original shape and also from that under compression). The shorter is the interval between the gel point and the start of rotation, the greater is the number of free links, and thus, the stronger is the influence of centrifugal force on the polymer structure being formed. Such is the second mechanism of gravitational influence, the action of which is generally concealed by convection.

Figure 9 displays the results of an alternate effect of both of the mechanisms mentioned above: sedimentation and network deformation caused by centrifugal force (the specimen was put in rotation before the onset of reaction). It is seen that the mechanisms strengthen each other.

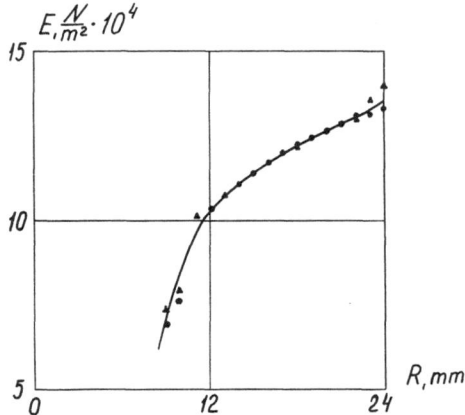

Figure 9. Elastic modulus distribution over a disk radius due to the alternate action of sedimentation and centrifugal force. The specimen was put in rotation before the onset of reaction. Thermal initiation.

Now let us recall for comparison, the results of the study of photopolymerization under the action of centrifugal force.[14] In contrast to a thermally initiated reaction, photopolymerization involves a stage in which convection, sedimentation and gel network deformation operate simultaneously. This is the stage of gel formation from the onset of reaction to the gel-point. The most significant of the above mechanisms is convection caused by a vertical density gradient due to the frontal character of photopolymerization. Such a gradient is unstable in the orthogonal field of centrifugal force. This convection intensifies sedimentation, as shown in Figure 10. A rapid accumulation of polymeric globules at the side cavity wall results in development of frontal polymerization at this surface. The combined action of the gravitational mechanisms defines a final polymer structure with properties depending linearly on the radius, as shown in curve 1,2 of Figure 11. It is known that gel formed by photoinitiation has pores of larger size than that produced by peroxide-initiated polymerization. This means that photogels are more elastic and sensitive to the action of external forces. Such an assumption is supported by the response of curves 3 and 4 in Figure 11, describing the mechanical properties of gel

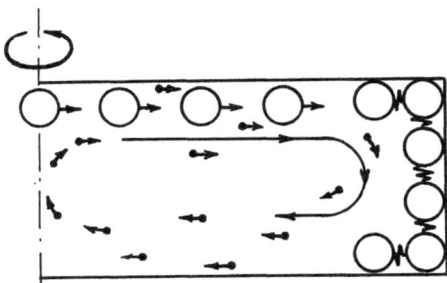

Figure 10. Schematic illustration of the influence of convection on gel formation

specimens put in rotation after the completion of polymerization in the whole volume. These curves have equal or greater slopes for E/E₀ than similar curves for samples obtained with thermal initiation of the reaction at the same stage of gel formation (Figure 7). Furthermore, in the overload zone (more than 180 g), the polymeric links partially break down, as reflected in chaotic behaviour of the elastic modulus at large values of the radius.

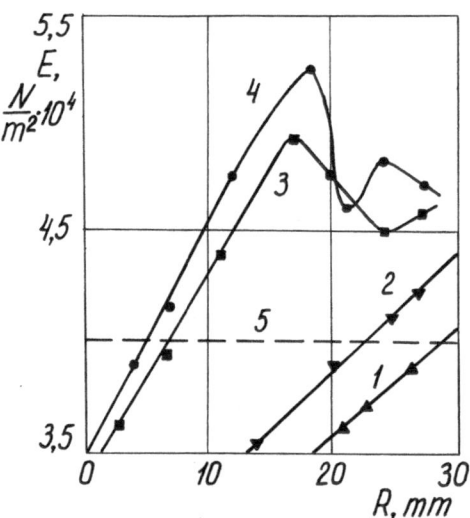

Figure 11. Radial variation of elastic modulus for matrices polymerized in the presence of centrifugal force. Time in minutes between the beginning of photoinitiation and the start of rotation:
curve 1 - 0 min, curve 2 - 6 min, curve 3 - 12 min, curve 4 - 18 min; curve 5 - without rotation.

CONCLUSIONS

1. In this paper we discussed the results of experimental investigation on the effect of a nonuniform force field, in particular, a centrifugal force, on structure formation of polyacrylamide gels under different initiation conditions.
2. The results suggest that there exist three mechanisms of gravitational nature that act during the whole process of polymerization. The first mechanism is sedimentation of isolated gel globules formed during the initial stage of the process. The second is buoyancy-driven convection caused by the dependence of density on temperature and conversion, which change during structure formation. The third can be defined as an immediate effect of the centrifugal force on the generation of a deformable polymer network.
3. Depending on the means of reaction initiation, these mechanisms operate either in an alternate or simultaneous manner enhancing one another. Their interaction results in the formation of a gel structure with a nearly linear dependence of properties on radial position.
4. It was found that the spatial distribution of properties largely depends on the behaviour of the polymerization process and at which stage the system is put into rotation.

To summarize, the investigations have demonstrated an extremely high gravitational sensitivity of polymerization and the final gel structure. This offers new possibilities for manufacturing controlled-property polymers by changing the intensity of body forces.

Acknowledgments

The authors highly appreciate the initiating efforts of L.L Regel to facilitate this research. The authors also wish to express sincere gratitude to V.V. Moshev for his valuable discussion of experiments, N.V. Pistsov for measuring the mechanical properties of the gel specimens, and S.N. Lysenko who carried out chemical investigations involved in the work. This work was executed with the financial support of the Russian State Scientific-Technical Program "Astronomy. Fundamental Space Research."

REFERENCES

1. L.L. Regel, M. Rodot, and W.R. Wilcox, eds., "Materials Processing in High Gravity," North-Holland, Amsterdam (1992). Also *J. Crystal Growth* 119 (1992).
2. L.L. Regel and W.R. Wilcox, eds., "Materials Processing in High Gravity," Plenum Press, New York (1994).
3. Sh. Abdurakhmanov, L. Bogatyreva, V. Briskman, M. Levkovich, et al., On polyacrylamide gel formation by photoinitiation under terrestrial and orbital conditions, *in*: "Numerical and Experimental Modeling of Hydrodynamic Phenomena under Weightlessness," Sverdlovsk (1988) (in Russian).
4. L. Bogatyreva, V. Briskman, M. Levkovich, et al., Gravitationally-sensitive mechanisms of polyacrylamide gel structurization, *Space Sci. Techn.* 4:43 (1989) (in Russian).
5. Sh. Abdurakhmanov, V. Babskii, L. Bogatyreva, V. Briskman, et al., Structure formation of polyacrylamide gel at photoinitiation under earth and orbital conditions, *in*: "Gagarin Scientific Readings on Astronautics and Aeronautics 1989," Nauka, Moscow (1990) (in Russian).
6. L. Bogatyreva, V. Briskman, K. Kostarev, V. Leontyev, M. Levkovich, T. Lyubimova, A. Mashinsky, G. Nechitailo, and P.G. Righetti, Heat/mass transfer mechanisms of the polymerization under terrestrial and microgravity conditions, *in*: "Proceedings of the VIII European Symposium on Material and Fluid Sciences in Microgravity," Brussels, ESA SP-333, Vol. 1 (1992).
7. C. Gelfi, P. de Besi, A. Alloni, P.G. Rigetti, T. Lyubimova, and V. Briskman, Kinetics of acrylamide photopolymerization as investigated by capillary zone electrophoresis, *J. Chromotog.* 598 (1992).
8. V. Briskman and K. Kostarev, Gravity sensitive mechanisms of polymerization and gel formation. Ways for controlling final material properties, *in*: "Proceedings of the International Aerospace Congress, IAC-94," Moscow, Russia (1994).
9. P.G. Righetti, A. Bossi, V. Giglio, A. Vailati, T. Lyubimova, and V. Briskman, Is gravity on our way? The case of polyacrilamide gel polymerization, *Electrophoresis* 15:1005(1994).
10. V. Briskman, K. Kostarev, V. Levtov, T. Lyubimova, A. Mashinsky, G. Nechitailo, and V. Romanov, Polymerization under different gravity conditions, *Microgravity Quart.* 5, 2: 59(1995).
11. V. Briskman, K. Kostarev, V. Levtov, V. Romanov, and T. Yudina, Comparative experimental research of polymerization on the "MIR" orbital station and on the Earth., AIAA.95-0263, 33rd Aerospace Sciences Meeting and Exhibit, Reno (1995).
12. V. Briskman, K. Kostarev, V. Moshev, L. Guseva, A. Mashinsky and G. Nechitailo, "On gravity dependence of polymerization, AIAA.96-0257, 34th Aerospace Sciences Meeting and Exhibit, Reno (1996).
13. V.B. Golubev, B.A. Korolev, K.G. Kostarev, and T.P. Lyubimova, On the role of buoyancy convection in the frontal polymerization processes, *in*: "Hydromechanics and Heat/Mass Transfer in Microgravity. First International Symposium," Perm-Moscow (1991).
14. V.A. Briskman, K.G. Kostarev and T.P. Lyubimova , Gel polymerization at high gravity, *in*: "Materials Processing in High Gravity," L.L. Regel and W.R. Wilcox, eds., Plenum Press, New York (1994).

POLYMER FILMS FOR SOLAR CELLS
GROWN DURING RAPID STIRRING

Kh.S. Karimov, Kh.M. Akhmedov, and A.M. Achourov

S.U. Umarov Physical Technical Institute
Dushanbe 734063, Tajikistan

INTRODUCTION

At present, renewable energy sources are used in many countries, and the efficiency of the utilization of all kinds of energy is increasing. This year is very important for utilization of renewable energy in the future. In September 1996, the World Solar Summit will take place in Zimbabwe, and the heads of states and governments will sign papers for development of renewable power engineering.

Utilization of renewable energy is especially important for Tajikistan. On one hand, Tajikistan has very little oil and gas. On the other hand, this country is very rich in renewable energy sources. For example, Tajikistan occupies the second place after Russia (among the republics of the former USSR) in hydraulic power. It is technically possible in Tajikistan to utilize 144 mlrd.kW.hr of hydraulic power, while only 10% of this is being produced now.[1]

Tajikistan, as a southern country, is rich in solar energy. There are about 300 sunny days a year. The power of solar radiation reaches 1 kW/m^2, and the average operational time of solar panels is 2700 hours per year.[2] Our republic can completely provide itself with sufficient energy if we collect and transform solar energy from about 0.01% of our territory. This energy will be equal to 30 mlrd.kW.hr per year.

At present, the fabrication of solar cells with high efficiency and low cost is the main problem. At our Physical Technical Institute we are dealing with organic solar cells. Work on solar cells based on organic semiconductors is important due to the great variety of their structural and physico-chemical properties. According to Chamberlain, a feasible efficiency for organic solar cells can practically be 10%, while the theoretically possible value is 23%.[3]

During the last 16 years, the efficiency of organic solar cells has increased about 20 times, from 0.25% to 4%.[4,5] For example, photovoltaic cells on the base of polythiophene (Al/polythiophene/Au) have an efficiency of 4%.[6] The efficiency of some solar cells based on organic semiconductors is given in Reference 5.

It is known that the efficiency of organic solar cells is limited mainly by the low electrical conductivity of organic semiconductors.[4] For example, in spite of poly-Nepoxipropylcarbazol (PEPC) with 3 wt% of 2,4,7-trinitrofluorenone (TNF) being a highly photosensitive organic semiconductor, solar cells based on it have a very low efficiency (~ 0.001%).[5] The electrical conductivity of organic semiconductor films strongly depends on the conditions of their precipitation. An increase in electrical conductivity of polymer organic semiconductors is an important goal. Therefore, it is reasonable to grow polymer films by different technologies, including high gravity. This paper presents the results of the growth of PEPC-TNF polymer films with rapid stirring.

EXPERIMENTAL METHODS

The organic semiconductor PEPC-TNF was the object of this study. Figure 1 shows the molecular structure of PEPC-TNF (n was 4 to 6). The molecular weight of PEPC was about 1000 atomic units of mass. Polymer films of PEPC-TNF were grown from a solution with 1-3 wt% of the polymer in the organic solvent (for example, acetonitrile, tetrahydrofurane, etc.).

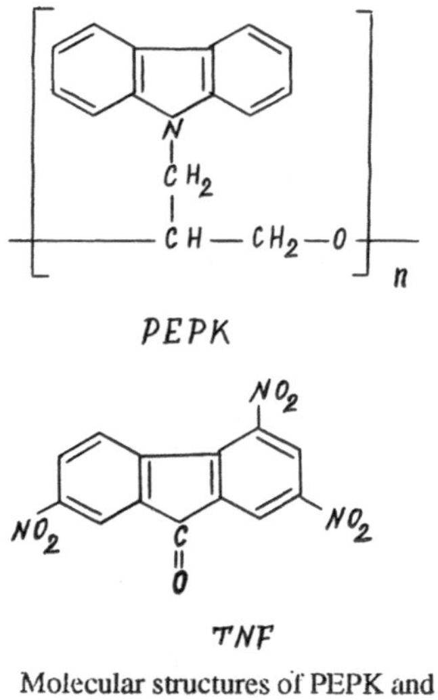

Figure 1. Molecular structures of PEPK and TNF.

Figure 2 shows our experimental setup for the growth of polymer films with stirring. This setup consisted of a cylindrical glass vessel (1), with lid (2) and base (3). A cylindrical metal substrate (4) without base and elastic support (5) were installed inside vessel (1). The diameter of substrate (4) was 15 cm and its height was 10 cm. The substrate (4) could be lifted by the lifting mechanism that consisted of pivots (6), lift (7), reducer (8), and electric motor (9). The glass mixer (10), which rotated the solution in the horizontal plane, was installed vertically along the vessel and the substrate axis. The upper end of mixer (10) was mechanically connected to an electric motor, which provided rotation at an angular speed of 1400 rpm. The aforementioned elements of the setup were installed on support (12). A glass pipe (13) and tap (14) also were installed at the bottom of vessel (1).

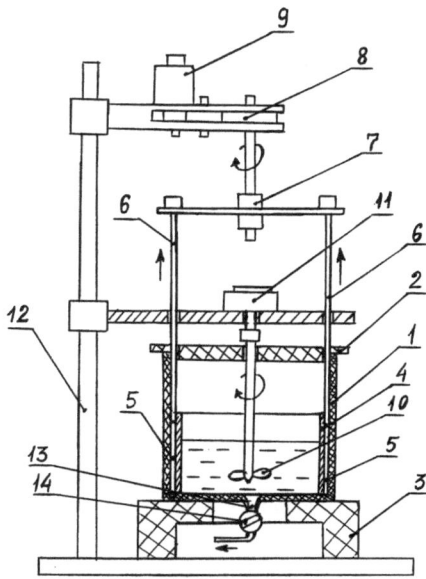

Figure 2. Experimental arrangement for growth of polymer films.

The experimental setup worked in the following way. The polymer solution was poured into vessel 1. It should be noted that the solution must wet the substrate 4. The electric motor 11 was started and the mixer 10 stirred the solution. After half an hour, electric motor 9 was started and substrate 4 was lifted at a speed of 1-10 mm/min. When the lower end of the substrate 4 was higher than the solution level, electric motor 11 was stopped and the solution was removed through tap 14. The precipitated polymer film on the substrate surface was dried at room temperature for 24 hours. Following this procedure a thin film of 1-2 mm thick was present.

The films were also grown on substrate 4 by another method. At the beginning of the process, when the solution was rotated, the substrate 4 was not lifted but instead the solution was removed very slowly through tap 14. In this case, it was very difficult to regulate the speed of the solution level decrease.

Reference samples were obtained by precipitation without stirring of the solution.

EXPERIMENTAL RESULTS

In order to measure the transverse electrical conductivity in the middle part of the inner surface of substrate 4, an electrical contact of 1 cm^2 was made on the polymer film with graphite conductive paste. On the opposite side of substrate 4, the polymer film was removed and the substrate was completely covered by graphite paste. The electrical resistance of the polymer films was measured by a measuring bridge, and the conductivity was calculated from these measurements.

The DC transverse electrical conductivity of the rotationally-prepared thin films in the direction normal to the substrate surface is 2 to 4 times higher than that of the reference samples. The electrical conductivity of the reference sample was equal to $3*10^{-8}$ Ohm^{-1} cm^{-1}.

Probably as a result of the influence of stirring, the structure of the polymer films changed. They could be more homogeneous and well-oriented, resulting in the observed increase in electrical conductivity. The described procedure can be used for the commercial production of thin films with higher transverse electrical conductivity.

We plan to investigate the longitudinal conductivity of films along the vertical and horizontal axes, and the conductivity of films on a plane substrate placed in the solution horizontally. It may be fruitful to grow polymer films in the experimental setup with rotation of a substrate.

REFERENCES

1. "Energetic conceptions of the Republic of Tajikistan," Dushanbe, Barki Tojik (1995) p.16.
2. A.B. Avakian *et al.*, "Gidroenergeticheskie resursy," Moscow, Nauka (1967) p.600 (in Russian).
3. G.A. Chamberlain, Organic solar cells: a review, *Solar Cells* 8:47 (1983).
4. M.I. Fiodorov, Kh.M. Akhmedov, and Kh.S. Karimov, "Solnechnye elementy na osnove organicheskih poluprovodnikov. Obzornaya informatsiya," Dushanbe, Tajik NIINTI (1989) p.51.
5. Kh.M. Akhmedov, Kh.S. Karimov, and M.I. Fiodorov, Organic solar cells, *Heliotechnics* 178 (1995).
6. S. Glenis, G. Tourillon, G. Horowitz, *et al.*, Electrochemically grown polythiophene and poly (3-methylthiophene) organic photovoltaic cells, *Thin Solid Films* 111:93 (1984).

INFLUENCE OF GRAVITY ON PERIKINETIC COAGULATION

R. Folkersma, A.J.G. van Diemen, J. Laven, and H.N. Stein

Department of Chemical Engineering and Chemistry
Eindhoven University of Technology
5600 MB Eindhoven, The Netherlands

SUMMARY

Experimentally, a difference was found in coagulation rates of polystyrene, quartz and silica particles in aqueous solutions, when comparing microgravity conditions (sounding rocket), 1g-conditions and high-g conditions (centrifuge). In microgravity, coagulation was faster than at 1g; at 1g, coagulation was faster than at high g. The coagulation rate was determined by measuring light transmission as a function of time. This dependence of coagulation rate on gravity seems not to coincide with theoretical expectations. However, at 1g the coagulation rate of neutrally buoyant dispersions was also higher than that of dispersions with a density difference between disperse and continuous phases. In order to explain these unexpected results, we discuss a number of relevant phenomena involved in the perikinetic coagulation process: hydrodynamics, interaction potentials, surface roughness, double layer characteristics during flow, and convection

INTRODUCTION

Coagulation is the process of clustering of small particles that are dispersed in a liquid phase, e.g. under the influence of London-Van der Waals attraction. Aggegation of dispersed particles is a vital step in many practical and technological processes: structure of soil, water treatment, biotechnology, emulsion polymerization. A pronounced influence of gravity on aggrgation is not expected. A previous investigation[1] showed, however, that orthokinetic (shear induced) aggregation under microgravity conditions was faster than at 1g, especially at low shear rates. The aim of the present investigation is to determine the influence of gravity on perikinetic (Brownian) coagulation.

THEORY

During coagulation, three types of forces are important: Van der Waals attraction, electrostatic repulsion and hydrodynamic interactions. The attractive forces act only over short ranges, which implies that the particles have to come near one another to aggregate. This can be established by Brownian motion, for which Von Smoluchowski[2] derived the following equation for the number N_k of aggregates of k particles after coagulation time t, starting with N_0 single particles:

$$N_k = \frac{N_0(t/T)^{(k-1)}}{(1+t/T)^{(k+1)}} \tag{1}$$

EXPERIMENTAL METHODS

The method used in this study to measure the coagulation rate was to record the light transmission. The relation between the decrease of turbidity and the decrease in the number of particles is given by:

$$1/E(dE/dt)_{t-0} = 1/\tau_0(d\tau/dt)_{t-0} = (-2+2(y+2)/3)f^{2/3}1/N_0(dN/dt) \tag{2}$$

where E is the extinction, τ is turbidity, f is the volume correction factor, $y = d\ln\tau/d\ln\lambda$, and λ is the wavelength in the medium. The value of $d\ln E/dt$ was calculated from the slope of the transmission vs. time curve.

Coagulation experiments were performed at 1g, high-g and micro-g conditions. Polystyrene (PS) was used as the dispersed phase in water (0.5 M NaCl). The volume fraction of the particles was 0.0001. The diameter of the PS particles was ≈ 2 μm. The coagulation rate was measured as a function of $\Delta\rho$ (density difference) and gravity. For the 1g experiments, the density of water was matched (or nearly matched) to that of PS by addition of sucrose, methanol or D_2O. At high g, the acceleration was varied from 1g to 7g by means of a centrifuge of the AMC in Amsterdam (Figure 1). For micro-g, experiments were performed in a sounding rocket at Kiruna, Sweden. Here, over 6 minutes of low gravity conditions were produced.

RESULTS AND DISCUSSION

Figure 2 shows the 1g results. The coagulation rate increased as the density of the liquid approached that of the dispersed phase. Figure 3 shows that the coagulation rate decreased in going from 1g to 2g, especially for dispersions with a small density difference. Figure 4 compares light intensity versus time curves from 1g and low g experiments. The coagulation rate was faster at low g. This retarding effect of increasing gravity on coagulation is unexpected.[3] The following possible causes are being considered:

Hydrodynamics. A hydrodynamic calculation showed,[4] although for considerably larger Reynolds numbers Re, a slowing of the approach of two particles when they are settling. The non-linear terms in the Navier-Stokes equations may remain important when two particles are close at small Re.

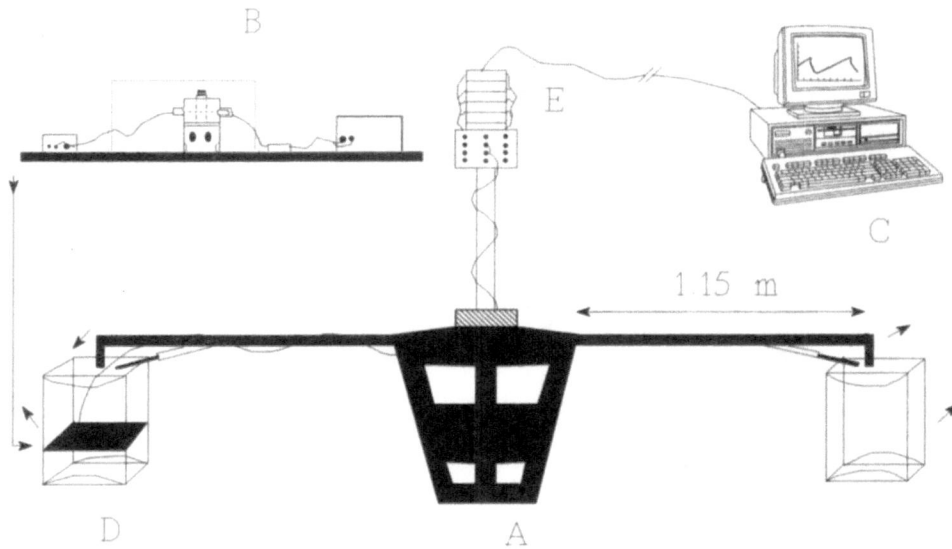

Figure 1. Schematic diagram of the Amsterdam centrifuge experiment. A: Rotating drive system. B: Coagulation apparatus. C: Computer and measuring system. D: Swing basket. E: Contact slip rings.

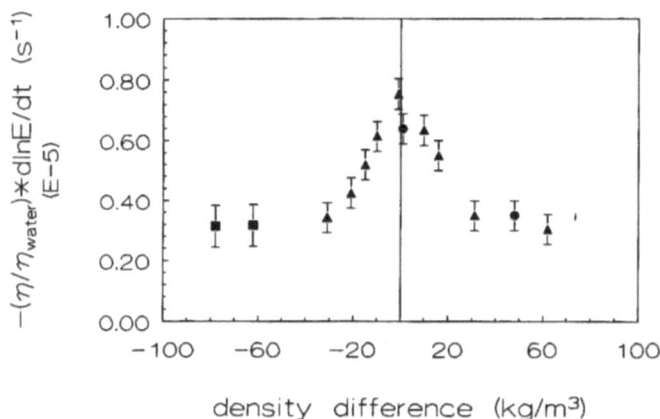

Figure 2. Results of 1g experiments. Viscosity-corrected, initial rate of change of extinction versus density difference between dispersed phase and polystyrene particles.
▲ methanol/water; ● D_2O/H_2O; ■ sucrose/water.

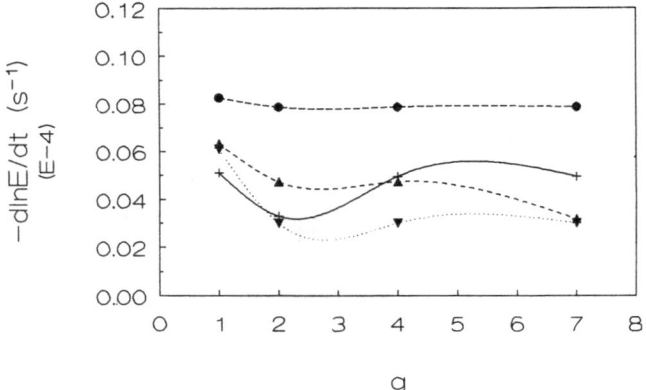

Figure 3. Results of centrifuge experiments. Initial rate of change of extinction as a function of acceleration, for different values of density difference $\Delta\rho$: + -31 kg/m^3; ▲ -16 kg/m^3; ● +1 kg/m^3; ▼ +17 kg/m^3.

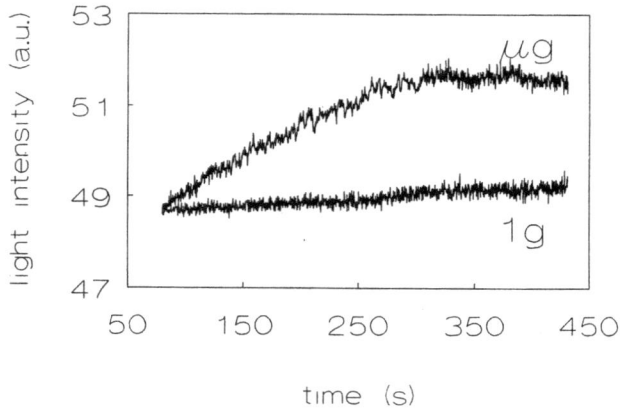

Figure 4. Light transmittance versus time for 1 g and low g experiments.

We are now employing a three-dimensional finite-element method to calculate the drag and lift coefficients of the spheres. In these calculations, the non-linear terms in the Navier-Stokes equation are not neglected, even at small Re.

Interaction potentials. Coagulation in the secondary minimum may be not strong enough to lead to lasting contact between the particles, when settling. Additionally the surface roughness of the particles may play a role. Calculation of attractive and repulsive energies were performed (example in Figure 5) using the exact results for repulsion at larger values of the surface potential (interpolating between the values of Verwey and Overbeek) and Clayfield's[5] formula for retarded London-van der Waals attraction.

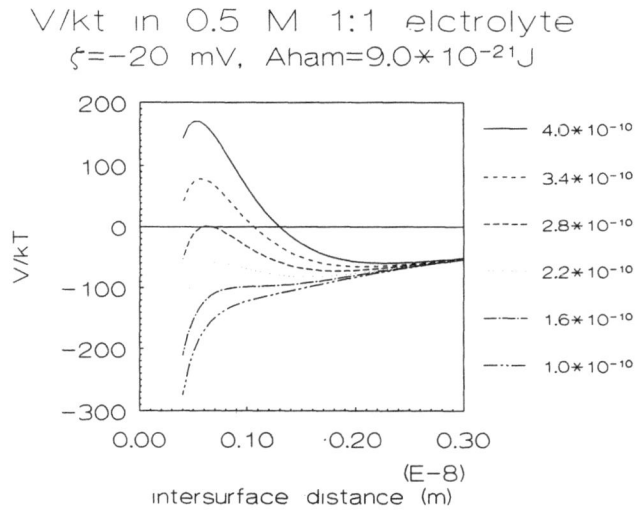

Figure 5. Calculated interaction potential for different thicknesses m of the Δ-layer.

Influence of flow on the charge distribution between two surfaces. When two particles approach each other, the liquid in the gap between them is squeezed out. This is accompanied by a pressure gradient in the gap. We consider this phenomenon here with regard to the generation of a streaming potential. The present investigation is intended to investigate whether this streaming potential may influence the pressure in the gap, such as to retard coagulation, and whether this may be influenced by gravity. Additionally, flow may distort the shape of the double layer, leading to dipole-dipole and higher order electrostatic interactions between the particles.

Convection in the vessel. In order to calculate the shear rates expected to occur in a "stagnant" dispersion subjected to temperature gradients and an acceleration field, we calculated the flow rates. In the present treatment, we started from a circular convection loop. In combination with the calculated interaction potentials, we are planning to calculate whether convection can disrupt the formation of doublets of particles.

REFERENCES

1. L.L.M. Krutzer, The influence of flow type and gravity on orthokinetic coagualation. Ph.D. Thesis, Eindhoven University of Technology (1993).
2. M. von Smoluchowski, *Z. Phys. Chem* 92:129 (1917).
3. J. Happel and H. Brenner, "Low Reynolds Number Hydrodynamics," Martinus Nijhoff Publisher, Dordrecht (1965); Chapter 6.
4. I. Kim, S. Elghobashi, and W.A. Sirignano, *J. Fluid Mech.* 246:465-488 (1993).
5. E.J. Clayfield, E.C. Lumb, and P.H. Mackey, *J. Coll. Int. Sci.* 37:382-389 (1971).

INFLUENCE OF CENTRIFUGATION ON COAGULATION
OF COLLOIDAL DISPERSIONS OF TEFLON™

J. Simmons,[1] L.L. Regel,[1] W.R. Wilcox,[1] and R. Partch[2]

[1]International Center for Gravity Materials Science and Applications
[2]Chemistry Department
Clarkson University
Potsdam, New York 13699-5814

ABSTRACT

The influence of centrifugation on Teflon® colloidal dispersions was studied using a table-top centrifuge and the High Inertia Rotating Behemoth (HIRB). Experiments at about 700 g resulted in a sticky, porous film of Teflon much different from solids created by coagulation using agitation at 1 g. Acceleration at 4 g without stirring caused coagulation only at the air/dispersion interface. Simultaneous stirring and acceleration gave a much different result than agitating the dispersion at 1 g.

INTRODUCTION

Colloidal dispersions have important applications in industry, biology and food technology. Colloid particles remain dispersed due to electrostatic repulsion between the particles.[1,2] Surfactants can be adsorbed onto the surface of colloid particles to increase the interparticle repulsion and make the dispersion more stable. Coagulation occurs when the repulsive potential between the particles is overcome, and involves Brownian motion, convection, sedimentation and creaming of the agglomerating particles.[3] Coagulation of colloidal dispersions of polytetrafluoroethylene (PTFE) can be induced by mechanical agitation. This phenomenon has been attributed to irreversible contact of the particles by air, because PTFE is poorly wet by the liquid (i.e., the contact angle is very high).

The study of the stability of colloids and their coagulation is of great importance to colloid science and applications. One aspect of this science that has scarcely been investigated is the effect of gravity. PTFE was chosen for the present study since it is a unique polymer used in many consumer and industrial applications. Despite being hydrophobic, aqueous dispersions of PTFE can be made stable by the addition of several types of surfactants.[4] PTFE dispersions are usually coagulated by some form of mechanical agitation or by addition of salts or acids. The objective of this project was to determine how centrifugation influences coagulation of PTFE colloidal dispersions and to examine if differences occur in the coagulated product.

EXPERIMENTAL METHODS

Two Teflon PTFE dispersions from E.I. du Pont de Nemours and Company were used in this study: one with 6 weight percent Triton X-100 surfactant added (called "Teflon 30") and one containing only surfactant remaining from the emulsion polymerization (simply called "Teflon"). The Teflon 30 dispersion was very stable and remained dispersed even with strong agitation, while the Teflon dispersion could easily be coagulated by shaking. The particles in both dispersions were about 0.2 μm in diameter.

One set of experiments was performed on Teflon 30 using a Sorval Superspeed Centrifuge run for 15 minutes at a rotation rate of 3000 rpm, which is equivalent to an acceleration of about 700 g. Centrifugation of 20 ml of the dispersion resulted in a pasty solid sedimented against the bottom of the centrifuge tube. Many of the particles remained dispersed in the supernatant, which was gently decanted. The sedimented solid was easily redispersed by stirring with 20 ml of deionized water. The resulting dispersion was centrifuged under the same conditions. All of the redispersed particles sedimented in the second centrifugation, leaving a clear supernatant. This time, more stirring was required to redisperse the sediment in deionized water. This dispersion was centrifuged. When an attempt was made to redisperse this sedimented solid, a small amount of it could not be dispersed. This solid was collected and the centrifugation was repeated. Again, an attempt to redisperse the sedimented solid resulted in some solid that could not be redispersed and was collected. This process was repeated two more times until no more sediment could be redispersed by stirring in distilled water.

Centrifugation of 10 ml of the low-surfactant Teflon dispersion containing also resulted in a sedimented solid at the bottom of the centrifuge tube. When an attempt was made to redisperse this sample in 10 ml of deionized water, most of the solid could not be redispersed. This solid was collected and the new dispersion centrifuged under the same conditions. A large part of the resulting sediment could not be redispersed and was collected. After a third centrifugation, all of the material was collected as a non-dispersible solid.

Two studies are in progress on Clarkson's HIRB centrifuge. (Details and results will appear later.[5]) The experiments on HIRB are performed in a swing bucket that is attached to the 1.5 meter radius arm of the centrifuge. The swing bucket is attached to the arm by a hinge that allows it to swing outward with increasing rotation rate, thereby aligning itself with the resultant acceleration. The objective of one study is to determine the effect of centrifugation on the dispersion with less surfactant. A centrifuge tube like those used in the desk-top centrifuge was secured inside the swing bucket and covered so that the wind created by the centrifuge would not affect the experiment. A 10 ml sample of the dispersion was used for the tests. Thus far, experiments have been done at 2 g and 4 g.

For the other study, a laboratory mixer with a rotation rate of 350 rpm was mounted on the top of HIRB's swing bucket. A glass mixing cell with a diameter of 5.5 cm and a height of 11.5 cm was attached to the bottom plate of the swing bucket. The mixing impeller was a 45° pitched-blade turbine with a diameter of 3.7 cm. The four turbine blades were 1.2 cm in length. Figure 1 shows a schematic drawing of the mixing apparatus in the swing bucket. Thus far, experiments have used 80 ml of a mixture made of 20 ml of the low-surfactant Teflon dispersion and 60 ml of deionized water. Coagulation was determined after stirring the dispersion for two or three hours at 1 g or 4 g.

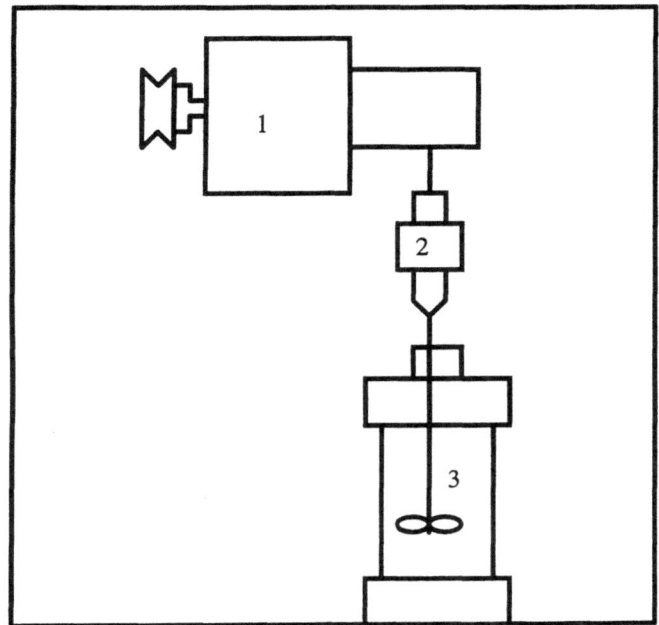

Figure 1. Experimental setup for stirring experiments on HIRB centrifuge. (1) Stirring motor and gear box; (2) Chuck to secure impeller shaft; (3) Mixing cell.

RESULTS

The non-dispersible solids created by centrifugation at ~700 g of the two dispersions were quite different. The solids created from the dispersion with surfactant (Teflon 30) were dense and so strong that separation into pieces required them to be torn apart with a spatula. They did not float in water. The solids created from the low-surfactant Teflon dispersion were less dense, with water and air captured within the structure. Compression of the solid caused water to flow out of it. These solids floated and were not wet by water. They sticked to surfaces they contacted, such as the spatula used to collect them and the sides of the vials used to store them. X-ray diffraction showed that all of these solids were Teflon. Scanning electron microscopy of samples from these two experiments revealed that the Teflon particles were fused together differently than a Teflon solid created by coagulation by agitation. Figure 2 shows non-dispersible solid obtained after four centrifugations of the Teflon 30 dispersion. Figure 3 shows solid created by one centrifugation of the low-surfactant dispersion. Figure 4 shows the product of the same dispersion coagulated by shaking at 1 g.

Centrifugation at 4 g of 10 ml samples of undiluted low-surfactant resulted in coagulation only at the air/dispersion interface. After both 5 and 7.5 hours the product covered almost the entire air/dispersion interface area and had the same dry weight, 0.22 gram. No solid was found in the rest of the dispersion. After 2 hours at 4 g, only 0.13 gram had coagulated, all at the air/dispersion interface. At 2 g, no coagulation was observed after 1.25 or 4 hours.

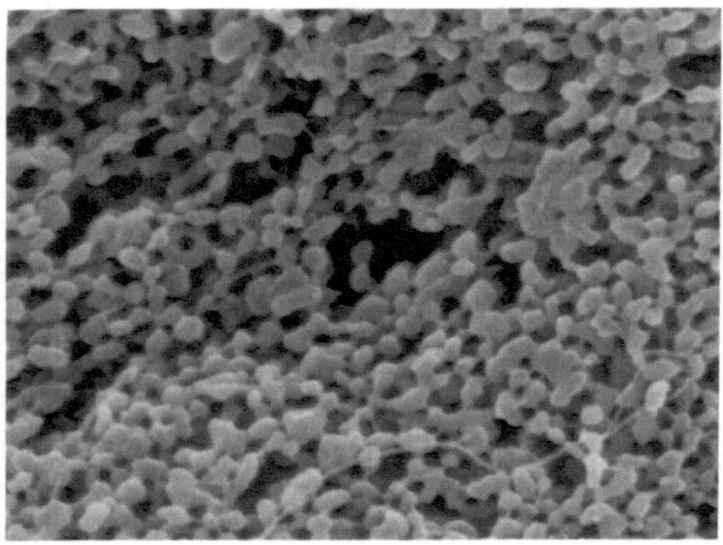

Figure 2. Scanning electron micrograph of solid created by centrifugation of the dispersion containing surfactant (Teflon 30) at 705 g. Original magnification of 15,000. Note 1 μm long white line

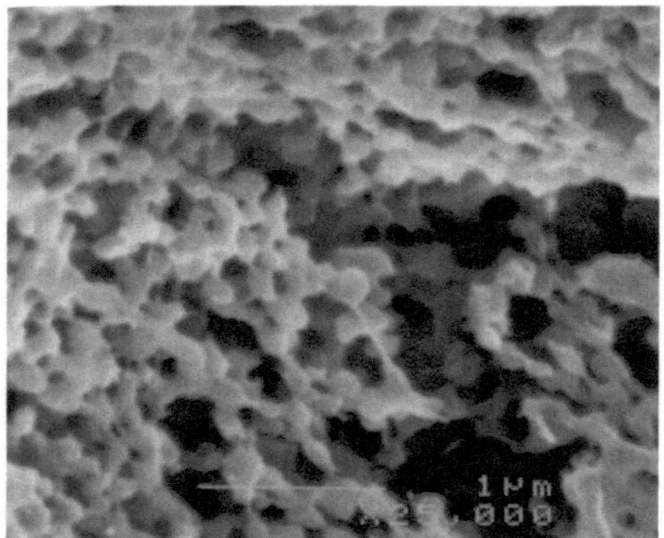

Figure 3. Scanning electron micrograph of solid created by centrifugation of Teflon dispersion with little surfactant at ~700 g. Original magnification of 25,000. Note 1 μm bar.

Figure 4. Scanning electron micrograph of solid created by coagulation of the low-surfactant dispersion by manually shaking at 1g. Original magnification of 25,000. Note 1 μm bar.

Stirring at 1g for 2 hours resulted in about 75% of the air/liquid surface being covered by flat, oval-shaped agglomerates with diameters of about 2-5 mm and a total dry weight of 0.15 gram. These non-wettable agglomerates had a dense surrounding layer supporting a less dense, very thin middle section. Attached to the impeller were many porous spherical agglomerates with diameters of 1-2 mm, and some agglomerates similar to those found at the air/water surface. A ring of coagulated material surrounded the impeller shaft where the air/water interface had been located during mixing. The total dry weight of the coagulation on the impeller was 0.16 gram. No coagulate was found on the bottom of the mixing cell. Stirring at 1 g for 3 hours yielded the same results, except that the agglomerates at the surface of the liquid had diameters of about 4 - 7 mm and a dry weight of 0.17 gram. The material coagulated on the impeller also had a dry weight of 0.17 gram.

An acceleration of 4 g significantly affected the coagulation caused by stirring. After 2 hours there was much less coagulate (0.03 gram) at the surface of the liquid, with less defined and weaker agglomerates than at 1 g. The coagulate (0.06 gram) on the impeller was very soft and not shaped like the well-defined spheres obtained at 1 g. Unlike the experiments without centrifugation, significant amounts (0.24 gram) of coagulated solid were found on the bottom of the mixing cell, including one piece over 1 cm in length. Results were similar for runs at 3 hours, except that more coagulation occurred than for 2-hour runs. There was 0.07 gram at the air-water interface, 0.14 gram on the impeller and 0.27 gram on the bottom of the mixing cell.

Acknowledgment

This research is partially supported by the National Science Foundation under grant DMR-9414304. We are grateful to du Pont for donating the Teflon dispersions used in this research.

REFERENCES

1. D. H. Everett, "Basic Principles of Colloid Science," The Royal Society of Chemistry, London (1988).
2. J. Lyklema, "Fundamentals of Interface and Colloid Science," Academic Press, London (1991).
3. J. Overbeek, Recent developments in the understanding of colloid stability, *in:* "Colloid and Interface Science," Academic Press, New York (1977).
4. S. Kratohvil and E. Matijevic, Stability of colloidal Teflon dispersions in the presence of surfactants, electrolytes, and macromolecules," *J. Colloid Interface Sci.* 57:104 (1976).
5. J. Simmons, "The Influence of Centrifugation on the Coagulation of Polytetrafluoroethylene Colloidal Dispersions," M.S. Thesis, Clarkson University, Potsdam, New York (1997).

SECONDARY FINITE AMPLITUDE FLOW IN A ROTATING PIPE AND ITS INSTABILITY

S.Ya. Gertsenshtein, N.V. Nikitin, and A.N. Sukhorukov

Moscow State University
Moscow 119899, Russia

ABSTRACT

We consider here the solution of two problems concerning the flow of a rotating fluid. In the first problem, we investigate the nonlinear stability of a viscous incompressible flow in a circular pipe rotating about its own axis. We solve the initial boundary value problem for the unsteady three-dimensional Navier-Stokes equations by the Bubnov-Galerkin method.[1-5] A series of methodological investigations is made. The nonlinear evolution of the periodic self-oscillating regimes is studied, and their characteristic stabilization times, amplitudes, and other integral and fluctuation characteristics found. The secondary instability of these finite-amplitude wave motions is examined. It is established that the secondary instability is initially weak and linear in character; the corresponding growth times are approximately an order greater than for the primary perturbations. There is the possibility of a sharp, explosive restructuring of the motion when the secondary perturbations reach a certain critical amplitude. A "survival curve"[5] is constructed, which makes it possible to determine the preferred perturbation, distinguishable from the rest if the initial perturbation amplitudes are equal, and the critical amplitude values starting from which the other perturbations may prevail even over the preferred one. The range of these surviving perturbations is obtained. It is shown that as a result of the nonlinear interaction of several perturbations at low levels of supercriticality, a periodic motion in the form of a single traveling wave is generated.

The second problem we investigate is the shear instability of a rotating vertical stratified column of fluid.

FIRST PROBLEM: METHODS

The problem is investigated using the unsteady three-dimensional Navier-Stokes equations. The velocity vector V_1 and the pressure p_1 are represented in the form of the sums $V_1 = V_0 + V$ and $p_1 = p_0 + p$ of the values for the main flow. In a rotating frame,

$\mathbf{V}_0 = (1-r^2)\mathbf{k_z} - \omega r \mathbf{k_\theta}$ and $p_0 = \omega^2 r^2/2 - 4z/R$, whose stability is being investigated via perturbations to $\mathbf{V}$ and p in:

$$\frac{\partial \mathbf{V}}{\partial t} + (\mathbf{V}_0\nabla)\mathbf{V} + (\mathbf{V}\nabla)\mathbf{V}_0 + (\mathbf{V}\nabla)\mathbf{V} = -\nabla p + \frac{1}{R}\Delta\mathbf{V}; \qquad div\mathbf{V} = 0 \qquad (1)$$

where $R = U_0 r_0 v^{-1}$ is the Reynolds number, $\omega = -\Omega r_0 U_0^{-1}$, U_0 is the maximum longitudinal velocity of the main flow, r_0 is the pipe radius, v is the kinematic viscosity of the liquid, Ω is the angular rate of rotation of the pipe, r, θ, and z are the dimensionless cylindrical coordinates, and $\mathbf{k_r}$ and $\mathbf{k_\theta}$ are the corresponding unit vectors. The perturbations are assumed to be periodic in z:

$$\mathbf{V}\left(r,\theta,z+\frac{2\pi}{\alpha},t\right) = \mathbf{V}(r,\theta,z,t), \qquad p\left(r,\theta,z+\frac{2\pi}{\alpha},t\right) = p(r,\theta,z,t) \qquad (2)$$

where $2\pi/\alpha$ is the interval of periodicity and α is the wave number. The usual non-slip conditions are imposed at the walls.

The approximate solution of the problem is found by the Galerkin method:

$$\mathbf{V}(r,\theta,z,t) = \sum_{k=-K}^{K}\sum_{n=-N}^{N} \mathbf{v}^{(k\alpha,n)}(r,t)e^{i(k\alpha z+n\theta)} =$$

$$\sum_{k=-K}^{K}\sum_{n=-N}^{N}\sum_{m=1}^{M} a_m^{(k\alpha,n)}(t)\varphi_m^{(k\alpha,n)}(r)e^{i(k\alpha z+n\theta)} \equiv \sum_{k=-K}^{K}\sum_{n=-N}^{N}\sum_{m=1}^{M} a_m^{(k\alpha,n)}(t)\Psi_m^{(k\alpha,n)}(r,\theta,z) \qquad (3)$$

where $\mathbf{v}^{(-k\alpha,-n)} = \overline{\mathbf{v}}^{(k\alpha,n)}$ (the overbar denotes the complex conjugate), and $\Psi_m^{(\beta,n)}$ are the vector eigen functions for stability of a liquid at rest in the pipe, as given by:[1,2,6]

$$\Psi_m^{(\beta,n)}(r,\theta,z) \equiv \varphi_m^{(\beta,n)}(r)^{i(\beta z+n\theta)}, \quad \Delta\Psi_m^{(\beta,n)} - \nabla p_m^{(\beta,n)} = -\mu_m^{(\beta,n)}\Psi_m^{(\beta,n)},$$

$$div\Psi_m^{(\beta,n)} = 0, \quad \Psi_m^{(\beta,n)}\Big|_{r=1} = 0$$

$$\int_0^1 \left\langle \Psi_m^{(k\alpha,n)}\Psi_l^{(j\alpha,s)}\right\rangle rdr = \delta_{ml}\delta_{kj}\delta_{ns}, \qquad \int_0^1 \left\langle \Delta\Psi_m^{(k\alpha,n)}\overline{\Psi}_l^{(j\alpha,s)}\right\rangle rdr = -\mu_m^{(k\alpha,n)}\delta_{ml}\delta_{kj}\delta_{ns} \qquad (4)$$

$$\int_0^1 \left\langle \nabla p\overline{\Psi}_l^{(j\alpha,s)}\right\rangle rdr = 0, \quad <f> \equiv \frac{\alpha}{4\pi^2}\int_0^{\frac{2\pi}{\alpha}}dz\int_0^{2\pi} fd\theta$$

Here, the angular brackets denote averaging over the homogeneous variables θ and z.

The method of solving this problem is described in Reference 6. The equations for the unknown $a_m^{(k\alpha,n)}$ (t) are obtained after carrying out the standard orthogonalization procedure:

$$\frac{da_m^{(k\alpha,n)}}{dt} + \frac{1}{R}\mu_m^{(k\alpha,n)}a_m^{(k\alpha,n)} = -\int_0^1 \left\langle [(\mathbf{V}_0\nabla)\mathbf{V} + (\mathbf{V}\nabla)\mathbf{V}_0 + (\mathbf{V}\nabla)\mathbf{V}]\overline{\Psi}_m^{(k\alpha,n)}\right\rangle rdr \qquad (5)$$

The integral on the right side of equation 5 is determined approximately in accordance with the Gaussian quadrature formula. The system of ordinary differential equations represented by 5 is solved on a computer by the Kutta-Merson method with automatic step selection and accuracy control.

RESULTS FOR PROBLEM ONE

Instability of the Main Flow

Our numerical experiments were concentrated on the range of parameters over which instability of the main flow with respect to perturbations of various types is observed. Thus, for example, at the values R = 200 and ω = 0.5, corresponding to the main series of calculations, we have linear instability with respect to three-dimensional wave disturbances of the form:

$$\mathbf{V}(r,\theta,z,t) = \text{Re}\left\{w(r,t)e^{i(\alpha z+n\theta)}\right\} \tag{6}$$

with azimuth numbers n=1 (spiral waves), 2, and 3.[7-9] In this case, the perturbation with maximum growth corresponds to the wave vector $\{\alpha, n\}=\{0.65, 1\}$.

Initially, we investigated "single-wave" periodic motions developing in the flow as a result of the evolution of linearly growing perturbations. These regimes can be obtained by giving the initial perturbation in the form of a solitary wave, $\mathbf{V}(r,\theta,z,0) = \text{Re}\left\{w(r)e^{i(\alpha z+n\theta)}\right\}$, with a wave vector from the region of linear instability and a certain velocity distribution w(r). In this case, the motion at t > 0 is described as a sum of waves with multiple wave vectors $\{k\alpha, kn\}$, and k = 0,1,2,...., and equation 3 takes the form:

$$\mathbf{V}(r,\theta,z,t) - \mathbf{v}^{(0,0)}(r,t) + 2\,\text{Re}\left[\sum_{k=1}^{K}\mathbf{v}^{(k\alpha,kn)}(r,t)e^{ik(\alpha z+n\theta)}\right] =$$

$$\sum_{m=1}^{M}a_m^{(0,0)}(t)\varphi_m^{(0,0)}(r) + 2\,\text{Re}\left[\sum_{k=1}^{K}\sum_{m=1}^{M}a_m^{(k\alpha,kn)}(t)\varphi_m^{(k\alpha,kn)}(r)e^{ik(\alpha z+n\theta)}\right] \tag{7}$$

The values of K and M and the number of nodes L in the Gaussian quadrature formula were varied from 1 to 3, from 2 to 24, and from 2 to 32, respectively. Methodological calculations confirmed the good convergence of this method. Thus, when R=200 and ω=0.5, the results of the calculations no longer show any qualitative change starting from K=1, M=4, and L=6. The main calculations were carried out for K=2, M=3 and L=12.

We note that the contribution of the higher harmonics is unimportant. For example, in the steady-state self-oscillating regime obtained for K=4 (α=0.5, n=1), the kinetic fluctuation energy, corresponding to the second (k=2), third (k=3), and fourth (k=4) harmonics is 15%, 2.2%, and 0.33% of the main wave energy, respectively. Varying the K parameter has little effect either on the integral or the fluctuational characteristics of the stabilized motion.

Calculations for the initial perturbation having the form of a solitary wave (Equation 6) with wave vector $\aleph_1=\{\alpha, n\}$ from the region of linear instability (R=200, ω=0.5) showed that there develops in the flow a periodic secondary motion in the form of a traveling wave with an exponential energy distribution over harmonics with wave vectors $\aleph_k=k\aleph_1$, k=1,2,...,K. In this case, the average flow is characterized by values of the liquid flow rate and angular momentum about the pipe axis that are reduced relative to the starting steady flow. The profile of the average longitudinal velocity component $<u>=1-r^2+v^{(0,0)}k_z$ is fuller, and the rotational component $\langle w \rangle = -\alpha r + v^{(0,0)} k_0$ and its gradient in the neighborhood of the pipe axis are somewhat reduced. With distance from the axis (at $r > 0.4$), the gradient increases somewhat. In the vicinity of the flow "core," there is a tendency for a calm zone to form.

The decrease in the angular momentum about the pipe axis is associated with the emergence in the transition phase of a tangential component of the shear stress at the wall, which tends to increase the rate of rotation of the pipe. In the limiting regime the wall friction returns to the initial value.

In the limiting regime the radial distribution of the root-mean-square fluctuation velocity:

$$I_v(r,t) = \left\langle \left| \mathbf{V}(r,\theta,z,t) - \mathbf{v}^{(0,0)}(r,t) \right|^2 \right\rangle^{\frac{1}{2}} \equiv \left[2 \sum_{k=1}^{K} \left| \mathbf{v}^{(k\alpha, kn)}(r,t) \right|^2 \right]^{\frac{1}{2}} \tag{8}$$

is constant almost everywhere, except for a relatively narrow wall zone. In this case, the maximum intensity of the radial, tangential, and longitudinal velocity components:

$$I_v(r,t) = \left[2 \sum_{k=1}^{K} \left| v_k(r,t) \right|^2 \right]^{\frac{1}{2}} \tag{9}$$

$$I_w(r,t) = \left[2 \sum_{k=1}^{K} \left| w_k(r,t) \right|^2 \right]^{\frac{1}{2}} \tag{10}$$

$$I_u(r,t) = \left[2 \sum_{k=1}^{K} \left| u_k(r,t) \right|^2 \right]^{\frac{1}{2}} \tag{11}$$

is attained on the pipe axis at $r = 0$ for the first two components, and in the neighborhood of $r = 0.5$ for the third. Here v_k, w_k, and u_k are the radial, tangential, and longitudinal components of the vector $v^{(k\alpha,\ kn)}$, respectively. The presence of a local minimum on the $I_w(r)$ curve is associated with the fact that the tangential component of the fluctuation velocity changes sign at $r \approx 0.5$. It is interesting to note that the maximum values of the various fluctuation components of the velocity differ only slightly; e.g., for $\alpha = 0.65$ and $n = 1$ the maxima of their root-mean-square values are 6-7% of the axial velocity of the main flow.

As in References 3 and 4, the characteristic "nonlinear" formation times of the limiting regimes in question are approximately equal to the corresponding Reynolds numbers.

Curiously, the phase velocities for fairly large-scale perturbations (when $\alpha < 0.3$) are negative. However, as in the linear case,[9] in the frame of reference rotating with the pipe the propagation velocity of the waves is always positive. For stable finite-amplitude regimes ($0.38 < \alpha < 0.95$ and $n = 1$), as α increases the propagation velocity grows from 0.58 to 0.63. As compared with the linear case, a certain decrease in phase velocity by approximately 0.02 - 0.03 is observed.

We note that the most intense perturbations are the long-wave finite-amplitude perturbations, whose scale is approximately twice as great as that of the perturbation corresponding to a minimum of the liquid flow rate.

Secondary Stability of Single-Wave, Finite-Amplitude Periodic Flows

Considerable interest attaches to the study of the secondary stability of the "single-wave" finite-amplitude periodic flows described above. In these investigations, we used the following representation of the velocity vector:

$$\mathbf{V}(r,\theta,z,t) = \mathbf{v}^{(0,0)}(r,t) + 2\,\mathrm{Re}\{\sum_{k=1}^{2}[\mathbf{v}^{(k\alpha_1,kn_1)}(r,t)e^{ik(\alpha_1 z + n_1\theta)} +$$

$$+\mathbf{v}^{(k\alpha_2,kn_2)}(r,t)e^{ik(\alpha_2 z + n_2\theta)}] + \mathbf{v}^{(\alpha_1+\alpha_2,n_1+n_2)}(r,t)e^{i[(\alpha_1+\alpha_2)z+(n_1+n_2)\theta]} + \qquad (12)$$

$$+\mathbf{v}^{(\alpha_1-\alpha_2,n_1-n_2)}(r,t)e^{i[(\alpha_1-\alpha_2)z+(n_1-n_2)\theta]}\}$$

where $\{\alpha_1; n_1\} = \aleph_1$ and $\{\alpha_2; n_2\} = \aleph_2$ are the wave vectors of the primary and secondary perturbations, respectively. The initial distributions, $v^{(k\alpha_1,kn_1)}(r,0)$, $k = 0,1,2$, were determined from the solution of the problem in the "single-wave" representation (Equations 6 and 7). The secondary perturbation with t = 0 was assigned the following form:

$$v^{(\alpha_2,n_2)}(r,0) = A_2\varphi_1^{(\alpha_2,n_2)}(r), \quad 10^{-6} \leq \overline{A}_2 \leq 10^{-2} \qquad (13)$$

These calculations demonstrated the instability of single-wave self-oscillating regimes with azimuth numbers n = 2 and 3 for all α, and moreover, of the "spiral" waves (n = 2) in the long-wave $\alpha < 0.38$ and short-wave $\alpha > 0.95$ parts of the spectrum. The periodic flows corresponding to these wave vectors disintegrate, in particular, upon the superposition of small ($A_2 \approx 10^{-5}$) secondary perturbations with wave vector $\{0.5, 1\}$. Thus, in the region of the parameters in question it possible to observe only "spiral" secondary flows with a wavelength of approximately 3 to 8 pipe diameters.

It is interesting to note that the secondary instability is much less intense than the linear instability of the initial rotating Hagen-Poiseuille flow. Thus the development of secondary perturbations with wave vector $\{0.5, 1\}$ and initial amplitude $A_2 = 10^{-5}$, superimposed on a steady-state self-oscillating motion with wave vector $\{0.5, 2\}$, is initially exponential in character with a growth rate one order less than when the same perturbations develop against the background of the main flow. At the same time, when the secondary perturbations reach a certain level the motion is rapidly restructured. For the secondary perturbations in question, this level is about 10% of the self-oscillation energy level; under these conditions the initial periodic flow is sharply damped and a new one develops.

Interaction of Secondarily Stable Perturbations

Within the framework of representation in equation 12, we investigated the interaction of two "secondarily" stable perturbations of the same type with initial conditions:

$$\mathbf{v}^{(\alpha_1;1)}(r,0) = A_1\varphi_1^{(\alpha_1;1)}; \quad \mathbf{v}^{(\alpha_2;1)}(r,0) = A_2\varphi_1^{(\alpha_2;1)} \qquad (14)$$

We showed that if the initial amplitudes are sufficiently small and equal ($A_1 = A_2 \approx 10^{-4}$), the perturbation with the greater growth rate under the linear theory will "survive." However, in the case of different initial amplitudes, the end state is determined by the initial state of the system. For each perturbation with wave vector $\{\alpha_1, 1\}$ and amplitude A_1 it is possible to construct a relation between the critical amplitude A and the wave number α, a so-called survival curve,[5] such that for values of A_2 below the curve the wave corresponding to $\{\alpha_1,$

1} survives, while for values of A_2 exceeding $A(\alpha_2)$ the wave corresponding to $\{\alpha_2, 1\}$ survives. In particular this curve shows that perturbations close to those with maximum growth under the linear theory are the most preferred.

It is interesting to note that the interaction of perturbations with sufficiently large initial amplitudes may lead to the survival of waves with smaller growth rates than with the linear theory. Thus, when $\aleph_1 = \{0.5; 2\}$, $\aleph_2 = \{1; 1\}$, and $A_1 = A_2 = 10^{-2}$, the periodic flow formed corresponds to $\aleph_1$, despite the fact that the growth rate of the second wave (within the framework of the linear theory) has a somewhat greater value.

We note that for a relatively low level of supercriticality in all the calculation variants considered, the interaction of several (up to four) waves from the region of linear instability leads to the survival of one of them and the formation of a secondary self-oscillating regime of the traveling wave type.

SECOND PROBLEM

Our second problem is the shear instability of a rotating vertical stratified column of fluid. The linear problem is solved in the frame of the Euler equations:

$$\frac{\partial \mathbf{V}}{\partial t} + (\mathbf{V}_0 \nabla)\mathbf{V} + (\mathbf{V}\nabla)\mathbf{V}_0 = -\frac{1}{\rho}\nabla p, \quad div\mathbf{V} = 0 \tag{15}$$

The full velocity is:

$$\overline{\mathbf{V}}(r,\theta,z) = \mathbf{V}_0\{u_0, v_0, w_0\} + \varepsilon\mathbf{V}\{u, v, w\} \tag{16}$$

The basic flow, which we will investigate for its stability is:

$$u_0 = 0, \quad v_0 = v_0(r), \quad w_0 = const, \quad p_0 = p_0(r), \quad \frac{1}{2}v_0^2 = -\frac{1}{\rho}p_{0r} \tag{17}$$

We will find perturbations of the form:

$$\mathbf{V} = \tilde{\mathbf{V}}(r)e^{i(kz+m\theta-\beta t)} \tag{18}$$

This free layer model has been calculated for the following parameters:

1) $\mathbf{V}_{01} = a_1 r, \quad 0 \le r \le r_1, \quad r_1 = 1, \quad \rho = \rho_1$

2) $\mathbf{V}_{02} = a_2 r + (a_1 - a_2)\dfrac{1}{r}, \quad r_1 \le r \le r_2, \quad \rho = \rho_2$

3) $\mathbf{V}_{03} = (a_2 r_2^2 + (a_1 - a_2))\dfrac{1}{r}, \quad r_2 \le r, \quad \rho = \rho_3$

The boundary conditions were derived by assuming that the perturbations are damped at infinity and from velocity and pressure continuity at $r = r_1$ and $r = r_2$.

A set of calculations was made to find the dependence of the growth coefficient β_i and phase velocity $c_r = \beta_r / m_l$ on the non-dimensional wave number $m_l = m (r_2 - r_1) / r_1$. We investigated the dependence of these values on the basic flow profile ($\mathbf{V}_{01}$, $\mathbf{V}_{02}$, $\mathbf{V}_{03}$) and the densities (ρ_1, ρ_2, ρ_3).

The growth coefficient β_i depends mainly on the velocity gradient on the second layer ($\mathbf{V}_{02}(r)$). The phase velocity c_r depends on the shape of the velocity profile and in particular on the position of the flex point.

In order to investigate the influence of the density difference, let us set $\beta_2 = 1$. If $d\mathbf{V}_{02}/dr > -2$, then β_i decreases with decreasing ρ_1 and *vice versa*. When $d\mathbf{V}_{02}/dr < -2$, the growth coefficient increases. The value of β_i always increases with decreasing ρ_3.

One can find the dependence of the phase velocity c_r on densities by substituting flow with variable density for the corresponding (with β_i) flow with constant density.

REFERENCES

1. G.I. Petrov, Application of the Galerkin method to the problem of stability of a viscous flow, *Prikl. Mat. Meth.* 4:3 (1940).
2. O.A. Ladyzhenskaya, "Mathematical Problems of the Dynamics of a Viscous Incompressible Fluid," [in Russian], Nauka, Moscow (1970).
3. A.E. Orszag and A.T. Patera, Secondary instability of wall-bounded shear flows, *J. Fluid. Mech.* 128:347 (1983).
4. B.L. Rojdestvenskii and V.G. Primak, "Chislennoe modelirovanie dvumernoi turbulentnosti v ploskom kanale," Preprint N20, M.V. Keldysh Institute of Applied Mathematics, USSR Academy of Science, Moscow (1981).
5. S.Ya. Gertsenshtein, E.B. Radichev, and V.M. Schmidt, Konechno-amplitudnye konvektivnye dvijeniya v sloe rastvora c tverdimy granitsami, *Dokl. Acad. Nauk USSR* 266:1330 (1982).
6. H. Salwen and C.E. Grosch, The stability of Poiseuille flow in a pipe circular cross section, *J. Fluid Mech.* 54:93 (1972).
7. T.J. Pedley, On the instability of viscous flow in a rapidly rotating pipe, *J. Fluid Mech.* 35:97 (1969).
8. P.A. Mackrodt, Stability of Hagen-Poiseuille flow with superimposed rigid rotation, *J. Fluid Mech.* 73:153 (1976).
9. F.W. Cotton and H. Salwen, Linear stability of rotating Hagen-Poiseuille flow, *J. Fluid Mech.* 108:101 (1981).

CORIOLIS EFFECT ON TRANSPORT PHENOMENA
IN ROTATING SYSTEMS: A NUMERICAL STUDY

Daniel T. Valentine and Craig C. Jahnke

Mechanical and Aeronautical Engineering Dept.
Clarkson University, Potsdam, NY 13699-5725

ABSTRACT

The results of a numerical investigation of the Coriolis effect in flows through rotating-rectangular conduits are presented. The secondary flows that are induced are described from the viewpoint of a vorticity-production mechanism. In the centrifugation of fully-developed flows through rectangular passages, as the rate of rotation increases, the secondary flows induced by the Coriolis effect undergo a number of interesting transitions that have an impact on transport phenomena. The main purposes of this investigation are to describe the Coriolis effect in terms of vorticity transport and to determine the evolution of steady solutions as the relevant parameters describing their dynamics are varied. In this study we found a region of nonunique solutions over a range of Reynolds numbers. We also found that the onset of roll cells is associated with a turning point bifurcation.

INTRODUCTION

At Clarkson University there is a relatively new centrifuge that is devoted to experimental investigations of centrifugation in materials science. It is the principal laboratory facility of the International Center for Gravity Materials Science and Applications (ICGMSA). As pointed out by Regel (Director of ICGMSA) and Wilcox (Co-Director),[1] one area of experimental research being considered for this facility is research on forced convection. One of the questions of interest to the flow-systems engineering and aerospace science communities is: What effect does rotating a piping system have on the flow through the system? Regel and Wilcox raised this question to the first author almost a year before this workshop (the 3rd International Workshop on Materials Processing at High Gravity). This paper is a formal response to their interesting question.

The effects of centrifugation on materials processing have been investigated and reviewed by Regel et al.[2,3] The mechanics of fluids in rotating environments is not only of interest to the materials processing community, it is also of interest to NASA and other space agencies. Space

Shuttle experiments have already been performed with a rotating microscope. Biological experiments in a centrifuge are being planned for the International Space Station. It is likely that long manned missions to Mars will require rotation of the spacecraft in order to avoid physiological deterioration of the astronauts. The rotation of a spacecraft orbiting the earth (usually one revolution with each orbital cycle) is another example in which a rotating environment can impact flows in fluid mechanical systems on board the craft; see, e.g., Yuferev.[4]

Transport phenomena in flow-through systems that would be influenced by the Coriolis force include the pressure drop in flow through tubes, the onset of turbulence, the division of flow in piping networks, heat transfer to the tube wall, and mass transfer both to the tube wall and within the flow. Multi-phase flows, as in the circulation of blood, would also be influenced by the Coriolis force. This could, in turn, influence many physiological processes, such as blood pressure, clotting, arteriosclerosis, cell movement into capillaries, respiration, and digestion. To begin to develop an understanding of the influence of the Coriolis force on flows through tubes of rotating systems, a computational investigation was undertaken to determine the influence of the Coriolis force on laminar flows in straight sections of pipe.

Rotating fluid systems viewed from reference frames attached to the rigid boundaries of the system are subject to Coriolis and centrifugal forces. We need to determine the implications on the fluid mechanics of mounting a piping system on a centrifuge. To begin to examine the Coriolis effect on the fluid mechanics, we considered the problem of fully-developed laminar flow in a rectangular pipe rotating about an axis perpendicular to the bulk-flow direction. This is the simpliest problem that can be studied to examine the Coriolis effect. We examined low to moderate Reynolds number flows over a wide range of Rossby numbers because of their practical importance in material processing flows and in physiological flows on rotating platforms.

There is already a developing body of literature on flows in rotating pipes. Barua[5] described the secondary flow in a slowly rotating circular pipe with fully-developed laminar flow. The axis of rotation in Barua's study was perpendicular to the centerline of the pipe. The secondary flow consists of a counter-rotating double-vortex configuration similar to that which occurs in a stationary curved pipe. This secondary flow causes a rise in the friction factor. Benton[6] computed the effect of Earth's rotation on flows through moderately sized pipes at sea level. Although quite small, he found that pipes parallel to the equator are influenced the most by Earth's rotation. Benton and Boyer[7] examined laminar flow in a rapidly rotating duct of arbitrary cross section. In this case the flow in the interior of the duct is approximately geostrophic.

There were a number of investigations over the last 20 years[8,9,10] that dealt with turbulence modeling problems in rotating channel flows. The published investigations utilized k-ϵ models with curvature and Coriolis corrections. Limited successes were reported. Earlier than the computational studies[11] there were experimental investigations in the turbulent flow regimes that provided data to compare with later computational models. Wagner and Velkoff [12] investigated experimentally the entrance region of turbulent flow in a rotating, rectangular duct. In the present study we do not deal with turbulent flows. Instead, we investigate laminar flows, because data in the laminar flow regime are limited at best. The flows are generally laminar for most biological flows, for coolant loop design problems for the Mars mission, and for material processing problems. In addition, the secondary flow features, i.e., the scales of motion which are of the order of the size of the flow domain, are observed to occur in both laminar and turbulent flows in rotating channels. Hence, investigating laminar flows is useful in helping to interpret similar phenomena in turbulent flow; see, e.g., Lezius and Johnston.[13]

Speziale,[14,15] and Speziale and Thangam[16] investigated the moderate rotation case for fully-developed laminar flows in rectangular channels. The assumption of fully-developed flow is a significant and important simplification that facilitates computational analysis considerably;

this is elaborated in the section of this paper entitled Mathematical Model. Speziale et al., and ourselves computed the secondary flows over a range of rotation rates, Reynolds numbers and two cross-section aspect ratios. For weak rotation rates the secondary flow is comprised of a double-vortex. As rotation rate exceeds a certain critical value, roll-cell vortices are induced near the center of the duct adjacent to the wall nearest the axis of rotation (i.e., adjacent the "low-pressure wall"). When the rotation rate exceeds a second critical value, the roll-cell vortices are suppressed and the flow assumes a Taylor-Proudman configuration (geostrophic flow) in the interior of the duct.

The number of roll cells that may appear depends on the aspect ratio of the channel. Six cells were predicted for relatively large aspect ratios, i.e., for an 8×1 channel cross section, where the "8" side is parallel to the axis of rotation. This is not surprising since cellular structures that occur tend to be nearly circular in shape, with the smallest dimension of the cross-section controlling the size of the possible cells. The roll cells have been related to a hydrodynamic instability mechanism; this conjecture is based on a comparison of the numerical predictions of critical conditions for the onset of roll-cell instability in channels reported by Speziale and Thangam with the infinite aspect ratio predictions based on linear stability theory reported by Hart[17] and Lezius and Johnston[3]. The numerical predictions of the critical conditions for the onset of roll-cell instability compared favorably with the predictions from linear stability theory.

Transport of heat in rotating channels with laminar throughflow has recently been examined by Fann, Yang and Mochizuki.[18] They investigated numerically the developing region from the inlet to up to 20 hydraulic diameters from the entrance. The flow was radially outward from the axis of rotation. Tekriwal[19] numerically studied turbulent heat transfer in rotating ducts that simulate the cooling-flow channels built into turbine blades. Previous studies have also been conducted on the transport of heat in rotating channels. By comparison, there is a dearth of information on laminar and transitional flows of homogeneous fluids in rotating channels.

Other work on the circular pipe was reported by Ishigaki.[20,21] He investigated the analogy between the flows through rotating pipes and curved pipes. For mild curvature and low rotation rates at equivalent conditions the Nusselt numbers and the friction factors for the two flows are essentially the same. This is because the secondary flows are quite similar for the two flows at an equivalent Dean number. For circular pipes under these conditions, roll cells and Dean vortices do not occur. Speziale[14] hinted at the notion of an analogy for situations where roll cells and Dean vortices occur. However, in comparing roll-cell vortices with Dean vortices, both of which arise from apparent instability mechanisms, we find that they do not occur on the same wall. In fact, the roll cells occur on the inner (or "low pressure") wall while the Dean vortices, which were investigated by Cheng et al.,[22] occur on the outer wall. Hence the analogy cannot be fully complete. On the other hand, it is not surprising that friction and heat would be the same for similar-strength secondary vortices at conditions without roll cells or Dean vortices.

The paper by Speziale[14] was the first numerical simulation of fully-developed laminar flow in rotating, rectangular pipes. His study was for an aspect ratio of two, with the pipe rotating about an axis parallel to the longer side. The results he reported were preliminary because he used a relatively course grid. Subsequently, Speziale[15] reviewed and reported some improvements to his original study. The present investigation revisits this problem to clarify some of the details and to extend the previously published results. To provide a clearer description of the Coriolis effect on this flow, we describe it in terms of a vorticity-production mechanism. Previously, only streamline patterns were reported. In the present work we also present the vorticity field to shed additional light on the physics. In addition, we examine the bifurcation diagram of steady states that we computed over a range of Reynolds numbers to study the qualitative nature of the changes in the secondary flow pattern when roll cells appear near the center of the wall closest to the axis of rotation.

The next section of this paper describes the mathematical problem solved. This is followed by a presentation and discussion of the computational results.

MATHEMATICAL MODEL

The problem we examined is illustrated in figure 1. It is the fully-developed laminar flow through a rectangular pipe rotating at a constant angular speed, Ω, about the y axis. The bulk flow is moving in the z direction. The secondary flow, (u,v), in the (x,y) plane (which is a cross section of the pipe) is sketched in the figure. This secondary flow is the basic flow structure at relatively low Reynolds numbers before any instabilities modify the pattern. It is this secondary flow, which is induced by the Coriolis force, and its modifications at moderate Reynolds numbers that is of primary interest in the present study.

The equations of incompressible flow in a reference frame rotating at Λ in a gravitational field g not necessarily aligned with Λ are described next. The Navier-Stokes, continuity and temperature equations of the flow of an incompressible, Boussinesq fluid are:

$$\rho(\frac{\partial U}{\partial t} + U \cdot \nabla U) = - \nabla p - 2\rho\Lambda \times U - \rho\Lambda \times (\Lambda \times r) + \rho g + \mu\nabla^2 U, \qquad (1)$$

$$\nabla \cdot U = 0, \qquad (2)$$

$$\frac{\partial \theta}{\partial t} + U \cdot \nabla\theta = \alpha\nabla^2\theta, \qquad (3)$$

respectively. In this system of equations $U = (u,v,w)$ is the velocity vector field in the rotating frame $x = (x,y,z)$. The parameter ρ is the mass density, p is the pressure, μ is the dynamic viscosity, α is the thermal diffusivity, and the magnitude of r is the radial distance from the axis of rotation.

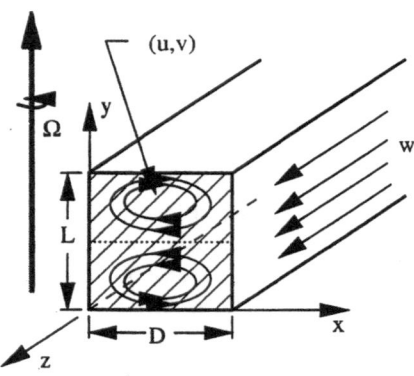

Figure 1. Illustration of the rotating-flow system examined.

For a Boussinesq fluid the temperature, T, is related to the density in the following way:

$$\rho = \rho_o - \rho_o \beta (T - T_o). \tag{4}$$

Rearranging this relationship, we may write:

$$\frac{\rho}{\rho_o} = 1 - \beta\theta, \tag{5}$$

where $\theta \equiv T - T_o$, which is the difference between the temperature and a reference temperature, T_o. The mass density at the reference temperature is ρ_o. The Boussinesq approximation implies:

$$\beta\theta \ll 1. \tag{6}$$

Dividing equation (1) by ρ_o, substituting for ρ/ρ_o , and invoking the Boussinesq approximation, we obtain:

$$\frac{\partial U}{\partial t} + U \cdot \nabla U = - \frac{1}{\rho_o} \nabla P - 2\Lambda \times U + \frac{\mu}{\rho_o} \nabla^2 U + B, \tag{7}$$

where $P \equiv p - \tfrac{1}{2} \rho_o (\Lambda \times r) \cdot (\Lambda \times r) + \rho_o \, g \, z_g$ is the reduced pressure, z_g , is the coordinate direction opposite to the direction of g, and $B \equiv \beta\theta[\Lambda \times (\Lambda \times r) - g]$ is the buoyancy-force term. Equations (7), (2) and (3) are the model equations for nonhomogeneous flows of a Boussinesq fluid. Natural convection in liquids is generally assumed to be modeled by this set of equations.

In this paper we wish to discuss the Coriolis effect in terms of vorticity transport. Thus, we introduce the vorticity vector,

$$\omega \equiv \nabla \times U. \tag{8}$$

Taking the curl of equation (7) and rearranging terms, we get the vorticity transport equation:

$$\frac{\partial \omega}{\partial t} + U \cdot \nabla \omega = \omega \cdot \nabla U + 2\Lambda \cdot \nabla U + \nabla \times B + \frac{\mu}{\rho_o} \nabla^2 \omega. \tag{9}$$

The vector field ω is the vorticity of the flow in the rotating reference frame. Its components are:

$$\omega = (\xi, \eta, \zeta) = i(\frac{\partial w}{\partial y} - \frac{\partial v}{\partial z}) + j(\frac{\partial u}{\partial z} - \frac{\partial w}{\partial x}) + k(\frac{\partial v}{\partial x} - \frac{\partial u}{\partial y}). \tag{10}$$

The production of vorticity due to the Coriolis effect is $2\Lambda \cdot \nabla U$. The other terms on the right hand side of equation (9) are the effects of stretching and turning of vorticity ($\omega \cdot \nabla U$), buoyancy ($\nabla \times B$) and diffusion ($\nu \nabla^2 \omega$, where $\nu = \mu/\rho_o$) production and transport of ω.

If $\Lambda = \Omega j$ and $g = - gj$, then we may write (9) as follows:

$$\frac{\partial \omega}{\partial t} + U \cdot \nabla \omega = \omega \cdot \nabla U + 2\Lambda \cdot \frac{\partial U}{\partial y} + \nabla \times B + \nu \nabla^2 \omega. \tag{11}$$

where

$$\nabla \times B = i(-\beta z \Omega^2 \frac{\partial \theta}{\partial y} - \beta g \frac{\partial \theta}{\partial z}) + j(-\beta x \Omega^2 \frac{\partial \theta}{\partial z} - \beta z \Omega^2 \frac{\partial \theta}{\partial x}) + k(\beta g \frac{\partial \theta}{\partial x} + \beta x \Omega^2 \frac{\partial \theta}{\partial y}).$$

The equation for $\nabla \times B$ given above illustrates that the buoyancy effect associated with Earth's gravity induces vorticity if the horizontal derivatives of θ (i.e., $\partial \theta/\partial x + \partial \theta/\partial z$) are finite. The buoyancy-effect associated with centrifugation is such that vorticity is produced if the gradient of θ (i.e., $\nabla \theta$) is finite. In addition, the x and y components of $\nabla \times B$ explicitly depend on z whether θ is or is not a function of z. Thus, the assumption of fully-developed flow cannot be invoked in the analysis of flows in rotating pipes if buoyancy effects are included. In the present study, we consider the flow of a homogeneous fluid ($\beta \theta = 0$) and fully-developed flow. In this way we isolated the Coriolis effect. Thus, we formulated the simpliest problem of a flow-through rotating system that is subject to the Coriolis effect.

Recall that for fully-developed flow u, v, w, and ζ are functions of x, y only; thus, these parameters are independent of z and of the direction of the bulk flow through the pipe. For this case, the equations above reduce to:

$$\frac{\partial \zeta}{\partial t} + u\frac{\partial \zeta}{\partial x} + v\frac{\partial \zeta}{\partial y} = 2\Omega\frac{\partial w}{\partial y} + \nu(\frac{\partial^2 \zeta}{\partial x^2} + \frac{\partial^2 \zeta}{\partial y^2}), \tag{12}$$

$$\frac{\partial w}{\partial t} + u\frac{\partial w}{\partial x} + v\frac{\partial w}{\partial y} = - \frac{1}{\rho_o}\frac{\partial P}{\partial z} + 2\Omega u + \nu(\frac{\partial^2 w}{\partial x^2} + \frac{\partial^2 w}{\partial y^2}), \tag{13}$$

$$\frac{\partial u}{\partial x} + \frac{\partial v}{\partial y} = 0, \tag{14}$$

$$\zeta = \frac{\partial v}{\partial x} - \frac{\partial u}{\partial y}, \tag{15}$$

where $P \equiv p - \rho_o \Omega^2 (x^2 + z^2)/2 + \rho_o gy$ is the reduced pressure. In dimensionless form these equations may be written as follows:

$$\frac{\partial \zeta}{\partial t} + \frac{\partial u\zeta}{\partial x} + \frac{\partial v\zeta}{\partial y} = \frac{1}{Ro}\frac{\partial w}{\partial y} + \frac{1}{Re}(\frac{\partial^2 \zeta}{\partial x^2} + \frac{\partial^2 \zeta}{\partial y^2}), \tag{16}$$

286

$$\frac{\partial w}{\partial t} + \frac{\partial uw}{\partial x} + \frac{\partial vw}{\partial y} = C + \frac{u}{Ro} + \frac{1}{Re} \left(\frac{\partial^2 w}{\partial x^2} + \frac{\partial^2 w}{\partial y^2}\right), \tag{17}$$

$$\frac{\partial^2 \psi}{\partial x^2} + \frac{\partial^2 \psi}{\partial y^2} = \zeta, \tag{18}$$

$$u = -\frac{\partial \psi}{\partial y}, \quad v = \frac{\partial \psi}{\partial x}. \tag{19}$$

The nondimensional dynamic parameters for the problem are defined as:

$$C = \frac{GD}{\rho W_o^2}, \quad Re = \frac{W_o D}{\nu}, \quad Ro = \frac{W_o}{2\Omega D},$$

where $G = -1/\rho_o \partial P/\partial z$, W_o is the characteristic through-flow velocity and D is the width of the pipe; see figure 1. The aspect ratio of the pipe cross section is $A = L/D$, where L is the dimension in the y direction. Thus, there are four parameters that describe the flow, viz, C, Re, Ro, and A.

Equations (16), (17), (18), and (19) were solved numerically subject to the boundary conditions $u = v = w = \psi = 0$ imposed on the entire wall of the pipe. The initial conditions for the initial-value problems are $u = v = \psi = \zeta = 0$, $w = w_l$, imposed everywhere at t = 0, where w_l is the fully-developed laminar flow profile when the rotation is zero. At $t = 0^+$, the rotation of the system is commenced instantaneously. It is set to $\Lambda = \Omega j$, where Ω is a constant. The evolution of the transient solutions were computed by applying the ETUDE method developed by Valentine[23,24] and applied to solve rotating flow problems by Valentine and Jahnke[25] and Jahnke and Valentine.[26] ETUDE is an explicit, transportive-upwind finite difference method that is first order in time and second order in space. In the present study the x × y grid selected was a uniform 41 × 81 grid. The nondimensional time step was $\Delta t = 0.0001$.

In addition to studying transient solutions, we also investigated steady solutions. These were obtained by setting the time derivatives in the system of equations to zero, applying second order in space finite differences (on uniform 41 × 81 and nonuniform 51 × 91 grids — the nonuniform grid was used to improve the resolution of the boundary layers and as a check of the results obtained with the uniform grid) and an arc-length continuation method. This approach has been used previously by the authors to solve rotating flow problems.[25,26] The fact that two numerical methods produced essentially the same steady states for several test cases, leads to the conclusion that the predictions reported in the present paper are correct. Both numerical methods are fully described in the papers cited and, hence, will not be described here.

RESULTS AND DISCUSSION

The first case we shall discuss is for C = 0.1346, Ro = 0.6, Re = 279 and A = 2. This Rossby number is greater than the critical condition for roll-cell instability given by Lezius and Johnston[13] for the $A = \infty$ problem. This Reynolds number is also found to be higher than the critical value for A = 2; this is because we predicted that roll cells occur for this case. The details of this case are described next.

At t = 0 the flow is fully-developed with u = v = ψ = ζ = 0 and w = w_I, where w_I is the fully-developed laminar-flow profile for Ω = 0 illustrated in figure 2. It is symmetric about the central axis of the rectangular cross section with left/right and top/bottom symmetries. At t = 0^+ the rotation is turnedon instantaneously with Λ = Ωj. The frame of reference is fixed to the boundaries of the pipe. Hence, when the pipe begins to rotate we would expect the peak in w to shift towards the y axis of rotation for positive (or right-hand) rotation about y. This is indeed the situation, as illustrated in figure 3(a) for t = 5. The secondary-flow streamlines in the cross-sectional plane, (x,y), are illustrated in figure 3(b). The corresponding z component of vorticity, ζ, is illustrated in figure 3(c). At t = 0, and t = 5, the gradient of the through flow, i.e., ∂w/∂y, is finite because of the no-slip conditions imposed along the top and bottom walls. This component of the w gradient provides the source of ζ due to the Coriolis effect in its transport equation, equation (16). Since ζ is produced, a secondary flow results due to the coupling between ζ, ψ, u and v; see equations (16), (18) and (19). It is this secondary flow that pushes the peak of w towards the x = 0 wall. Hence, the peak shifts simultaneously with the production of vorticity, ζ, by the Coriolis effect.

At t = 15 a pair of vortices is generated that is associated with an instability of the shear layer along the x = 0 wall, where ∂w/∂x is unstable. These vortices are illustrated in figure 4. At t = 20 the size of the roll cells are approaching a size close to their steady size as shown figure 5. At t = 30 they begin to spread vertically along the x = 0 wall; see figure 6. The steady-state solution for this case is summarized in figure 7; the stability of this solution has yet to be determined. The approach to steady-state conditions is summarized in the phase portrait (or Lissajou figure) in figure 8(a); it is a plot of ψ(0.25,0.25) versus ψ(0.5,0.125) over the same period of time as for ψ(0.5, 0.25) in figure 8(b). Both parts of figure 8 illustrate the oscillatory approach to the fixed point in phase space (i.e., the approach to steady state).

In figures 3 through 6, the development of two pairs of vortical structures are observed. The top and bottom vortices are caused by the top and bottom duct walls. For the A = ∞ case, the end-wall vortices do not occur. Since the A = ∞ case cannot be tested in the laboratory, the end wall vortices always occur. Speziale and Thangam[16] computed the A = 8 case and end-wall vortices occurred as expected. Hart[17] also discussed this phenomenon. The end-wall vortices

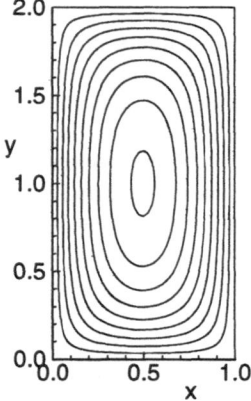

Figure 2. Steady-state solution of the w component of velocity for Re = 279 and Ω = 0; 0 ≤ w ≤ 0.5, Δw = 0.05.

arise for all rotation rates, Ω, because $\partial w/\partial y$ is always finite unless there is no flow through the pipe. Thus, the end-wall vortices are not a result of an instability phenomenon. They are a direct result of the Coriolis effect. The roll-cell vortices have been identified as being associated with the stability of the shear layer along the $x = 0$ wall because they occur near the condition when the linear stability analysis of the $A = \infty$ case suggests they would. Previous investigators compared the onset of roll-cells in the confined-boundary cases (i.e., $A < \infty$) with the linear stability theory for the $A = \infty$ case. The comparisons reported were reasonably good and, hence, it was concluded that the occurrence of roll cells is associated with the instability of the shear, $\partial w/\partial x$, along the $x = 0$ wall near the center of the tube. In the following discussion we provide additional clarification, from an alternative viewpoint, of the nature of the onset of roll cells in the $A = 2$ (confined flow) case. The viewpoint taken is to examine two of three sets of steady flows via bifurcation diagrams.

We examined three sets of steady flows. In the first set we looked at $Re = 30$, $C = 0.1346$ and $A = 2.0$. We raised the rotation rate, i.e., we decreased the Rossby number over the range $5.3 > Ro > 0.0008$. At all of these conditions the secondary flow consisted of the end-wall pair of vortices only. No roll cells were predicted because $Re = 30$ is well below the critical condition for their occurrence. Figure 9 illustrates this secondary flow for $Ro = 0.574$. As Ω is increased, the secondary flow strength increases. Since C is constant (and, thus, the energy supply is fixed), as Ro decreases we would expect the flow through the system to decrease. This is indeed the case as illustrated in figure 10.

The next two sets of computations were for $Ro = 0.5$ and 0.6, respectively, $A = 2$, $C = 0.1346$, and a range of Re from 1.0 to 330. The bifurcation diagrams for the two sets of cases are illustrated in figure 11. The onset of the roll cells is associated with the turning point bifurcation at $Re \approx 263$. When $\psi(2,42) = \psi(1/40,41/80)$ crosses zero, the roll cells are detected. As shown in the figure, the $Ro = 0.6$ set goes through a series of turning points before it continues towards larger values of Re. The set of $Ro = 0.5$ cases goes through a second turning point at $Re \approx 200$ before it continues toward larger values of Re. The region $200 < Re < 300$ is a region where multiple solutions of the Navier-Stokes equations are predicted. The steady-state solutions immediately below the first turning point in figure 11 are unstable; the change in stability of steady solutions from one side to the other side of a

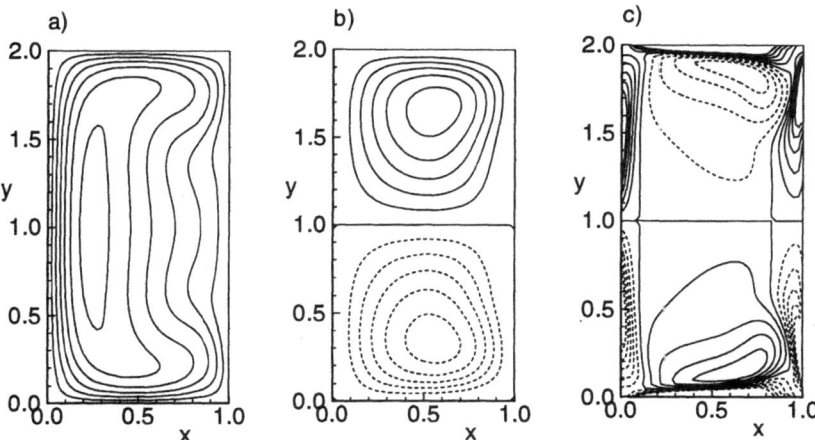

Figure 3. Flow at $t = 5$ for $Re = 279$, $Ro = 0.6$: (a) w, $0 \le w \le 3.6$, $\Delta w = 0.6$. (b) ψ, $-0.2 \le \psi \le 0.2$, $\Delta \psi = 0.04$. (c) ζ, $-10 \le \zeta \le 10$, $\Delta \zeta = 2.0$. Solid lines are positive contours. Dashed lines are negative contours.

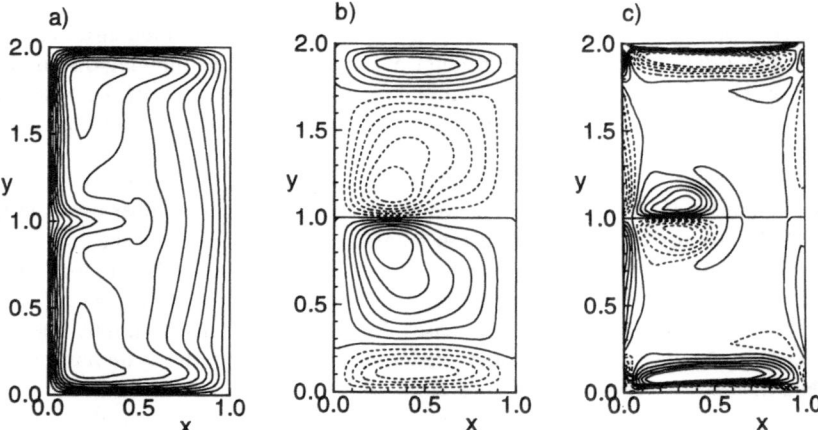

Figure 4. Flow at t = 15 for Re = 279, Ro = 0.6. (a) w, $0 \leq w \leq 2.4$, $\Delta w = 0.3$. (b) ψ, $-0.048 \leq \psi \leq 0.048$, $\Delta \psi = 0.008$. (c) ζ, $-6.25 \leq \zeta \leq 6.25$, $\Delta \zeta = 1.25$. Solid lines are positive contours. Dashed lines are negative contours.

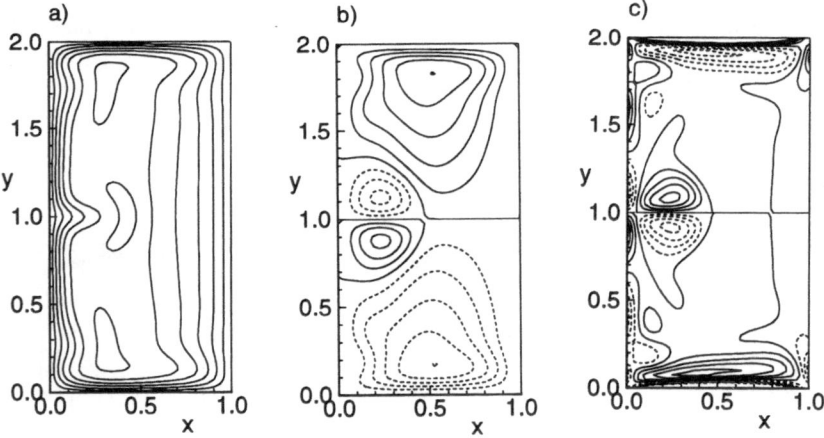

Figure 5. Flow at t = 20 for Re = 279, Ro = 0.6: (a) w, $0 \leq w \leq 1.8$, $\Delta w = 0.3$. (b) ψ, $-0.04 \leq \psi \leq 0.04$, $\Delta \psi = 0.008$. (c) ζ, $-6.25 \leq \zeta \leq 6.25$, $\Delta \zeta = 1.25$. Solid lines are positive contours. Dashed lines are negative contours.

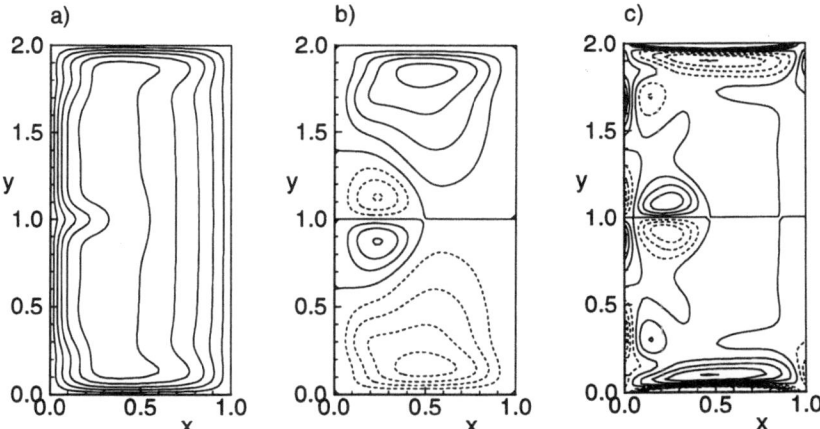

Figure 6. Flow at t = 30 for Re = 279, Ro = 0.6: (a) w, $0 \leq w \leq 1.5$, $\Delta w = 0.3$. (b) ψ, $-0.032 \leq \psi \ 0.032$, $\Delta \psi = 0.008$. (c) ζ, $-6.25 \leq \zeta \leq 6.25$, $\Delta \zeta = 1.25$. Solid lines are positive contours. Dashed lines are negative contours.

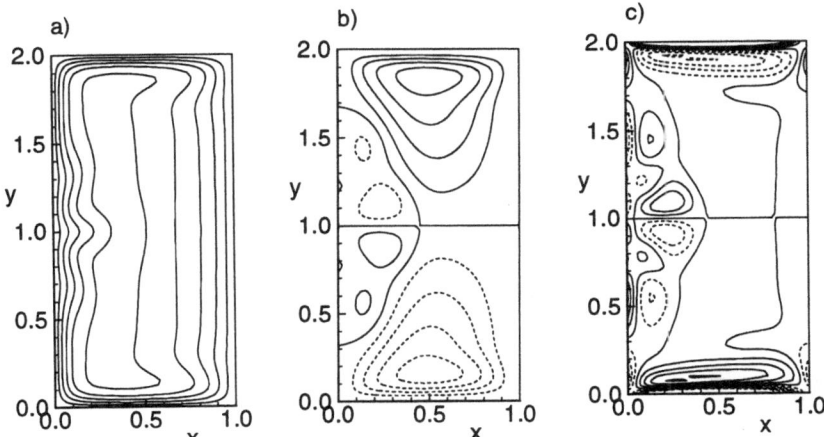

Figure 7. Flow at steady state for Re = 279, Ro = 0.6: (a) w, $0 \leq w \leq 1.5$, $\Delta w = 0.3$. (b) ψ, $-0.032 \leq \psi \leq 0.032$, $\Delta \psi = 0.008$. (c) ζ, $-6.25 \leq \zeta \leq 6.25$, $\Delta \zeta = 1.25$. Solid lines are positive contours. Dashed lines are negative contours.

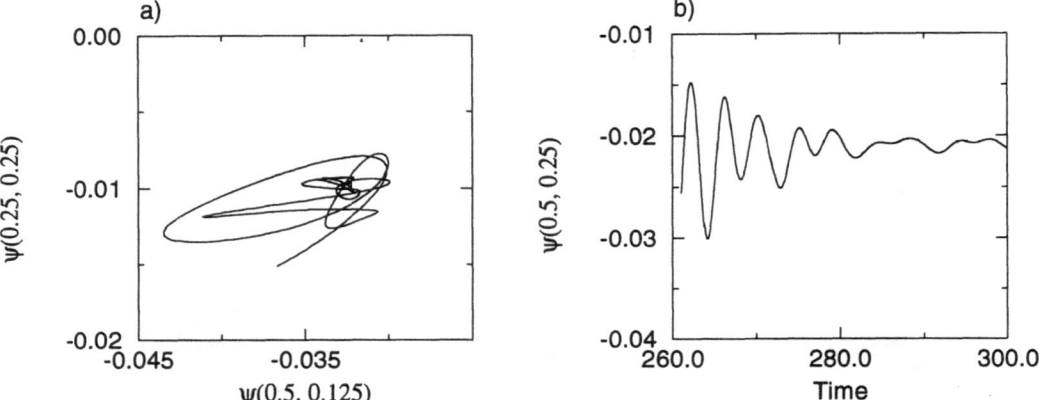

Figure 8. The (a) phase plot and (b) the time evolution for Re = 279 and Ro = 0.6.

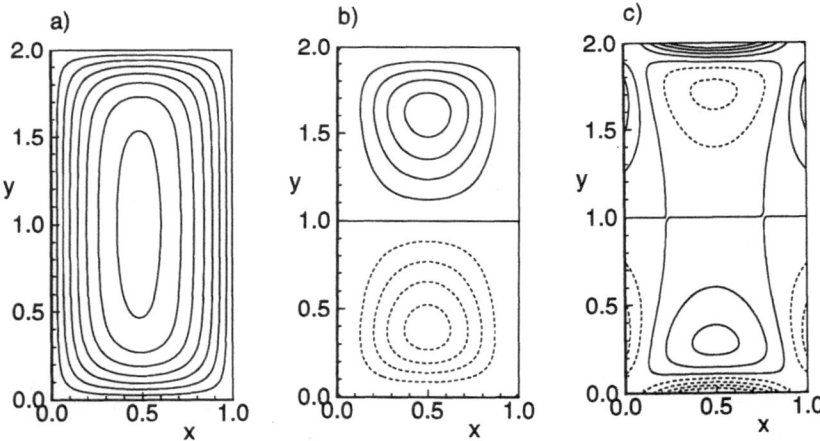

Figure 9. Flow at steady state for Re = 30, Ro = 0.574: (a) w, $0 \leq w \leq 0.4$, $\Delta w = 0.05$. (b) ψ, $-0.012 \leq \psi \leq 0.012$. (c) ζ, $-0.8 \leq \zeta \leq 0.8$. Solid lines are positive contours. Dashed lines are negative contours.

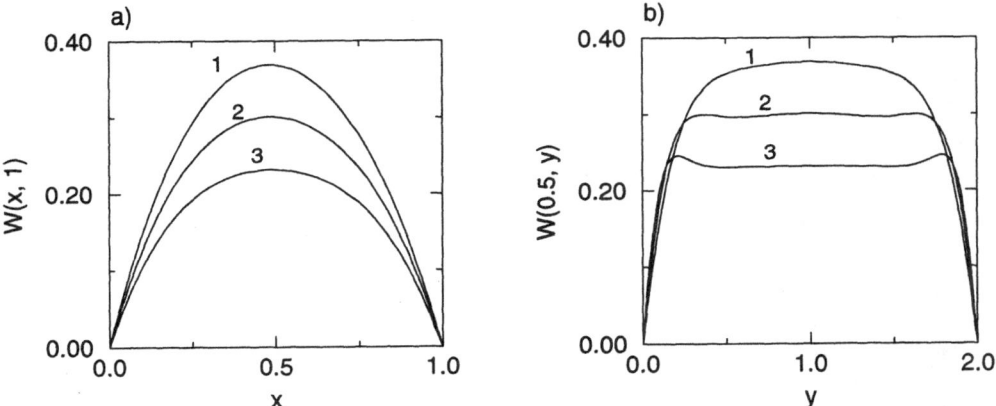

Figure 10. Profiles of the w component of the velocity for (1) Ro = 0.57, (2) Ro = 0.28, and (3) Ro = 0.09.

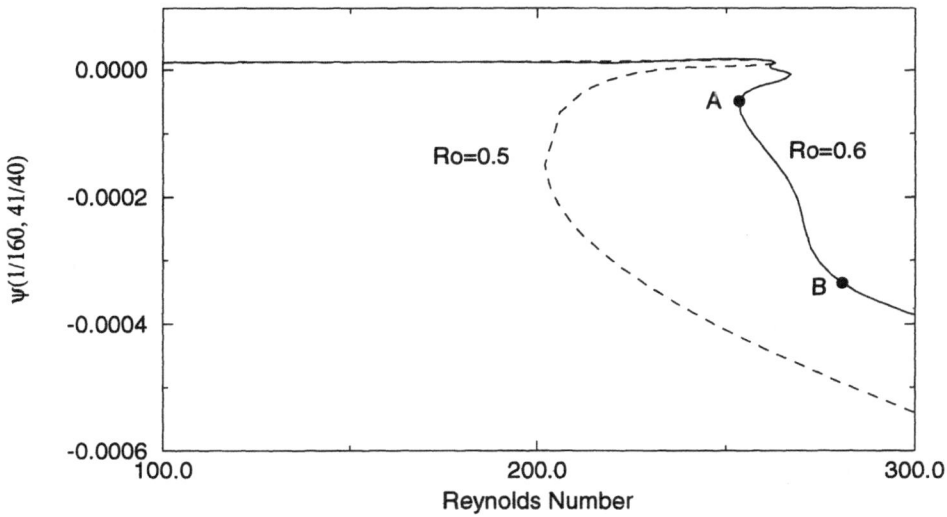

Figure 11. Bifurcation diagram for Ro = 0.5 and Ro = 0.6.

bifurcation point (in this case the turning point) is a theoretically established fact; see, e.g., Guckenheimer and Holmes.[27] The stability of solutions following the second and other turning points has yet to be established. Since multiple solutions exist over a relatively wide range of Re, it is not surprising that determining steady solutions in this range of Ro would be difficult to compute when using time-dependent methods with relatively course grids, as was done in the preliminary study reported by Speziale.[14] The stable solutions obtained would depend on the perturbations introduced by numerical error. Our solutions were checked by refining the grid in the boundary layers and comparing results from both time-dependent and steady-state computational methods of finer grids than the grids use by Speziale.[14,16] Confirmation of our predictions was, thus, attained.

The solutions for points A and B on the bifurcation diagram for Ro = 0.6 are illustrated in figure 12 and figure 7, respectively. Some details of the flow field for the latter are given in figure 13. For comparison purposes, figure 14 presents the Ro = 0.5, Re = 279 case. Note the change in flow pattern of the roll cells. What can we conclude? We can conclude that for Re = 279, A = 2, C = 0.1346 and Ro = 0.5 or 0.6, roll-cell vortices exist. They occur as a natural evolution of solutions from Re=1, where they do not exist, to Re = 279, where they do exist. Their occurrence is associated with a turning point bifurcation that occurs at approximately Re = 263. Between Re = 200 and 260 for Ro = 0.5 there is a region where multiple solutions of the Navier-Stokes equations exist. For Ro = 0.6 this range is reduced; however, the number of possible solutions is increased as illustrated by the increase in number of the turning points predicted.

Further work is underway investigating the details of this newly discovered region of multiple solutions. The main points of the present paper were to clarify the results of previously reported numerical predictions and to illustrate the complexity of the fluid mechanics in rotating pipe flows that are caused by the Coriolis effect.

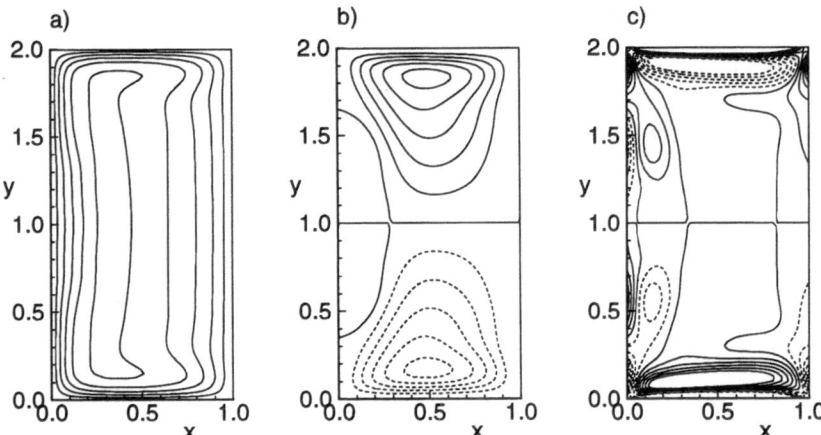

Figure 12. Flow at steady state for Re = 253.4, Ro = 0.60 (point A on figure 11): (a) w, 0 ≤ w ≤ 1.5, Δw = 0.3. (b) ψ, -0.035 ≤ ψ ≤ 0.035, $\Delta\psi$ = 0.007. (c) ζ, -3 ≤ ζ ≤ 3, $\Delta\psi$ = 0.6. Solid lines are positive contours. Dashed lines are negative contours.

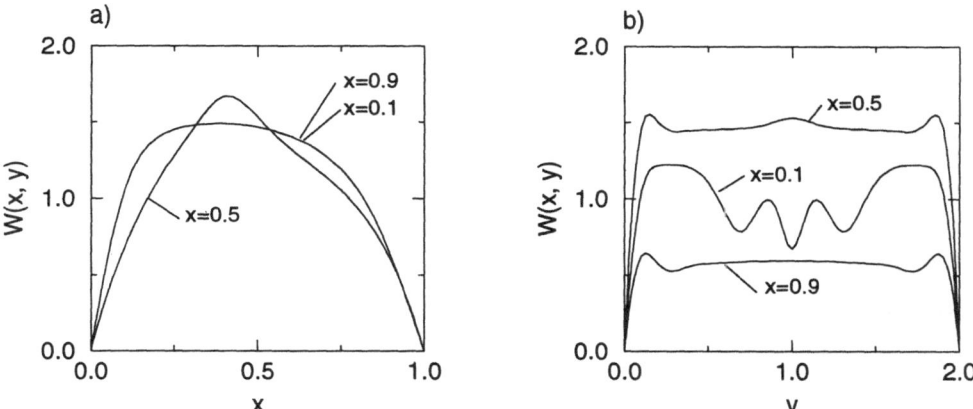

Figure 13. Profiles of the w component of the velocity for Re = 279, Ro = 0.6.

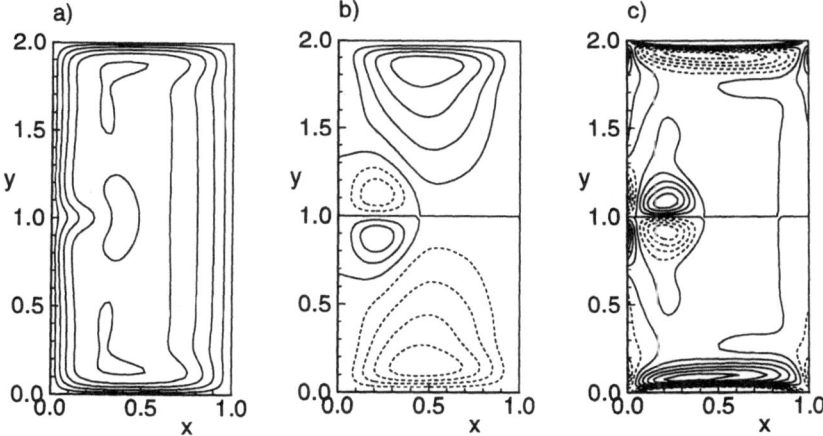

Figure 14. Flow at steady state for Re = 279, Ro = 0.50: (a) w, $0 \leq w \leq 1.5$, $\Delta w = 0.3$. (b) ψ, $-0.03 \leq \psi \leq 0.03$, $\Delta \psi$ = 0.0075. (c) ζ, $-5 \leq \zeta \leq 5$, $\Delta \psi = 1$. Solid lines are positive contours. Dashed lines are negative contours.

ACKNOWLEDGEMENTS

One of the authors (D.T.V) acknowledges the support of NASA under the contract OSP-5039.

REFERENCES

1. L.L. Regel and W.R. Wilcox, Influence of centrifugation on transport phenomena, Clarkson University (1996).

2. L.L. Regel, M. Rodot, and W.R. Wilcox, eds, "Material Processing in High Gravity, Proceedings of the First International Workshop on Material Processing in High Gravity," North-Holland, Amsterdam (1992). Also volume 119 of the *Journal of Crystal Growth*.

3. L.L. Regel and W.R. Wilcox, eds, "Materials Processing in High Gravity," Plenum Publishing Corporation, New York (1994).

4. V.S. Yuferev, The coriolis force---a factor that must be taken into account when growing crystals from a melt under weightless conditions, *Tech. Phys. Lett.* 20:97-98 (1994).

5. S.N. Barua, Secondary flow in a rotating straight pipe, *Proc. Roy. Soc. Lond.* A227:133-139 (1954).

6. G.S. Benton, The effect of the Earth's rotation on laminar flow in pipes, ASME *J. Applied Mech.* 23:123-127 (1956).

7. G.S. Benton and D. Boyer, Flow through a rapidly rotating conduit of arbitrary cross-section. *J. Fluid Mech.* 26:69-79 (1966).

8. B.E. Launder and D.P. Tselepidakis, Application of a new second-moment closure to turbulent channel flow rotating in orthogonal mode, *Int. J. Heat Fluid Flow* 15:2 (1994); Chem. Abstracts 120:195017 (1994).

9. A.K. Majumdar, V.S. Pratap and D.B. Spalding, Numerical computation of flow in rotating ducts, ASME *J. Fluids Engrg.* 99:148-153 (1977).

10. J.H. Howard, S.V. Patankar and R.M. Bordynuik, Flow prediction in rotating ducts using Coriolis-modified turbulence models, ASME *J. Fluids Engrg.* 102:456-461 (1980).

11. J.P. Johnston, R.M. Halleen and D.K. Lezius, Effects of spanwise rotation on the structure of two-dimensional fully developed turbulent channel flow, *J. Fluid Mech.* 56:533-557 (1972).

12. R.E. Wagner and H.R. Velkoff, Measurements of secondary flows in a rotating duct, ASME *J. Engrg. Power* 95:261-270 (1972).

13. D.K. Lezius and J.P. Johnston, Roll-cell instabilities in rotating laminar and turbulent channel flow, *J. Fluid Mech.* 77:153-175 (1976).

14. C.G. Speziale, Numerical study of viscous flow in rotating rectangular ducts, *J. Fluid Mech.* 122:251-271 (1982).

15. C.G. Speziale, Numerical Solution of Rotating Internal Flows, American Mathematical Society, *Lectures in Applied Mathematics* 22:261-287 (1985).

16. C.G. Speziale and S. Thangam, Numerical study of secondary flows and roll-cell instabilities in rotating channel flow, *J. Fluid Mech.*130:377-395 (1983).

17. J.E. Hart, Instability and secondary motion in a rotating channel flow, *J. Fluid Mech.* 45:341-351 (1971).

18. S. Fann, W.J. Yang and S. Mochizuki, Transport phenomena at the entrance regions of rotating heated channels with laminar throughflow, ASME *J. Heat Transfer* 116:239-242 (1994).

19. P. Tekriwal, Heat transfer predictions with extended k-ε turbulence model in radial cooling ducts rotating in orthogonal mode, ASME *J. Heat Transfer* 116:369-380 (1994).

20. H. Ishigaki, Analogy between laminar flows in curved pipes and orthogonally rotating pipes, *J. Fluid Mech.* 268:133-145 (1994).

21. H. Ishigaki, Analogy between turbulent flows in curved pipes and orthogonally rotating pipes, *J. Fluid Mech.* 307:1-10 (1996).

22. K.C. Cheng, R.C. Lin and J.W. Ou, Fully-developed laminar flow in curved rectangular channels, ASME *J. Fluids Engrg.* 98:41-48 (1976).

23. D.T. Valentine, Control-volume finite difference schemes to solve convection-diffusion problems, ASME *Computers in Engrg.* 3:111-117 (1988).

24. D.T. Valentine, Decay of confined, two-dimensional, spatially-periodic arrays of vortices: A numerical investigation, *Intnl. J. Num. Methods Fluids* 21:155-180 (1995).

25. D.T. Valentine and C.C. Jahnke, Flows induced in a cylinder with both end walls rotating, *Phys. Fluids* 6:2702-2710 (1994).

26. C.C. Jahnke and D.T. Valentine, Boundary layer separation in a rotating container, *Phys. Fluids* 8:1408-1414 (1996).

27. J. Guckenheimer and P. Holmes, "Nonlinear Oscillations, Dynamical Systems, and Bifurcations of Vector Fields," Springer-Verlag, NY (1983).

INDEX